MANUFACTURING PROCESSES

BY

MYRON L. BEGEMAN

PROFESSOR OF MECHANICAL ENGINEERING,
SUPERINTENDENT OF ENGINEERING SHOP
LABORATORIES, THE UNIVERSITY OF TEXAS

SECOND EDITION

NEW YORK: JOHN WILEY & SONS, INC.
LONDON: CHAPMAN & HALL, LIMITED

PRINTED IN THE UNITED STATES OF AMERICA

MANUFACTURING
PROCESSES

PREFACE TO THE SECOND EDITION

During the past five years American industry has demonstrated its ability to produce at rates beyond all previous conception. Spurred on by necessity and aided by intensive research, many new materials, machine tools, and processes were developed. This situation brings about a need for revising a text on this subject.

A knowledge of present manufacturing processes is of extreme importance to engineers engaged in industry. With the development of many new processes a product can frequently be made several ways, and this presents a problem to the designer and production engineer. The possibilities and limitations of the various processes and machine tools should be understood if low factory costs are to be attained. The purpose of this text continues to be the training of young engineers in the technical fundamentals of important manufacturing processes, engineering materials, and the modern machine tools necessary for processing these materials.

Principal among the changes made in this revision are new chapters on Special Casting Methods, Powder Metallurgy, Hot Forming of Metals, and Cold Forming of Metals. In addition the chapter on Plastic Molding has been rewritten as well as much of the material in the chapter on Welding and Allied Processes. All chapters have some new material particularly in regard to newly designed machines and the tools they use. Careful attention has been given to the illustrations, many of which are new.

The co-operation of many industrial firms is acknowledged as well as the helpful suggestions received from my colleagues and others who have been using the text. I wish also to acknowledge Mr. J. R. Holmes's work in preparing various line drawings; Dr. Joseph Jones's assistance in reading the manuscript; Dr. E. A. Murray's suggestions on the subject of plastics; and Mr. J. R. Morrill's assistance in revising the chapter on welding.

MYRON L. BEGEMAN

Austin, Texas
January 2, 1947

CONTENTS

LIST OF TABLES

CHAPTER 1

FOUNDRY PRACTICE

From early historic time, castings have been fundamental to man's progress. With the gradual advancement of civilization to the present industrial age, castings have constituted a basic foundation for the development of mechanical processes. The field for cast parts is increasing constantly, because scientific research has brought about applications and adaptations which hitherto were not considered within the scope of the castings industry.

Casting is the process of pouring molten metal into a mold and allowing it to solidify. By this process, intricate parts can be given strength and rigidity frequently not obtainable by any other method. While all metals can thus be formed to shape, iron is especially adapted to this method, because of its fluidity, its small shrinkage, and the ease with which its properties can be controlled.

The *mold*, into which the metal is poured, is made of some heat-resisting material. Sand is most often used, as it is easily packed to shape, is somewhat porous, and resists high temperatures. However, permanent molds of metal can be used for small castings, particularly those of nonferrous composition. In die casting metal molds are used exclusively.

Sand molds are filled by pouring the molten metal into an opening at the top of the mold. Properly constructed passages allow the metal to flow to all parts of the mold by gravity. Permanent molds can be filled the same way; but in die casting, the practice is to force the metal into the mold under pressure, either by compressed air or by an operating plunger. In centrifugal casting, the metal is introduced into the mold by gravity, additional pressure being obtained by centrifugal force.

Tools and Equipment for Molding

Small or medium-sized castings are made in a *flask* — a box-shaped container without top or bottom. It is made in two parts, held in alignment by dowel pins. The top part is called the *cope* and the lower part the *drag*. If the flask is made in three parts the center is called a *cheek*. These flasks can be made either of wood or metals. Wood is the cheapest material, and wooden flasks can be quickly made, but they have the disadvantage of wearing out rapidly and of being destroyed by contact with hot metal. Metal flasks of steel, cast iron, magnesium, or aluminum alloys are widely used in production work because of their rigidity and permanence.

One type of flask often used in the production of small castings and in machine molding is the *snap flask*. This flask can be removed from the mold by releasing the latches at one corner and immediately used again for making additional molds. A steel slip jacket is placed over the mold to hold it together and to prevent breaking from pressure of the molten metal as it enters the mold. These jackets assist in holding the halves of the mold in proper alignment and eliminate difficulties caused by shifting of the cope. Being of steel, they are not seriously damaged by occasional contact with hot metal.

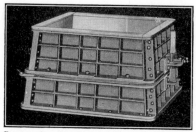

Courtesy American Foundry Equipment Company.

Fig. 1. Side View of Standard-Weight Dowmetal or Aluminum Tapered Flask.

The flask shown in Figure 1 is an all-metal taper strip flask. The inside surfaces of these flasks are accurately ground to a smooth finish and have an inside taper of 5 degrees. The entire flask lifts off the mold without the necessity of unlatching the corner hinges. Metal jackets for the molds, made on the same dimensional form as used for assembling the flasks, preserve the mold in the same shape as it was in the flask. These flasks are made either of aluminum or of Dowmetal.

A *molding board* and a *bottom board* complete the flask. The molding board is a smooth board on which the flask and patterns are placed when the mold is started. It should be perfectly flat and well reinforced with cleats on the bottom. When the mold is turned over, the function of this board is ended; the mold is placed on a similar board, called a bottom board, which acts as a support for the mold until it is poured.

Before any metal is poured into a mold, it is necessary that the flask be clamped in some way to prevent the buoyant effect of the molten metal from lifting up the cope. Small molds are usually held down by flat cast-iron weights placed on the top of the molds. Larger flasks are held together by clamps placed on the sides or ends — either U-shaped clamps, held tight by driving wooden wedges under the end, or clamps which can be quickly adjusted to fit the height of the flask.

A *gagger* is a small L-shaped metal accessory used in floor molds to help support hanging bodies of sand in the cope. It is used only in large molds having crossbars. The gagger is first coated with a clay wash and then placed next to one of the crossbars. The lower end should be close to the pattern, and the upper end should extend to the top of the mold.

The hand tools of a molder are few and need little explanation. A brief description of the most important tools is given here:

Riddle. A riddle of a standard mesh screen is used to remove lumps or foreign particles from the sand. Both hand and power riddles are available, the latter being used where large volumes of sand are involved. A power riddle which gives the sand a rapid vibrating movement is shown in Figure 2.

Rammer. Hand rammers are used to pack the sand in the mold. One edge of the rammer, called the peen end, is wedge-shaped; the other, called the butt end, is flat. Floor rammers are similar in construction but have long handles. Pneumatic rammers are used in large molds, saving considerable labor and time.

Bellows. Standard hand-operated bellows are used for blowing loose sand from the cavities and surface of the mold.

Trowel. Small trowels of various shapes are used for finishing and repairing mold cavities as well as for smoothing over the parting surface of the mold. The usual trowel is rectangular in shape and has either a round or square end.

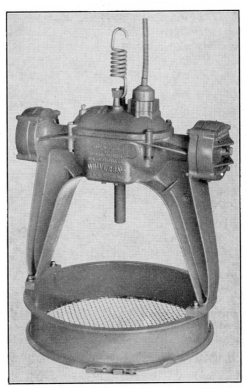

Courtesy Foundry Supplies Manufacturing Company.

FIG. 2.　Universal Free-Wheeling Power Riddle.

Slick. The principal hand tool for repairing molds is called a slick. It is a small double-ended tool having a flat on one end and a spoon on the other. This tool is also made in a variety of other shapes.

Lifter.   Lifters are used for smoothing and cleaning out depressions in the mold.   They are made of thin sections of steel of various widths and lengths with one end bent at right angles.   A combination slick and lifter is known as a Yankee lifter.

Swab.   This tool is used for moistening the sand around the edge before the pattern is removed.   A simple swab is a small brush having long hemp fibers; a bulb swab has a rubber bulb to hold the water and a soft hair brush at the open end.   Pressure on the bulb forces the water out through the brush.

Draw spike or screw.   The draw spike is a pointed steel rod, with a loop at one end.   It is driven into a wood pattern to hold the pattern when it is withdrawn from the sand.   The draw screw is similar in shape but threaded on the end to engage metal patterns.

Vent wire.   This wire has a sharp point and is used to punch holes through the sand after it has been rammed.   In this manner the mold is provided with vents which carry off the steam and gases generated by the hot metal in contact with the sand.

Gate cutter.   This tool is a U-shaped piece of thin metal used for cutting a shallow trough in the mold to act as a passage for the hot metal.

Molding Processes

Molding practice in the ordinary commercial foundry may be classified according to (1) the *processes* used in making the mold or (2) the *type of material* of which the mold is made.   Under the first heading we have the following divisions:

1. Bench molding.   This type of molding is for small work, done on a bench of a height convenient to the molder.

2. Floor molding.   As the size of castings (with resultant difficulty in handling) increases, the work is done on the foundry floor.   This type of molding is used for practically all medium- and large-sized castings.

3. Pit molding.   Extremely large castings are frequently molded in a pit instead of a flask.   The pit acts as the drag part of the flask, and a separate cope is used above it.   The sides of the pit are brick-lined, and on the bottom there is a thick layer of cinders with connecting vent pipes to the floor level.   Since pit molds can resist pressures developed by the hot gases, this practice saves greatly on pattern expenses.   Figure 3 shows a large pit mold partially completed.

4. Machine molding.   Machines have been developed to do a number of the operations that the molder ordinarily does by hand. Such operations as ramming the sand, rolling the mold over, forming the gate, and drawing the pattern can be done by these machines much better and more efficiently than by hand.   So far, no machine has been developed that is completely automatic. A discussion of these machines appears later in this chapter.

Courtesy Steel Founders' Society of America.

FIG. 3. Large Pit Mold Partially Completed.

Molds classified as to the materials commonly used in making are:

1. Green-sand molds. This most common method, consisting of forming the mold from damp molding sand, is used in most of the processes previously described. Figure 4 illustrates the procedure for making this type of mold.

2. Skin-dried molds. Two general methods are used in preparing the skin-dried molds. In one, the sand around the pattern, to a depth of about 1/2 inch, is mixed with a binder so that when it is dried it will leave a hard surface on the mold. The remainder of the mold is made up of ordinary green sand. The other method is to make the entire mold of green sand and then coat its surface with a spray or wash which hardens when heat is applied. Sprays used for this purpose include linseed oil, molasses water, gelatinized starch, and similar liquid solutions. In both methods the mold must be dried either by air or by a torch to harden the surface and drive out excess moisture.

3. Dry-sand molds. These molds are made entirely from fairly coarse molding sand mixed with a binding material similar to these already mentioned. Since such molds must be oven-baked before being used, the flasks are of metal. Such a mold holds its shape when poured and is free from gas troubles due to moisture. Both the skin-dried and dry-sand molds are widely used in steel foundries.

4. Loam molds. Loam molds, like pit molds, are used for large work. The mold is first built up with bricks or large iron parts. These parts are then plastered over with a thick loam mortar, the shape of the mold being obtained with sweeps or skeleton patterns. The mold is then allowed to dry thoroughly so that it can resist the heavy rush of molten metal. Such molds take a long time to make and are not extensively used.

5. Metal molds. Metal molds have their principal use in the die casting of low-melting-temperature alloys. Castings are accurately shaped with a smooth finish, thus eliminating much machine work. The disadvantages of metal molds are their expense of manufacture, their gradual deterioration by oxidation of surface in contact with the hot metal, their inability to contract along with the cooling metal, and the severe chilling effect they have upon some alloys. Some metal molds are also hard to vent.

Molding Sand

Good molding sand found along river beds is uniform in structure because of the polishing action of the river currents. Here and along lakes is found most of the sand suitable for molding work. Although the principal constituent of sand is silica, or quartz, many other elements in small quantities are found with it. Silica sand is suitable for molding purposes because of its ability to withstand a high temperature without decomposition. If the percentage of other elements, such as limestone, soda, and magnesia, becomes too high, the melting temperature of the sand is reduced, and there may be some tendency for the sand to fuse with the molten metal. The proportion of these ingredients should not exceed 2 to 3%.

Pure sand is not suitable in itself for molding, since it lacks binding qualities. The latter can be obtained by adding around 8 to 15% of clay to the sand. Clay is made up principally of alumina, which is also highly refractory. It is formed by decomposing feldspar by a hydration process. Clay, when added to the sand and dampened by water, forms a mixture which becomes very cohesive and is easily shaped into molds.

The size of the sand grains will depend on the type of work to be molded. For a small and intricate casting the use of a fine sand is desirable so that all the details of the mold will be brought out sharply. Such sand may be slightly higher in clay content, since the thin sections of metal will chill rapidly and there will be very little gas to escape through the mold. As the size of the castings increases, the sand particles likewise should be coarser to permit the ready escape of the gases that are generated in the mold. Also, because of the prolonged high temperature, this sand should be high in

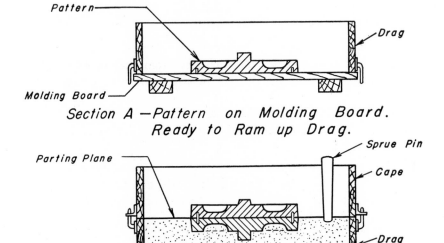

Section A —Pattern on Molding Board.
Ready to Ram up Drag.

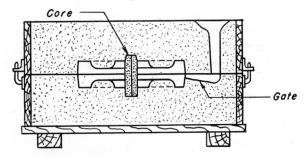

Section B —Drag Rolled over and Pattern
Assembled Ready to Ram Cope.

Section C —Mold Complete with Dry
Sand Core in Place.

FIG. 4.   Procedure for Making Mold.

silica and free from other contaminating elements which lower its melting temperature.

Good molding sand should be:

1. Refractory — to resist the high temperatures of the molten iron without fusing.

2. Cohesive, when moistened — to provide sufficient bond to hold together.

3. Porous or permeable — to permit the escape of gases and steam formed in the mold.

When a mold is made in two or more parts, it is necessary to provide some means of keeping the sand from sticking together at the parting surface. This is accomplished by dusting or spreading the surface with pure silica sand that is very fine and free from clay. This is called *parting sand*. Materials such as brick dust or burnt core sand are sometimes used, but pure beach sand is preferable.

Molding Procedure

In order to understand fully the use of patterns, we must first become somewhat familiar with foundry practice. Assume, for example, that we wish to make a cast-iron gear blank. The procedure for molding this part is illustrated in Figure 4. The mold for this blank is made in the usual flask, which consists of two parts. The two parts are held in a definite relation to one another by means of pins on either side of the drag which fit into openings in angle clips fastened to the sides of the cope.

The first step in making a mold is to place the pattern on a molding board, which fits the flask being used. Next, the drag is placed on the board with the pins down, as shown in section *A* of Figure 4. Molding sand, which has previously been tempered, is then riddled in to cover the pattern. This sand should be pressed around the pattern with the fingers; then the drag should be completely filled. The sand is then firmly packed in the drag by means of a hand rammer. In ramming the sand around the sides of the flask, the peen end should be used first, additional sand being placed into the drag as the sand is packed down. The inside area of the drag is then packed down with the butt end of the rammer. The amount of ramming necessary can be determined only by experience. Obviously, if the mold is not sufficiently rammed, it will not hold together when handled or when the molten iron strikes it. On the other hand, if it is rammed too hard, it will not permit the steam and gas to escape when the molten iron comes into the mold.

After the ramming has been finished, the excess sand is leveled off with a straight bar known as a *strike rod*. In order to insure the escape of the gases when the casting is poured, small vent holes are made through the sand to within a fraction of an inch of the pattern.

The completed lower half of the mold is now ready to be turned over so that the cope may be placed in position and the mold finished.

A little sand is sprinkled over the mold, and a bottom board placed on top. This board should be moved back and forth several times to insure an even bearing over the mold. The drag is then rolled over and the molding board removed, exposing the pattern. The surface of the sand is first smoothed over with a trowel and is then covered with a fine coating of dry parting sand. This is done to prevent the sand in the cope from sticking to the sand in the drag when the mold is separated to remove the pattern.

The cope is next placed on the drag, as shown in section *B* of Figure 4, the pins on either side holding it in proper position. In order to provide a place for the iron to enter the mold, a tapered pin

Courtesy American Foundry Equipment Company.

FIG. 5.   Molder Blowing Out Mold for Dowmetal Flask Side.

known as a *sprue pin* is placed approximately an inch to one side of the pattern. The operations of filling, ramming, and venting of the cope proceed in the same manner as in the drag.

The mold is now complete except for the removal of the pattern and the sprue pin. The latter is first withdrawn and a funnel-shaped opening is scooped out at the top so as to have a reasonably large opening into which to pour the iron. Next, the cope half of the flask is carefully lifted off and set to one side. Before the pattern is withdrawn, the sand around the edge of the pattern should be moistened with a swab so that the edges of the mold will hold firmly together when the pattern is withdrawn. To loosen the pattern, a draw spike is driven into it and rapped lightly in all directions. The pattern can then be withdrawn by lifting up on the draw spike.

Finally, before the mold is closed again, a small passage known as a gate must be cut from the mold at the bottom of the sprue opening. The completed mold is shown in section *C*, Figure 4. This passage is shallowest at the mold, so that after the iron has been poured the metal in the gate may be broken off close to the casting. In order to insure a better surface on the casting, the mold is usually faced with powdered graphite. When the iron comes in contact with this material, it causes it to burn and form a thin film on the surface of the mold which assists in keeping the iron from penetrating the sand. Figure 5 shows a molder blowing out an open mold preparatory to placing the cope back on the drag. In some cases, where snap flasks are used, the flask may be removed and a steel jacket placed around the mold. This permits using the same flask for other molds and eliminates the danger of having the hot iron burn the flask. Before pouring the iron into the mold, a weight should be put on top to keep it in place and to eliminate any tendency of the liquid iron to separate the cope and drag.

Sand-Conditioning Equipment

Properly conditioned sand is an important factor in obtaining good castings. Conditioning is difficult to do by hand, since the moisture content must be controlled, and the binder material should be uniformly distributed around the sand grains. The most desirable sand for molding has a minimum of binder and moisture. A typical mixer for preparing the sand is shown in Figure 6. This mixer consists of a circular pan, in which is mounted a combination of plows and mullers driven by a vertical shaft. This arrangement gives a shoveling action to the sand, turning it over on itself and lining it up in front of the mullers, which give an intensive kneading and rubbing action. The result is a thorough distribution of the sand grains with the bonding material. After the sand is mixed, it is discharged through a door in the bottom of the pan. Both green sand and core sand may be prepared in this manner.

Courtesy National Engineering Company.

FIG. 6.   No. 3 Mixer for Mixing and Conditioning Foundry Sand.

A representative sand-reclamation and conditioning unit is shown in Figure 7. As the molds are shaken out, the sand falls through a grate onto a belt conveyor which carries it over a magnetic separator. It is then discharged onto a bucket elevator, from which it goes through a hexagonal revolving screen to remove any refuse. The screen is enclosed for air exhaust of dust and fines. This important process of reclaiming and reconditioning of the sand permits control of the required permeability without the necessity of adding large amounts of new sand.

The advantages of such sand conditioning for all classes of foundries have been demonstrated many times. Some of these advantages are economy of new sand and binder, close control, uniformity of sand condition, and low cost in preparing the sand; likewise, all the sand, or a considerable portion of it, is taken off the floor, releasing this space for molding and other facilities. Such a system greatly

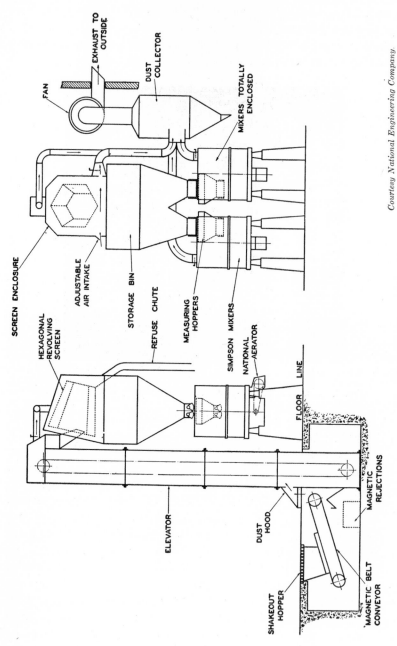

Courtesy National Engineering Company.

Fig. 7. Sand-Preparing Equipment.

improves the general operating conditions of the foundry. The units with overhead sand storage have high production of uniform high-quality castings, hold closely to tolerances and weights with a minimum of defects, and reduce cleaning labor to a minimum. The fact that all the sand in the system is mulled each time it is used, so that some is maintained virtually at facing grade, eliminates the preparation of special facing sands as such. Much of the work formerly required to be made in dry sand is made in green sand, in some cases skin-dried, with consequent economies of labor, handling, and baking. It also is possible to reduce the number of grades of sand used for different types of work, one grade of sand ordinarily being adequate when properly reconditioned.

Sandslingers

Uniform packing of the sand in molds is an important operation in the production of castings. Hand ramming is fairly satisfactory for small work, but not for medium- and large-sized molds. To accomplish this operation in a satisfactory manner, a mechanical device

Courtesy Steel Founders' Society of America.

FIG. 8.   Motive-Type Sandslinger in Operation.

known as a sandslinger has been developed. Figure **8** shows a motive-type sandslinger, which is a self-propelled unit operating on a narrow-gage track. The supply of sand is carried in a large tank of about 300 cubic feet capacity, which may be refilled at intervals by overhead handling equipment. A delivery belt feeding out of a hopper on the frame at the fixed end conveys the sand to the rotating impeller head. The impeller head, which is enclosed, contains a single rotating cup-shaped part which slings the sand into the mold. This part, rotating at high speed, slings over a thousand small buckets of sand a minute. The ramming capacity of this machine is **7** to 10 cubic feet, or 1000 pounds of sand per minute. The density of the packing can be controlled by the speed of the impeller head. For high production, machines of this type are available having a capacity of 4000 pounds of sand per minute.

Similar machines can also be obtained either with a tractor mounting or as a stationary unit. Tractor-type sandslingers travel along the sand piled on the floor and are used in foundries having no auxiliary sand-handling equipment. In addition to the ramming operation, these machines cut, riddle, and magnetically separate the sand from the scrap. The stationary machine is adapted to production work and must be served by sand preparation and conditioning equipment, as well as conveyors for removing the molds. Sandslinger machines greatly increase foundry production and insure the uniform ramming of molds so necessary for good castings.

Sand Testing

Periodic tests are necessary to determine the essential qualities of foundry sand. The properties change by contamination of foreign materials, by washing action in tempering, by the gradual change and distribution of grain size, and by continual subjection to high temperatures. Tests may be either chemical or mechanical; but aside from determining undesirable elements in the sand, the chemical tests are little used. Most mechanical tests are simple and do not require elaborate equipment.

Test for moisture content. Moisture content of foundry sands varies according to the type of molds being made and the kind of metal being poured. For a given condition there is a close range within which the moisture percentage should be held in order to produce satisfactory results. Any system of sand control should include a periodic check on moisture content, and complete records should be kept for future reference.

Many foundries rely on the judgment of the molders for the proper tempering of the sand, the usual test being to squeeze a handful of the sand into a lump, noting how it fractures or crumbles. Experienced molders can judge sand fairly well by this simple test, but this method has too much chance of error to be recommended. The most accurate method is that of drying out the sand and noting the weights before and after. In Figure 9 is shown a moisture teller which consists of a direct motor-driven fan; electric heating unit, base, and housing; and three sand pans. Fifty grams of tempered sand, accurately weighed, is placed in the shallow pan. The bottom of

Courtesy Harry W. Dietert Company.

FIG. 9.   Moisture Teller for Quick Moisture Tests of Foundry Sands.

the pan is made of Monel-metal filter cloth of 500 mesh to permit the air to escape. The switch for starting the motor can be set for the time necessary to dry the material. Air from the blower is heated to 230 F and blown through a diffuser which spreads the heated air over the sand. By weighing the sand after it cools and noting the difference in the initial and final readings the percentage of moisture can be determined.

Another method employing a sand rammer is based on the fact that the more moisture there is in the sand the closer it will be packed together. A gage on the indicator measures the compression and may also be calibrated to record moisture content direct if desired. Still another method measures the flow of current between two electrodes imbedded in the sand, and the moisture can be determined by comparing the reading with known impedance value for certain conditions.

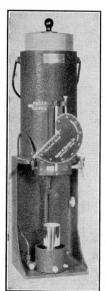

FIG. 10. Permeability Meter for Measuring Green and Dry AFA Permeability of Molding and Core Sands.

Permeability test. One of the essential qualities of molding sand is sufficient porosity to permit the escape of gases generated by the hot metal. This depends on several factors, including shape of sand grains, fineness, degree of packing, moisture content, and amount of binder present. Permeability is measured by the quantity of air which will pass through a given sample of sand in a prescribed time and under standard conditions. A permeability meter, meeting AFA Standard Testing Specifications, is shown in Figure 10. Additional pieces of equipment necessary for conducting the test are a sand rammer, balance, and weights.

The permeability meter consists of an aluminum casting in the form of a water tank and a base. Inside the tank floats a balanced air drum which is sealed at the bottom by the water. The air tube extending down to the specimen opens into the air space of the drum. The sand specimen is placed in the small cup at the base and is sealed with mercury. In taking a permeability reading, the air valve is opened and the dial rotated to water level in the pressure gage. Permeability is read direct for the AFA orifice-test method on the curved portion of the dial, and the straight scale is used for AFA stop-watch research method.

Green-sand permeability is determined by packing sand in a standard tube; dry sand or core permeability by clamping the specimen in a special tube on a rubber gasket and pouring mercury around it to seal the sides. The permeability of molds may also be determined by using an extension tube, which is pressed into the top of a mold. Permeability values obtained are actual volumes of air that the sands will pass. The manufacturer of this equipment gives the following values for different sands:

Nonferrous castings	5–19
Small iron castings	10–30
Iron castings 200 lb.	60
Large iron castings	100
Steel castings — dry sand	100–200

Mold and core hardness test. The mold hardness tester shown in Figure 11 operates on the same principle as a Brinell hardness tester.* A steel ball 0.2 inch in diameter is pressed into the surface of the mold, and the depth of penetration is indicated on the dial in thousandths of an inch. The reading will not change even with excessive pressure on the tester. Medium-rammed molds give a value around 50. Such a quick method of checking mold hardness is particularly useful in investigating mold uniformity and different machine settings.

Courtesy Harry W. Dietert Company.

FIG. 11. Mold-Hardness Tester for Measuring the Surface Hardness of Green-Sand Molds.

The hardness of cores is tested by a similar instrument using a carboloy cone point. The tester is drawn across the surface of the core in a manner similar to the fingernail-scratch test frequently used by foundrymen. A numerical figure is obtained which is a measure of the cone-point penetration into the core surface. This instrument is useful in controlling surface hardness of both dried molds and baked cores.

Clay-content test. The purpose of this test is to determine the percentage of clay in molding sands. The equipment necessary for this test consists of a drying oven, balance and weights, and a sand washer. A small quantity of sand is thoroughly dried out, and a sample of 50 grams is selected and placed in a wash bottle. To this sand is added 475 cc of distilled water and 25 cc of a 3% caustic soda solution. This mixture is then stirred 5 minutes in a rapid sand stirrer or 1 hour if a rotating sand washer is used. Sufficient water is then added to fill the bottle up to a level line marked on the bottle,

---

* The Brinell test consists of indenting the surface of a metal specimen with a 10 mm hardened-steel ball by means of a predetermined load.

and, after settling for about 10 minutes, the liquid is siphoned off.
The bottle is then refilled twice more and the siphoning operation
repeated, time being allowed for the sand to settle. The bottle is
finally placed in the oven, and, after the sand is dried out, a sample is
weighed. The percentage of clay is determined by the difference in
the initial and final weights of the sample.

Fineness test. This test, to determine the percentage distribution
of grain sizes in the sand, is performed on a dried-sand sample from
which all clay substance has been removed. A set of standard testing
sieves is used having U. S. Bureau of Standards meshes 6, 12, 20, 30,
40, 50, 70, 100, 140, 200, and 270. These sieves are stacked and
placed in one of the several types of motor-driven shakers. The
sand is placed on the coarsest sieve at the top, and, after 15 minutes
of vibration, the weight of the sand retained on each sieve is obtained
and converted to a percentage basis.

To obtain the AFA fineness number, each percentage is multiplied
by a factor as given in the following example. The fineness number
is obtained by adding all the resulting products and dividing the total
by the percentage of sand grain.

EXAMPLE OF AFA FINENESS CALCULATION

Mesh	Percentage Fineness	Multiplier	Product
6	0	3	0
12	0	5	0
20	0	10	0
30	2.0	20	40.0
40	2.5	30	75.0
50	3.0	40	120.0
70	6.0	50	300.0
100	20.0	70	1400.0
140	32.0	100	3200.0
200	12.0	140	1680.0
270	9.0	200	1800.0
Pan	4.0	300	1200.0
Totals	90.5		9615.0

$$\text{Grain fineness number} = \frac{9615}{90.5} = 106$$

This number is a useful means of comparing different sands for uses
in the foundry. A check of sands in use should be made every month.

Sand-strength test. Several strength tests have been devised to
test the holding power of various bonding materials in green and
dry sand. Compression tests are the most common, although tension,

shear, and transverse tests are sometimes used in strength investigations. Procedure varies according to the type of equipment used, but in general the tests are similar to those used for other materials. The fragile nature of sand requires special consideration in the handling and loading of test specimens.

A sand-strength machine with tensile core-strength accessory is shown in Figure 12. This machine consists of a frame, on which is mounted a pendulum weight and a pusher arm. It is motor-driven, although hand operation may be used if desired. In testing green strength under compression, a cylindrical specimen of sand is placed between the pusher arm and compression head of the pendulum weight. The compression load is applied by the motor at the rate of 7.5 pounds per 15 seconds. Pressure is continued until the specimen fails, at which time the compression value (indicated by a small magnetic bar) is read directly on the curved scale.

Courtesy Harry W. Dietert Company.

FIG. 12. Universal Sand-Strength Machine with Tensile Core-Strength Accessory.

To test core material in tension, the formed specimen is placed in the jaws of the machine at the upper part of the pusher arm. As the power is applied, the core pulls the weight up until failure occurs. Again a strength value is read directly on the curved scale.

The tests just described are the ones most commonly used in sand control. In addition to these, there are several others used to check various properties. New sand may be given a sintering test to determine whether or not it has a tendency to burn on to the metal at high temperatures. Chemical analysis is frequently necessary in order to check the composition of sand grains, since some elements greatly reduce the refractory qualities of the sand. Strength at high temperatures and expansion coefficients of different sands can be determined to check the action of the sand in contact with hot metal. The object of all these tests is to improve foundry operations and the quality of castings produced.

Molding Machines

Machines can eliminate much of the hard work of molding and at the same time produce better molds. Molding machines varying considerably in design and method of operation are named according to the manner in which the ramming operation is performed by squeezing, jolting, or some similar means.

The plain *jolt molding machine,* shown in Figure 13, is equipped with adjustable flask-lifting pins to permit the use of flasks of various

Courtesy The Taber Manufacturing Company.

FIG. 13.   Jolt-Molding Machine Equipped with Flask Lifting Pins.

sizes within the machine capacity. The illustration shows a match plate in place on the machine and a completed mold on the bench. The principle of this machine is very simple. The machine raises the flask a short distance and then lets it drop. As the flask is dropped suddenly, the momentum of the sand causes it to pack evenly about the pattern. Uniform ramming thus obtained gives added strength to the mold and reduces the possibility of swells, scabs, or runouts. Furthermore, castings produced under such uniform conditions will not vary in size or weight. The lifting pins on the machine engage the flask and raise it from the match plate after the mold is complete. Jolt machines quite obviously can take care of only one part of a flask at a time and are especially adapted to large work.

Squeezer machines press the sand in the flask to uniform density between the machine table and overhead plates. Small machines can be manually operated, but in most cases the pressure is applied by an air cylinder below the table.

Courtesy Milwaukee Molding Machine Company.

Fig. 14. Jolt-Squeeze Molding Machine.

Many molding machines have incorporated both the jolt and squeeze principle. A machine of the *jolt-squeezer type* is shown in Figure 14. To produce a mold on this machine, the flask is assembled with the match plate between the cope and drag, and the assembly is placed upside down on the machine table. Sand is shoveled into the drag and leveled off, and a bottom board is placed on top. The jolting action then rams the sand in the drag. The assembly is turned over and the cope filled with sand and leveled off. A pressure board with an opening for the sprue pin is placed on top of the flask, and the top platen of the machine is brought into position. By the application of pressure the flask is squeezed between the platen and table, which packs the sand in the cope to the proper density. After the pressure is released, the platen is swung out of the way. The cope is then lifted from the match plate while the plate is vibrated, after which the plate is removed from the drag. This machine

Courtesy The Osborne Manufacturing Company.

Fig. 15.   Jolt-Squeeze Strip-Pin Life Machines Installed in Foundry.

eliminates six separate hand operations: ramming, smoothing the parting surface, applying parting sand, swabbing around the patterns, rapping the pattern, and cutting the gate.

Some patterns are difficult to withdraw without cracking or otherwise damaging the mold. To eliminate this trouble, *stripper-plate machines* may be used. A stripper plate, having the same outline as the pattern, supports the sand in the flask as the pattern is withdrawn. After the flask is rammed, the operator opens the vibrator valve and at the same time opens the operating valve to " draw " position. This causes the lifting pins to rise, contacting the stripping plate and flask, and the mold is lifted slowly from the pattern. In some cases the mold remains stationary, and the patterns are lowered through the stripper plate. Several jolt–squeeze strip-pin lift machines installed with overhead sand supply are shown in Figure 15. The completed molds, in aluminum snap flasks, rest on the roller conveyor.

For large molds that are difficult to handle, machines have been developed which roll the mold over and draw the pattern from the mold. Figure 16 shows a *jolt-roll-over pattern-draw molding machine*

Courtesy The Osborne Manufacturing Company.

FIG. 16. Jolt-Roll-over Pattern-Draw Molding Machine Installed with Conveyor for Removing Molds.

set up for magnesium aircraft castings. The figure also illustrates the use of roller conveyors for handling the molds from the machine. In the operation of this machine, the flask, first packed with sand by jolting, is rolled over and the pattern then withdrawn. For quantity production two machines are required, one for making drags and the other for copes.

Oscillating molding machines are built on the principle that it is necessary to keep the sand in motion by a vibrating action while under pressure in order to produce uniformly permeable molds. Molding sand is made up of irregularly shaped grains having a coating of some bonding element and a small moisture content. Any packing action does not actually compress the sand, but only causes the grains to interlock and adhere with the assistance of the bonding material.

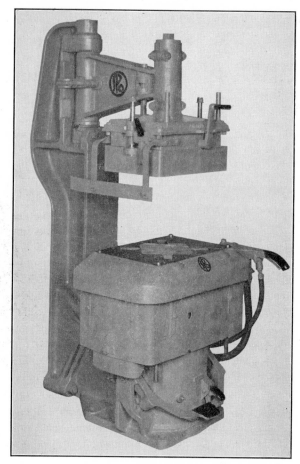

Courtesy Spo, Inc.

Fig. 17.   Oscillating Squeeze-Strip Molding Machine.

When the sand is vibrated under pressure, the packing action is speeded up, and uniform results are obtained.

The main feature of this type of machine is the oscillating table. Four vibrators are used beneath the table, at an angle, which direct rapid blows toward the outside of the mold. Sand disturbances are created throughout the mold which cause the sand to flow evenly around all portions of the pattern.

Figure 17 shows a standard *oscillating squeeze machine* which may be used as a stripping machine. This machine operates as follows:

Courtesy Spo, Inc.

FIG. 18. Oscillating Squeeze-Strip Roll-over Molding Machine. Patterns Mounted on Machine Ready for Operation.

Sand is placed in the mold and vibrated while the squeeze pressure is applied at the top. The vibration keeps the sand moving while the pressure speeds up the operation. For best results the moisture content should not exceed 5%, and a minimum of bonding materials should be used. The four pins shown are stripping pins, which are used to contact the frame of the flask during the stripping operation. This machine automatically strips the pattern from the mold on the down stroke of the table. No stripping plate is shown in this picture, although one may be used.

A similar machine of roll-over design is shown in Figure 18, with patterns mounted on the platen ready for operation. Flasks are put over these patterns, filled with sand, and leveled off with the strike bar mounted on the machine. The molds are then oscillated slightly to permit the sand to settle before putting the bottom boards in place. A bar is placed across both bottom boards; and during the operation of rolling over, the molds are held in place by the hook shown in the figure. The squeeze piston is then raised until it contacts the bottom board and the hook is released simultaneously with the squeezing operation. During this period the machine is both oscillated and squeezed. The figure shows clearly the floating springs which allow the table to vibrate freely while being oscillated. Near the end of the operation these springs are compressed and contact buttons engaged for the finish of the squeeze. Pressure-release valves are provided to bring all molds to a uniform density.

Molds made on these machines are uniform throughout, and all pattern variations are brought out in detail. It is also possible to ram up chills or cores without having them dislocated or broken during the operation. This method of molding is rapid, consuming no more time than that required for an ordinary squeeze operation.

Cores

When a casting is to have a cavity or recess in it, such as a hole for a bolt, some form of *core* must be introduced into the mold. A core is sometimes defined as " any projection of the sand into the mold." This projection may be formed by the pattern itself or made elsewhere and introduced into the mold after the pattern is withdrawn. Either internal or external surfaces of a casting can be formed by a core.

Cores may be classified under two headings: *green-sand cores* and *dry-sand cores*. Figure 19 shows various types of cores commonly used. Green-sand cores, as shown in Figure 19A, are those formed by the pattern itself and made from the same sand as the rest of the mold. This drawing shows how a flanged casting can be molded with the hole through the center " cored out " with green sand.

Dry-sand cores are those formed separately to be inserted after the pattern is withdrawn but before the mold is closed. They are usually made of sharp river sand, which is mixed with a binder and then baked to give the desired strength. The box in which they are formed to proper shape is called a *core box*. Dry-sand cores placed in a mold before it is closed are shown in Figure 20. In the

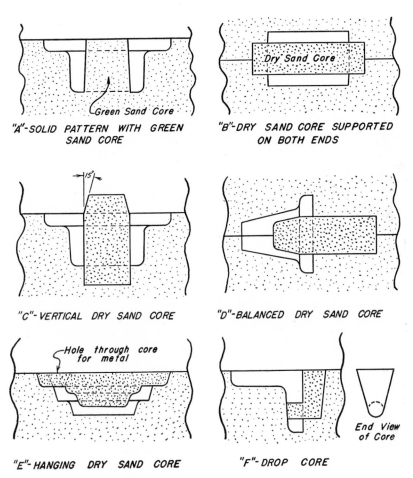

"A"-SOLID PATTERN WITH GREEN SAND CORE

"B"-DRY SAND CORE SUPPORTED ON BOTH ENDS

"C"-VERTICAL DRY SAND CORE

"D"-BALANCED DRY SAND CORE

"E"-HANGING DRY SAND CORE

"F"-DROP CORE

FIG. 19. Various Types of Cores Used in Connection with Patterns.

background of this illustration are the molding machines with the match plates from which the molds are made. In the foreground are the two halves of the flask resting on roller conveyors to facilitate handling. A large dry-sand core has been placed in the right half. The steel casting as it comes from the mold rests on the conveyor.

Several types of dry-sand cores are illustrated in Figure 19. At

B is the usual arrangement for supporting a core when molding a cylindrical bushing. The projections on each end of the cylindrical pattern are known as *core prints* and form the seats which support and hold the core in place. A vertical core is shown at *C*, the upper end of which requires considerable taper so as not to tear the sand in the cope when the flask is assembled. Cores which have to be supported only at one end must have the core print of sufficient length

Courtesy of Steel Founders' Society of America.

Fig. 20. Mold before Closing with Match-Plate Patterns and Coupled Casting.

to prevent the core from falling into the mold. Such a core, shown at *D*, is known as a *balanced core*. A core supported above and hanging into the mold is shown at *E*. This type usually requires a hole through the upper part to permit the metal to reach the mold. A *drop core*, shown at *F*, is required when the hole is not in line with the parting surface and must be formed at a lower level.

In general, green-sand cores should be used where possible to keep the pattern and casting cost to a minimum. Separate cores naturally increase the production cost. Core boxes must be made, and the cores must be formed separately, baked, and properly placed in the molds, all of which adds to the molding cost. However, more accurate holes can be made with dry-sand cores, for they give a better surface and are less likely to be washed away by the molten metal.

The designer of cores must remember that long, thin cores are very difficult to remove from the casting. If a core is surrounded by a heavy section of metal, the pressure of the liquid metal may be sufficient to force some of the metal into the core. This adds greatly to the cleaning costs.

In setting dry-sand cores into molds, adequate supports must be provided. Ordinarily, these supports are formed into the mold by the pattern; but, for large or intricate cores, additional supports in the form of *chaplets* (small metal shapes made of a low-melting alloy) are placed in the mold to give additional support to the core until the molten metal enters the mold and fuses the chaplets into the casting. The use of chaplets should be limited as much as possible because of the difficulty in securing proper fusion of the chaplet with the metal.

Essential Qualities for Dry-Sand Cores

A core must have sufficient *strength* to support itself. Naturally this strength depends upon the kind of sand and binding material used. Sharper grains of sand will bond together better and form a stronger core. It is also advisable to have sand of a uniform grain size to provide plenty of voids through which the escaping gases may pass. The size of the sand depends largely on the finish desired, but is also dependent on the size of work being cast.

Porosity or permeability is also an important consideration in the making of cores. As the hot metal pours over the cores, gases are generated by the heat's being in contact with the binding material, and provision must be made to carry away these gases. The size of the sand grain and its freedom from fines in between the grains largely determine its permeability. In addition to the natural porosity of the sand, it is usually advisable to vent the core as well. This may be done with an ordinary vent wire or, where two pieces go together to make up a core, the vent may be scraped out with the sharp corner of a trowel. Sometimes strings of wax are used, being placed in the core when it is first made up. During the process of baking, the wax is melted out, leaving a passage for the escaping gases.

To insure a smooth casting, the core must have a *smooth surface*. This smoothness depends largely on the grain size of the sand, although the surface may be improved by coating the core with a thin mixture of water and graphite. In the attempt to produce a smooth surface care must be exercised not to go too far, since permeability is lost as the fineness of the sand increases.

All cores must have sufficient *refractory* property to resist the action of the heat until the hot metal has found its place in the mold. Sand naturally is very refractory; so a binder must be selected that will stand the temperature required of the core. A thin coating of graphite or similar material adds considerably to its ability to withstand the intense heat momentarily. It must be kept in mind that it is not desirable to have the core remain hard after the metal has cooled: the binding material used should disintegrate or be burned out by the prolonged contact with the hot metal so that the core may be removed easily from the finished casting. This is also important from the standpoint of preventing shrinkage cracks during cooling.

Core Making

The first consideration in making dry-sand cores is to mix and prepare the sand properly. If the binder is dry, it should be thoroughly mixed with the sand before any moisture is added. In any event, the mixture must be homogeneous so that the core will be of uniform strength throughout. All ingredients should be measured out exactly, and care should be taken not to get the mixture too wet. Wet material sticks to the tools and core box, causes the cores to sag before they are baked, and tends to produce a hard core. Large foundries generally use some sort of mechanical mixer.

The core is formed by being rammed into a core box or by the use of sweeps. Fragile and medium-sized cores should be re-enforced with wires to give added strength to withstand deflection and the floating action of the metal. In large cores perforated pipes or arbors are used. In addition to giving the core strength, they also serve as a large vent.

When the cores are properly formed, they are placed on small metal plates which support them during the baking period. These plates are placed in the core oven and heated to a temperature ranging from 350 to 450 F. The actual process of baking consists of first driving off the volatile matter and moisture that is in the core and then allowing sufficient time for the binder to become oxidized and hard. The color of the cores on emerging from the oven is a good index of their condition. The usual color for oil binders is a nut brown, while flour or similar binding materials produce a light brown. A burnt core is very dark and crumbles when handled.

One of the most critical stages in core making is the fitting of the cores after baking, since many mistakes can be corrected at this point. The fitting consists of procuring two halves of a core, scraping or filing the contact surfaces so that a perfect fit is obtained, venting by cutting a trough in each half, and finally gluing them together. Broken edges or corners must be repaired by a patching sand made up of silica sand mixed with some binder such as glue or molasses. Various core binders can be used for gluing the halves together. These binders harden and hold the halves securely together after they have been dried out in the oven. A graphite or silica wash can be applied to the surface of the core to improve surface smoothness and add to its refractory property. After the drying process cores should again be inspected to insure a proper fit into the mold.

Binders and Core Mixtures

Among several types of binders used in making cores are those classified as oil binders. One of these, linseed oil, is the principal binder used in the making of small cores. The oil forms a film around the sand grain which hardens when oxidized by the action of the heat. Such cores should be baked between 350 and 425 F. A common mixture uses 40 parts of river sand and 1 part of linseed oil. An advantage of this core is that it does not absorb water readily and retains its strength in the mold for some time. A similar core oil having the following analysis has proved very successful: raw linseed oil, 42.5–45%, gum rosin, 27.5–30%, with the remainder kerosene which is water-white and acid-free. The gum rosin, although having some binding properties, is also used to prevent the thinned oil from draining to the bottom of the core on standing. About a pint of this oil is required for 100 pounds of core sand.

In another group of binders, soluble in water, we find wheat flour, dextrin, molasses, gelatinized starch, and many commercial preparations. The ratio of binder to sand in these mixtures is rather high, being 1 to 8 or more parts of sand. Frequently a small percentage of old sand is used in place of new sand. Such mixtures are all moistened with water to the proper dampness for working, a temperature of about 350 F being sufficient to harden the cores. One mixture using wheat flour contains 1 part of flour, 6 parts of sharp river sand, and 2 parts of molding sand. This should be mixed thoroughly and wet with a thin clay wash. Another cereal binder obtained from the corn-products industry, contains about 90% starch and 6% glucose.

It is available in powdered form and, when used in the same proportions as an oil binder, produces a weaker and softer core.

In addition, pulverized pitch or rosin may be used. During the baking these products melt and flow between the sand grains and when cooled form a very hard core. Both these binders melt below 350 F. Such cores are not very refractory on account of the rapid melting and consequent softening of the core when it comes in contact with the hot metal.

There are many other commercial binder preparations, the analyses of which are difficult to obtain, but most of them contain one or more of the afore-mentioned materials. Combinations of several binding materials into one preparation may result in an improved binder, as the desirable qualities of each component frequently supplement the other. Considerable experimental work is advisable in large foundries in order to work out mixtures and proportions which give best results in practice. No single binder can answer all the requirements in a diversified foundry.

Core-Making Machines

Pneumatic core-blowing machines offer a rapid means of producing cores in quantity production work. In this method, sand is blown into a core box at high velocity and under pressure. The resulting core is of uniform structure with a high degree of permeability.

A machine of this type is shown in Figure 21. The box on the top is a sand hopper which feeds the sand to the reservoir and blow plate beneath. This assembly moves back and forth from feeding to blowing position, and is open only in the filling position. The table below supports and clamps the core box in place and is provided with suitable vertical adjustment.

In operation, the core box is placed on the table and clamped in position. The table and box are then raised, sealing the space between the blow plate and top of the core box. Compressed air, at pressures ranging from 100 to 120 pounds per square inch, is introduced into the reservoir, forcing the sand through the small holes in the blow plate to the mold cavity. Suitable vents are built into the core box to permit the air to escape. These vents must be small enough to resist any flow of sand through them, as their sole purpose is to relieve the air from the box. Their location plays an important part in successful core making, since they are used to direct the flow of sand to the desired parts of the core. A core box, filled in a few seconds, is then removed from the machine. In order to keep the machine in continuous operation, duplicate core boxes should be provided. The function of the machine is only to fill the core boxes. Additional equipment, such as conveyors and mechanical drawing devices, greatly facilitates handling operations, reducing the cycle operating time.

The holes in the blow plate are countersunk and vary in size from 3/16 inch to 1/2 inch in diameter. The larger holes are placed

Courtesy The Osborn Manufacturing Company.

FIG. 21. Core-Blowing Machines Producing Cores for Plumbing Fixtures.

opposite the larger portions of the core and the smaller ones opposite small or restricted pockets that would not tend to fill from the main openings; thus, their size and position offer a means of controlling the flow of sand into the core-box cavities. No definite rule can be given as to number and size of holes, each installation being worked out in accordance with past experiences and best judgment.

Core boxes for this process should be made of metal. Cast iron, aluminum, and magnesium are the metals most commonly used. Cast-iron boxes, although having greater wearing qualities, are heavier and more difficult to machine than boxes of the lighter alloys. For medium- and large-sized boxes it is economical to use the lighter alloys. All boxes must fit properly to avoid excess air leakage at parting lines. In addition, the outer surfaces should be machined to fit squarely in the clamping fixture.

Sand used in this process need not differ from that of regular coremaking practice. It should, however, have good flowability and a minimum of moisture content. Sharp silica sand which has been thoroughly cleaned is recommended.

This type of equipment is especially adapted to production work

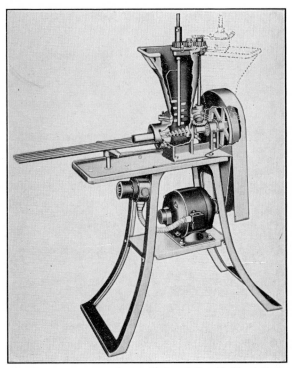

Courtesy Wadsworth Core Machine & Equipment Company.

FIG. 22. Section View of Rapid Core Machine for Extruding Stock Cores of Uniform Section.

and is limited to applications where the expense of metal core boxes is justified. It is rapid in its operation, and cores produced are true to form with excellent permeability.

A machine used in the making of stock cores 3/8 inch to 3 inches in diameter is shown in Figure 22. The upper part of the machine consists of a large hopper into which the core sand is placed. A mixer and feeding rod through the center of the hopper feeds a uniform amount of sand to the tapered spiral conveyor located horizontally below the trough. The rotation of this conveyor forces the sand through a die tube at a uniform speed and pressure. Formed cores emerge continuously on a grooved tray, all cores being vented through the center by an extension on the end of the spiral conveyor. Shapes controlled by the shape of the die opening are numerous, including round, square, oval, and hexagonal.

These machines require little skill in their operation. They produce cores rapidly, their output being around 10 feet of core per minute.

Core Ovens

Core baking is the process of hardening the core by drying out the binder in a controlled-temperature oven. A variety of core ovens is available, the choice depending on the nature and size of the cores as well as the quantity involved. These ovens are either *stationary* or *continuous*. The simplest of the stationary type is equipped with drawers which can be withdrawn for loading and unloading. A typical gas-fired oven of this type is shown in Figure 23. A plate at the rear of each drawer closes the opening to prevent the escape of heat when the drawer is withdrawn. This type of oven is suitable for small and medium-sized cores. Another type, shown in Figure 24, uses portable racks upon which the cores are loaded. These racks are moved about with hand-lift trucks or by overhead monorail trolleys and are convenient in that they may be placed next to the core maker's bench. For very large cores and molds the car-type oven is best. Work of this type is placed on the cars by hoists or cranes, and the car is rolled into the oven.

Continuous ovens are those in which the work is placed on some sort of a conveyor that moves it slowly through the oven. Loading and unloading is a continuous operation, the length of time for baking being controlled by the rate of movement of the conveyor. Vertical ovens of this type are widely used, since a large amount of work can be obtained from a relatively small amount of floor space. A suspended tray-type elevator which holds the tray in a horizontal position at all times is the usual means of conveying the cores. Such a core oven is shown in Figure 25. Cores are loaded onto the trays at one side of the oven, move vertically to the heated zone, and are removed as they come down on the opposite side. Horizontal ovens use the apron-type conveyor with steel slats.

Continuous ovens are especially adapted to drying out cores of approximately the same size, as the time through the oven is constant for a given run. Where there is considerable variation in size, cores

Courtesy The Foundry Equipment Company.

FIG. 23.　Coleman Rolling-Drawer Core Oven with Gas-Fired External Recirculating Heating System.

Courtesy The Foundry Equipment Company.

FIG. 24.　Battery of Six Coleman Double-End Transrack Core Ovens.

of nearly the same size are grouped together and baked in stationary ovens. The time must be carefully controlled so as not to over- or underbake. Most core furnaces use gas or oil as the fuel, and the temperature is controlled by indicating or recording thermometers.

Courtesy Steel Founders' Society of America.

FIG. 25. Vertical Core-Drying Oven with Suspended Trays to Support Cores.

REVIEW QUESTIONS

1. List the principal hand tools of a molder.

2. What hand-molding operations does a jolt–squeeze molding machine eliminate?

3. Classify and describe briefly the various processes of making molds.

4. For what are the following foundry tools or materials used: Slick, lifter, swab, gagger, chaplet, strike rod, cope, cheek, and drag?

5. Why should molds be faced, and what materials are used for that purpose?

6. How does molding sand differ from river sand?

7. What essential qualities should molding sand possess?

8. Distinguish between a squeeze molding machine and a jolt molding machine.

9. What is a stripping-plate machine, and for what type of molding is it adapted?

10. How are skin-dried molds made? Loam molds?

11. What qualities should dry-sand cores possess?

12. Give two compositions of core mixtures.

13. Is there any objection to using the same pattern for both iron and aluminum castings? State reasons.

14. Describe tests for determining moisture and clay contents of foundry sand.

15. How is the permeability of molding sand measured?

16. How does a sand mixer operate, and for what purpose is it used?

17. Show by sketch how a mold is made with a sweep pattern.

18. Name and sketch five kinds of cores.

19. List six binders used in core making.

20. Under what conditions would you recommend the use of metal molds?

21. Describe the various types of machines used for making cores.

22. Describe the operation of an oscillating-squeeze molding machine.

23. What is a sandslinger, and how does it operate?

BIBLIOGRAPHY

CAMPBELL, H. L., *Metal Castings*, John Wiley & Sons, 1936.

Cast Metals Handbook, 3d edition, American Foundrymen's Association.

LAING, J., and ROLFE, R. T., *A Manual of Foundry Practice*, Sherwood Press, 1934.

LINCOLN, R. F., "Fundamentals of Core Blowing," *The Foundry*, February, March, 1940.

MELMOTH, F. A., "Variables in Steel Foundry Practice," *Trans. AFA*, Vol. 36, p. 323, 1928.

Steel Castings Handbook, Steel Founders' Society of America, 1941.

STIMPSON, GRAY, GRENNAN, *Foundry Work*, American Technical Society, 1932.

WENDT, R. E., *Foundry Work*, 4th edition, McGraw-Hill Book Company, 1942.

CHAPTER 2

PATTERN WORK

The first step in making a casting is to prepare a model, known as a pattern, which differs in a number of respects from the resulting casting. These differences, known as pattern allowances, compensate for metal shrinkage, provide sufficient metal for machined surfaces, and facilitate molding. A thorough understanding of these allowances is necessary for successful pattern design and construction.

Most patterns are made of wood because of its cheapness and ability to be worked easily. Also, only a small percentage of patterns go into quantity production work, and therefore the majority do not need to be made of material that will stand hard usage in the foundry. Where durability and strength are required, patterns are made from metal, usually aluminum alloy, brass, or magnesium alloy. In large work, steel or cast-iron patterns may be preferred. For metal patterns, a wooden master pattern must first be made from which the metal pattern is cast.

Before a pattern is made, the pattern maker must visualize from the blueprint what the casting will look like when completed and how it can best be molded. This preliminary estimate is important, as the molding expense in the foundry depends to a great extent on proper pattern construction. After the molding procedure and the general form the pattern will take have been decided upon, a layout of the pattern as it will be built is made. Such layouts are, in general, reproductions of the detail on the drawings submitted laid out to full-size scale.

A flat smooth board of suitable size is selected for the layout and dressed so that at least two of the edges are smooth and square. The layout is made on this board with the square, shrink rule, marking gage, dividers, and a pocket knife. The usual practice is to make all lines in the layout with a sharp-pointed tool, as such practice tends toward narrow lines and insures greater accuracy than is obtained by using a soft lead pencil. All necessary pattern allowances are taken into account in the making of this layout, and, when complete, it serves as a full-sized detail from which the pattern maker may easily check his work.

39

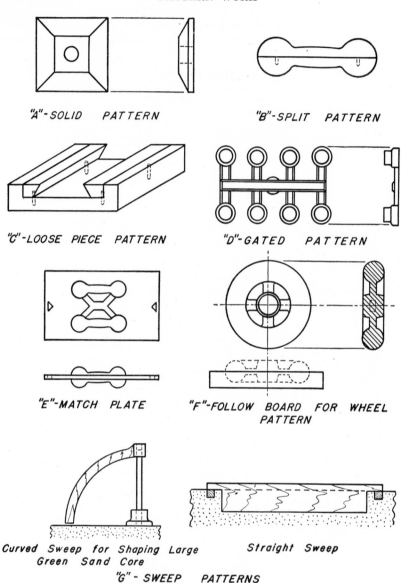

"A"-SOLID PATTERN   "B"-SPLIT PATTERN

"C"-LOOSE PIECE PATTERN   "D"-GATED PATTERN

"E"-MATCH PLATE   "F"-FOLLOW BOARD FOR WHEEL
                                  PATTERN

Curved Sweep for Shaping Large   Straight Sweep
    Green Sand Core
              "G" - SWEEP PATTERNS

Fig. 1. Types of Patterns.

Types of Patterns

In Figure 1 are shown seven types of pattern construction. The
simplest form is the solid or single-piece pattern shown at *A*. •Many
patterns cannot be made in a single piece because of the difficulties

encountered in molding them. To eliminate this difficulty, some patterns are made in two parts, as shown in the figure at *B*, so that half of the pattern will rest in the lower part of the mold and half in the upper part. The split in the pattern occurs at the parting line of the mold. At *C* is shown a pattern with two loose pieces which are necessary to facilitate withdrawing it from the mold. The method of constructing this pattern is discussed later in connection with Figure 3. In production work where many castings are required, *gated patterns*, as shown in *D*, may be used. Such patterns are made of metal to give them strength and to eliminate any warping tendency. The gates or runners for the molten metal are formed by the connecting parts between the individual patterns. *Match plates* provide a substantial mounting for patterns and are widely used in connection with machine molding. At *E* is shown such a plate, upon which are mounted the patterns for two small dumbbells. It consists of a flat metal or wooden plate, to which the patterns and gate are permanently fastened. On either end of such plates are holes to fit onto a standard flask. The *follow board*, which is shown at *F*, may be used either with single- or multiple-gated patterns. Patterns requiring follow boards are usually somewhat complicated and difficult to make as a split pattern. The board is routed out so that the pattern rests in it up to the desired parting line, and this board then acts as a molding board for the first molding operation. Many molds of regular shape may be constructed using *sweep* patterns. Two examples of sweeps are illustrated at *G* in the figure. The curved sweep might be used to form part of the mold for a large cast-iron kettle; the straight sweep for any type of groove or ridge. The principal advantage of this type of pattern is that it eliminates expensive pattern construction.

The type of pattern to be made for a given part depends largely on the judgment and experience of the pattern maker. Pattern cost and number of castings to be made also help govern this decision. When only a few castings are to be made, it is quite obvious that the pattern should be constructed in the cheapest manner possible. In such a case a single pattern of wood construction would best serve the purpose. Single wood patterns may also be used with economy in the production of large castings for two reasons: (1) Wood is a light material, and the pattern is easy to handle; (2) Large castings are usually cast singly in a mold, and a multiple or gated pattern would only increase molding and casting difficulties. In large castings there is a distinct saving in pattern cost if the sweep or skeleton type of pattern can be used. This type can well be used for large patterns having a uniform symmetrical section. Practically all high-production work on molding machines use the match-plate type of

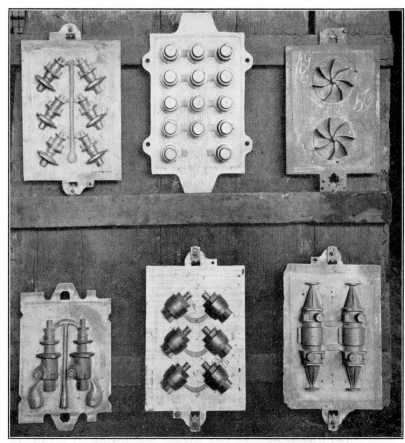

Courtesy The Osborne Manufacturing Company.

FIG. 2. Typical Production Match Plates.

pattern. Aside from the fact that several castings may be molded simultaneously with the pattern of this type, there are also numerous savings effected by the machine molding. Typical production match plates of lightweight metal are shown in Figure 2. Although expensive to make, such patterns will last a long time under severe use.

Pattern Allowances

The method of making a mold has been discussed in Chapter 1. The question naturally arises as to why a finished gear blank or any other object could not be used for making molds without the trouble and expense of making a special pattern. This is not practical for several reasons.

Shrinkage. When any metal cools it naturally shrinks in size. Hence, if the object or model itself were used for the pattern, the resulting casting would be slightly smaller than desired. To compensate for this possibility, a *shrink rule* must be used in laying out the measurements for the pattern. A shrink rule for cast iron is 1/8 inch longer per foot than a standard rule, which is the average shrinkage for cast iron. If the gear blank was planned to have an outside diameter of 6 inches when finished, the shrink rule in measuring it 6 inches would actually make it $6\frac{1}{16}$ inches in diameter, thus compensating for the shrinkage. Such a rule naturally saves a great deal of time which would otherwise be used in computing the proper shrinkage of the various dimensions. The shrinkage for brass varies with its composition but is usually close to 3/16 inch per foot. For steel the shrinkage is 1/4 inch per foot and for aluminum 5/32 inch per foot. It is necessary, therefore, that the pattern maker have several shrink rules available.

When metal patterns are to be cast from original patterns, double shrinkage must be allowed. For example, if the metal pattern is to be made of aluminum and the resulting castings of cast iron, the shrinkage on the original wood pattern would have to be 5/32 inch plus 1/8 or 9/32 inch per foot.

Draft. When a pattern is drawn from a mold, there is always some danger of tearing away the edges of the mold in contact with the pattern. This tendency is greatly decreased if the surfaces of the pattern, parallel to the direction it is being withdrawn, are given a slight taper. This tapering of the sides of the pattern, known as *draft*, is done to provide a slight clearance for the pattern as it is lifted up.

The amount of draft on exterior surfaces is about 1/8 to 1/4 inch per foot. On interior holes which are fairly small the draft should be larger; for such conditions it is usually around 3/4 inch per foot. These figures are influenced considerably by the size of the pattern and the method to be used in molding it. In allowing for draft, the usual practice is to add it to the pattern; that is, the top dimensions would be slightly larger than they would be if no draft were allowed. Draft should be kept to a minimum and an effort made to maintain uniform metal thickness.

Finish. When a draftsman draws up the details of a part to be made, each surface to be machined is indicated by a finish mark.

This mark, a small f drawn on the edge of the surface to be machined, indicates to the pattern maker that additional metal must be provided at this point so that there will be some metal to machine. In other words, surfaces of parts that have to be machined must be made thicker. The amount that is to be added to the pattern depends on the size and shape of the casting, but, in general, the allowance for small and average-size castings is 1/8 inch. When patterns are several feet long, this allowance must be increased because of the tendency of castings to warp in cooling. A pattern maker soon learns from experience the proper allowances that he must make for various conditions. It must be kept in mind that the term finish does not in any way apply to the sanding or finishing of the pattern itself.

Distortion. This allowance applies only to those castings of irregular shape which are distorted in the process of cooling. The distortion is a result of metal shrinkage and is influenced by the mold design and material. A casting in the form of a letter U will contract at the closed end upon cooling, while the open end will be held by the sand in fixed position. Hence the legs of the U pattern should converge slightly so that when the casting is made the sides will be parallel. Such an allowance depends upon the judgment and experience of the pattern maker as well as on a knowledge of the shrinkage characteristics of the metal.

Shake. When a pattern is rapped in the mold before it is withdrawn, the cavity in the mold is slightly increased. In an average size casting this increase in the size can be ignored, but, in large castings or ones that must fit together without machining, shake allowance should be considered. This is accomplished by making the pattern slightly smaller to compensate for the rapping of the mold. No figures can be given for this allowance, as it depends on a number of conditions and must be left to the judgment of the pattern maker.

Method of Constructing a Solid Pattern

The details of a cast-iron V block are shown in Figure 3A. The first step in making this pattern is to make a layout of the part, taking into account the various allowances. Such a layout is shown in part B of the figure. In making this layout the end view is drawn first, using a shrink rule. As the detail calls for " finish " all over, an additional amount of metal must be provided which is shown by the second outline of the V block on the layout. In providing for the draft, consideration must be given to the method of molding the pattern. This is shown at D in the figure. A slight taper is provided on all vertical surfaces of the pattern to facilitate its removal from the sand. The final outline on the layout board represents the actual size and shape of the pattern and is the outline that is used for constructing the pattern.

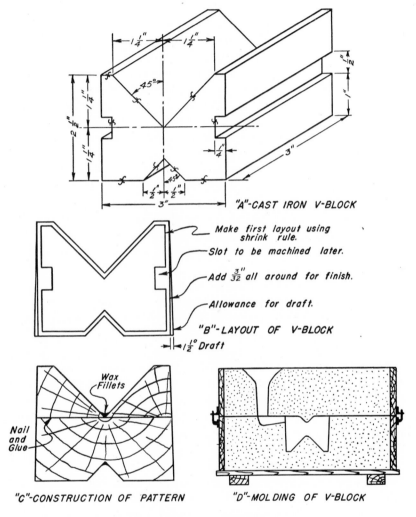

"A"-CAST IRON V-BLOCK

Make first layout using shrink rule.
Slot to be machined later.
Add $\frac{3}{32}$" all around for finish.
Allowance for draft.
"B"-LAYOUT OF V-BLOCK
$1\frac{1}{2}°$ Draft

Wax Fillets
Nail and Glue
"C"-CONSTRUCTION OF PATTERN   "D"-MOLDING OF V-BLOCK

FIG. 3.   Method of Constructing and Molding a Cast-Iron V Block.

The method of constructing the V-block pattern is shown at C. The three parts are nailed and glued together, and sharp interior corners are filleted to eliminate the tendency for metal shrinkage cracks to develop at such points. This same procedure should be followed in making any pattern which requires some degree of accuracy.

Fillets

A *fillet* is a concave connecting surface or the rounding out of a corner at two intersecting planes. In all castings sharp corners should be avoided for several reasons. Rounded corners and fillets assist materially in molding, as there is less tendency for the sand to break out when the pattern is drawn. The metal flows into the mold more easily, and there is less danger of its washing sand into the mold. The appearance of the casting is improved, and it is generally stronger, having fewer internal or shrinkage strains.

A casting in a mold cools on the outside first. As the cooling progresses to the center, the grains of the metal arrange themselves normal to the surface in a dendritic structure. In patterns which have sharp corners, there is a tendency for the metal at the corners to open up because of shrinkage. In patterns with rounded corners this tendency is eliminated, and a sound casting is the result.

Fillets are made of wood, leather, metal, or wax. In lathe work the fillets can usually be taken care of very well in the turning. Wood fillets may also be used in other shapes, but if they are made with a feather edge they are quite fragile. Wood fillets which are inserted in the corner are more durable but require considerably more labor to make and put in place.

For irregular shapes and patterns which are to be subject to considerable use in the foundry, leather fillets have been found to be very satisfactory. They are cut to the desired length, laid face down on a flat surface, and brushed over with glue. They are then put on the pattern and rubbed into place, a fillet tool of the proper curvature being used. Excess glue that is squeezed out at the edges should be wiped off immediately. On large fillets a few brads will help in holding the leather in place.

Wax fillets are used a great deal on small work, because they are cheap and easy to apply. The most convenient way to have this wax is in strips which have been extruded to proper shape. A strip is laid in a corner and formed into place by means of a fillet tool, the end of which has previously been heated. This melts the wax and forms it to the curve of the tool being used. All surface wax should be removed and the surface of the fillet smoothed over by being rubbed lightly with a rag moistened with gasoline. Wax strings of various diameters are easily made by placing some wax in a warm plunger and pressing it out through a hole of the desired diameter.

Sanding and Shellacking Patterns

Since the best possible finish is desired on a casting, it is important that the pattern itself be carefully finished so as to produce a smooth mold. No special allowance is made for the sanding that is required prior to shellacking or varnishing. All that is necessary is to remove the tool marks and other slight irregularities on the wood surface. The sanding operation should always be the last. No tool work should be done later, because the cutting edge would be spoiled by the small particles of sand imbedded in the wood.

Sandpaper is made in a variety of sizes, from very fine to coarse. The sizes are expressed numerically as: no. 000, no. 00, no. 0, no. 1/2, no. 1, no. $1\frac{1}{2}$, no. 2, no. $2\frac{1}{2}$, and no. 3. The last-mentioned figures are the coarse grades and are seldom used for pattern work, the best grades for this work being no. 0, no. 1/2, and no. 1. In finishing a pattern it is best to start with a fairly coarse grade, which will remove the surface irregularities faster than will a fine grade. When sanding on a flat surface, the paper should be backed up and held by means of a small wooden block. Best results are obtained by moving the block back and forth across the grain. On edge surfaces, sanding should be along the direction of its length to avoid curving the edges. When the surface appears to be perfectly smooth it is ready to receive its first coat of shellac.

Shellac seems to be the best material for finishing patterns. It fills the pores of the wood, gives a smooth finish, and leaves a surface that is impervious to moisture. *Shellac,* a resinous material formed by the insect *Coccus lacca,* is found on the branches of certain trees. These insects cover themselves with the sap from the trees, and the resulting product is a resinous scale on the branches. This material is gathered, crushed, and then washed thoroughly with hot water. The final product is melted and poured into thin sheets. The shellac for pattern work is obtained by dissolving this shellac gum with alcohol, the proportion being about 3 pounds of gum to 1 gallon of alcohol. In preparing it, an earthenware, glass, or aluminum container should be used, because most metal containers cause some oxidation, which discolors the shellac. The container should be as nearly air-tight as possible, to prevent evaporation of the alcohol.

Shellac can be applied either by spraying or by using a brush, although in most cases the latter method is used. It dries very quickly, and in a few hours the pattern can be handled. After one coat, the surface will appear quite rough owing to the raised grain of the wood. The pattern should then be sanded once more with a fine grade of sandpaper and a second coat given. Three coats are usually required to give a good surface.

Shellac in its natural state is orange-colored; however, various colors can be obtained by mixing coloring matter with it. Lampblack is used for black shellac. When pigment or lampblack is used, it should first be mixed with alcohol, after which the shellac is added and the mixture thoroughly stirred.

Many foundries have a color scheme of their own for indicating the kind of metal to be cast, core prints, and the like. The following color scheme for all foundry patterns and core boxes of wood construction is recommended by the American Standards Association and is in general use:

1. Surfaces to be left unfinished are to be painted black.
2. Surfaces to be machined are to be painted red.
3. Seats of and for loose pieces are to be marked by red stripes on a yellow background.
4. Core prints and seats for loose core prints are to be painted yellow.
5. Stop-offs are to be indicated by diagonal black stripes on a yellow base.*

Hand Tools Used in Pattern Work

The tools used in pattern work are essentially the same as those of the carpenter and cabinet maker. However, certain requirements of accuracy and special pattern allowances make the work of the pattern maker more detailed. It is assumed that those starting out in shop work have had some manual-training experience and have already a fundamental knowledge in the use of bench tools. Although much of the work in making patterns is done with machine tools, skill in the use of hand tools is an essential accomplishment for a pattern maker. Most patterns of medium and large size are built up from small pieces of wood, necessitating considerable fitting and assembly work. In general, it can be stated that pattern work requires more skill and accuracy in the use of hand tools than is demanded of other types of wood work.

* American Recommended Practice, B45.1 — 1932, American Standards Association.

Wood-Working Machines

In addition to the many hand tools required, there are a number of machine tools that are indispensable to the pattern maker. These tools are great time savers and are especially valuable in preparing stock and working large patterns. Care must be exercised in the operation of all such equipment because of the high speed involved and the rapidity with which the machine will cut. Chief among these machines are the band saw, circular saw, sander, jointer, planer, wood milling machine, combination tool grinder, and the wood lathe.

Band saw. The band saw is adapted to a large variety of work and may be used either for straight or curved cutting. Since many

Courtesy Yates-American Machine Company.

FIG. 4. Saw Table.

patterns are made up of irregular shapes, this machine is indispensable. The adjustable work table makes it possible to cut on an angle and bevel the work.

Circular saw. The circular saw, shown in Figure 4, is provided permanently with two arbors carrying a rip saw and a cross-cut saw, respectively. They are adjusted into cutting position by means of a hand wheel. The table is adjustable to any position up to 45 degrees and is provided with adjustable gages for controlling the width and angle of the cut. This machine is a great help in preparing stock for patterns and may also be used for grooving and making special joints.

Sander. The four main types of sanders used are the disk sander, the vertical spindle sander, the drum sander, and the belt sander.

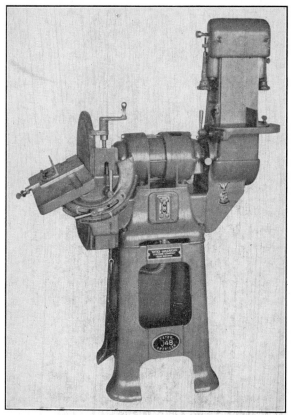

Courtesy Yates-American Machine Company.

FIG. 5. Disk and Belt Sander.

The first one mentioned (shown in Figure 5) is most commonly used and is usually made in combination with one of the other types. In front of the disk is a table for holding the work, capable of being adjusted to various angles although it normally rests at 90 degrees with the disk. The sand or garnet paper is easily replaced, and it cuts rapidly, leaving a very smooth finish. This machine is used to true up pieces and to give a smooth finish to the work. On small patterns the necessary draft may be put on the work by tilting the table slightly.

Jointer and planer. The work of the jointer and planer is quite similar, although each has a well-defined use which justifies its difference in design. Illustrations of these two machines are shown in

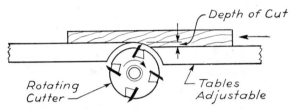

FIG. 6.   Illustrating Operation of a Jointer.

Figures 6 and 7. The rotating cutter on a jointer operates underneath the work and is supported between two parallel iron tables, one of which is capable of vertical adjustment on wedge parallels to control the thickness of the cut. As a board is pushed over the cutters, it is planed to a smooth surface on the bottom side. Likewise,

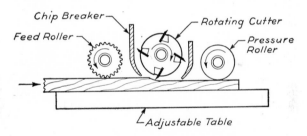

FIG. 7.   Planer or Single Surfacer.

by turning the board over, the other side could be dressed smooth; but the board would not necessarily be of uniform thickness. In order to have both sides of the board parallel and of a definite thickness, the last operation must be performed on a single-surface planer. This machine has a horizontal table upon which the board rests, and the rotating cutter is above the board. The previously smoothed surface of the board is placed on the table and the board fed to the cutter by means of a feed roller. A pressure roller is also placed just ahead of the cutter to keep the board in place. As the board passes through the planer, the rotating cutter finishes off the top side of the board parallel to the lower surface.

Wood milling machine. The most highly developed machine for pattern making, and one that has a wide variety of uses, is the wood milling machine. Such a machine is shown in Figure 8. It consists

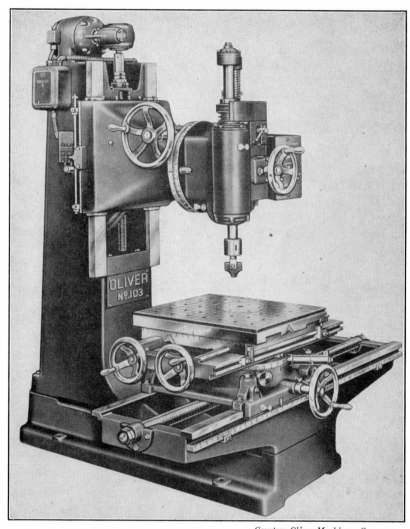

Courtesy Oliver Machinery Company.

FIG. 8.   Wood Milling Machine.

of a vertical rotating spindle, to which cutters may be attached, with an adjustable table beneath. The work is attached to the table and can be fed against the rotating cutter in any direction desired.

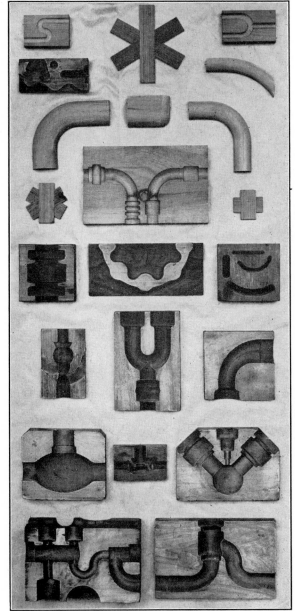

FIG. 9. Specimens of Work Done on " Oliver " Pattern Miller.

It is especially adapted to the working of large patterns and core boxes, but may also be used for gear cutting, making fillets, grooving, forming irregular shapes, and for many other intricate and special jobs. Special machines of this same type have been developed to take care of core-box work. The versatility of this machine is illustrated in Figure 9.

Combination tool grinder. The combination tool grinder, or oil grinder as it is sometimes called, is adapted for sharpening the vari-

Courtesy Yates-American Machine Company.

Fig. 10. Speed Lathe for Wood Turning.

ous kinds of wood tools. This machine has several grinding wheels of various grades as well as a conical-shaped stone for grinding curved tools. A special fixture is mounted in front of the wheels, on which the tool to be sharpened can be mounted and held in a definite position. The carriage holding the tool can be moved back and forth across the face of the stone. After the grinding operation on this machine, the tools must be finished on an oil stone to remove the wire edge and to give the tool a keen cutting edge.

Wood lathe. The wood-turning lathe shown in Figure 10 is a high-speed lathe, much simpler in construction than the metal-turning type. In the best lathes the headstock is connected directly to a motor having a variable-speed controller which gives a speed regulation of 600 to 3000 rpm. The tailstock is adjustable on the lathe bed according to the length of the stock to be turned. Inasmuch as the tools used in turning work are all held by hand, a plain adjustable tool rest is provided.

Two types of work can be done on a lathe, namely, turning between centers and face-plate turning. In turning between centers, the centers of the stock at the ends must first be located. The stock is then driven onto the live center with a wood mallet, and the dead center, which does not turn, is forced into the wood by a forward movement of the tailstock spindle. Both centers should be forced into the wood a sufficient distance to make a deep impression. A little wax or oil on the dead center will prevent squeaking or overheating through friction.

For turning short pieces that cannot be mounted between centers, it is necessary to use a face plate which screws onto the headstock spindle. A variety of sizes of these face plates is usually provided with each lathe, and the work to be turned is held to the face plate by means of wood screws. When the work is so large in diameter that it interferes with the bed of the lathe, the face plate is mounted on the outer end of the spindle. A floor stand for supporting the tools is provided for such work.

Face-Plate Turning

Face plates vary considerably in diameter. The small ones, usually called screw plates, are provided with only one screw hole through the center. As the diameter of the work increases, more screws are necessary to hold the work securely, and larger plates are used.

In order to insure having the axis of the pattern at right angles to the face plate, it is usually advisable to place an intermediate thickness of wood on the face plate between it and the pattern. This wooden plate should be faced off so that it is perpendicular to the axis on the lathe, and then the pattern stock should be attached to it with screws.

Assume that a pattern for the gear blank shown in Figure 11 is to be turned. A face plate, as shown at B, is used for a pattern of this size. The pattern stock should be prepared by cutting it out on a band saw slightly larger than the finished diameter and by smoothing off one side on the sander. The work should then be screwed to the wood face plate by one or more screws. With this mounting most of the outside diameter and one side can be finished. A template should be provided to check the size of the recess in the pattern.

After this much of the pattern is made, it should be removed from the face plate. The face plate should then be recessed, as shown at

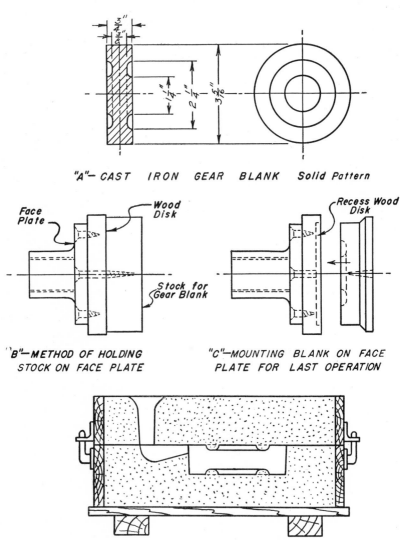

"A"— CAST IRON GEAR BLANK Solid Pattern

"B"—METHOD OF HOLDING STOCK ON FACE PLATE

"C"—MOUNTING BLANK ON FACE PLATE FOR LAST OPERATION

"D"— METHOD OF MAKING AND MOLDING A SOLID PATTERN GEAR BLANK

Fig. 11. Construction and Method of Molding Pattern for Solid Gear Blank.

C, so that the finished diameter of the gear blank fits in closely. This insures that the pattern will be fastened perfectly true and concentric with the previous turning. The gear blank may now be completed in the same manner as the first half. The method of molding the gear blank in the foundry is shown at D.

Patterns with Loose Pieces

In some cases patterns have to be made with projections or overhanging parts so that it is impossible to remove them from the sand,

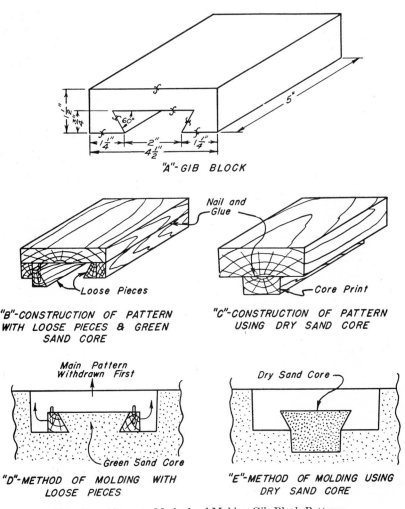

"A"- GIB BLOCK

"B"-CONSTRUCTION OF PATTERN
WITH LOOSE PIECES & GREEN
SAND CORE

"C"-CONSTRUCTION OF PATTERN
USING DRY SAND CORE

"D"-METHOD OF MOLDING WITH
LOOSE PIECES

"E"-METHOD OF MOLDING USING
DRY SAND CORE

FIG. 12. Alternate Methods of Making Gib-Block Patterns.

even though they are parted.   In such patterns the projections have
to be fastened loosely to the main pattern by means of wooden or
wire dowel pins.   When the mold is being made, such loose pieces
remain in the mold until after the pattern is withdrawn and are then
drawn out separately through the cavity formed by the main pattern.

The use of loose pieces is illustrated in the pattern for a gib casting
which fits over a dovetailed slide.   See Figure 12A for a detail of
this casting.

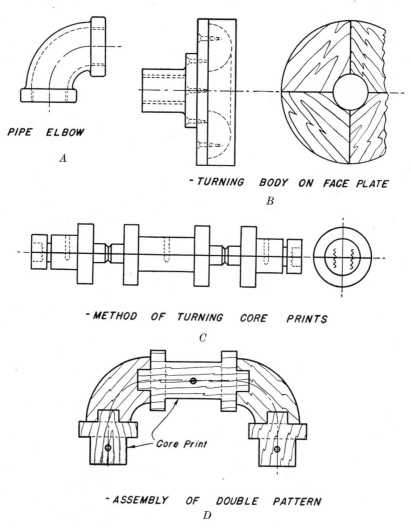

PIPE ELBOW

A

- TURNING   BODY ON FACE PLATE

B

- METHOD OF TURNING CORE PRINTS

C

Core Print

- ASSEMBLY OF DOUBLE PATTERN

D

FIG. 13.   Construction of Pattern for Pipe Elbow.

In starting out on this pattern it is necessary first to make a layout, with all allowances for draft, finish, and shrinkage. Next it should be decided how the pattern is to be molded. Two methods are possible, as illustrated at *D* and *E* of the figure. The first method requires two loose pieces to facilitate withdrawing the pattern from the sand. The main pattern is first withdrawn, leaving the two loose pieces in the sand. These pieces may then be withdrawn from the sand owing to the additional space occupied by the main part of the pattern. The pattern constructed in this manner is made up of five pieces, as shown in the figure at *B*.

The loose pieces may be eliminated by using a dry-sand core. If such construction is desired, the pattern would then be made as shown at *C*. In addition a core box would be necessary. This latter method is less economical because of the expense involved in making the core box and core.

Patterns for Pipe Fittings

The pipe elbow shown in Figure 13 is typical of many cast fittings. In production work, the patterns would be mounted on a match plate, but for a few castings a split pattern would be used.

The round part of the elbow is turned on a face plate, as shown in *B*. Four pieces of wood are prepared and screwed to the wood face plate and are then turned to shape, a template being used to check the curvature. The work is then sanded and removed from the face plate.

The enlarged parts at each end of elbow are turned separately with the core prints as illustrated at *C*. If only a single pattern is made, it will be necessary to have balanced core prints, but in the best construction a double pattern is made. A section of such a pattern after it is assembled is shown at *D*.

Pattern and Core Box for Jackscrew Base

The detail for this part is shown in Figure 14*A*. Such a pattern requires a rather long core print in order to balance and hold the dry-sand core in place. A layout is first made to take care of the various allowances and to provide a working detail for the pattern and core box. The completed pattern is illustrated at *B*. The body of the pattern is made up of two pieces of stock assembled as shown at *C*. Dowel pins are located and drilled before the turning starts. At completion of the turning the two ends held by corrugated fasteners are sawed off and the necessary draft sanded on the ends of the pattern.

The core box and its construction are shown at D and E. It is designed to mold half of the core at a time. After being baked, the

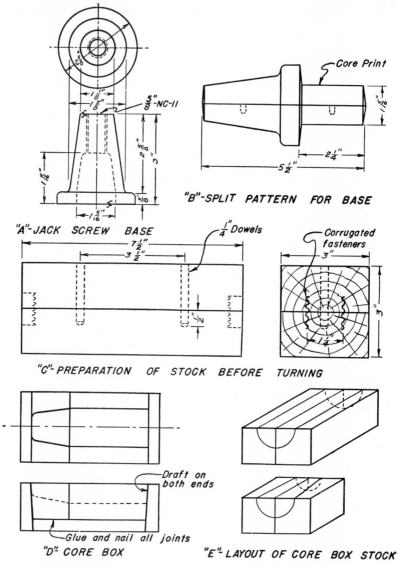

FIG. 14. Construction of Pattern for Jackscrew Base.

halves are assembled and glued together. They are then ready for use in a mold. The two blocks which make up the main part of the box must have their sides square and be carefully laid out as shown in the figure. The layout lines on each block represent the outline of the wood that is to be cut out. When this is done, the two blocks are assembled on a bottom board, and draft is provided at each end of the box by sanding. The box is completed when both end pieces are put on. Any corner that becomes an inside corner in the casting should be filleted.

Kinds of Wood Used in Pattern Work

Although patterns are made from a variety of woods, white pine is employed chiefly for them. This wood is straight grained, light, easy to work, and has little tendency to check or warp. When a more durable wood is necessary for fragile patterns or for hard use, mahogany is preferred. Other woods suitable for this work are cherry, beech, poplar, basswood, and maple. Maple is especially desirable for work on the lathe. Lumber from mature trees is best for pattern work, as the structure of such wood is more compact and less susceptible to shrinkage.

Since wood is the principal material used in pattern work as well as in many other manufacturing processes, a brief description of its properties, characteristics, and preparation is advisable.

Structure of Wood

Wood has a cellular structure. The cells are quite long and grow tightly together, forming a compact porous material. In most woods these cells constitute almost half the volume, which accounts for the comparatively light weight. This cellular structure also gives wood its ability to hold glue and paint. When nails or screws are forced into wood, the elongated cells naturally give some but hold the fastener securely in place. These characteristics all tend to make wood an excellent material for patterns.

The cross section of a log shown in Figure 15 is composed of concentric rings or layers which may readily be distinguished from each other. These rings, caused by a change in the size of the cells, are known as *annular rings*. In the spring, when the growth is more rapid, the cells are larger and sharply distinguished from the last cells formed. This makes it possible to determine the age of the tree, as the rings may easily be counted. Some history of the tree's growth may also be determined by close examination of the rings, since their spacing is wider during seasons of abundant moisture than in seasons of drouth. The rings are seen more easily in trees grown in temperate regions with well-defined seasons than in tropical lands where climatic changes are more gradual.

In addition to the annual ring cells, we may observe in practically all woods straight cells which run in a radial direction through the log. These cells, called *medullary rays*, serve to carry nourishment from the bark to the interior of the wood. They are quite small and in some woods are not discernible without the aid of a microscope.

As a tree grows, it continually forms new cells under the bark and at the tips of the roots and branches. The wood just under the bark is usually termed *sapwood* and is not nearly so dense as the interior wood, or *heartwood*. This is due to the fact that, as the tree

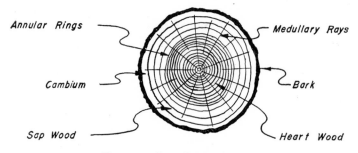

FIG. 15. Cross Section of Log.

grows older, the heartwood cells gradually become filled with resins and other materials. This causes the wood to become darker in color and more resistant to weathering. The percentage of sapwood to heartwood varies with different woods and in some cases may be as high as 50–50.

The type of grain in a wood is frequently important in determining the use. For pattern work a straight-grained wood such as white pine is usually preferred. The grain in this wood is fine in structure and runs uniformly in direction, making it an easy wood to work. Usually woods classified as fine-grained have their annual rings close together. The spacing of the rings is not the only determining factor, however, for many woods, such as oak and walnut, have large pores and yet have closely spaced rings. These woods are classified as *open-grained* woods, whereas such woods as white pine and gum are *close-grained*. Many other woods have peculiar grain structure, characterized by such names as *spiral*, *curly*, *wavy*, and *birds'-eye*. These peculiarities are highly desirable for many uses, such as furniture, because of the beauty they give to the wood. Special methods of sawing are used to bring out the various grain figures.

Seasoning of Lumber

Before lumber can be used commercially it must be dried or seasoned — that is, the moisture must be taken out of the pores or cells of the wood. This may be done by either natural or artificial means. Frequently both methods are used, the lumber being first *air-seasoned* for a period of time and then *kiln-dried*.

Air drying consists of piling the lumber in such a manner that air can circulate around all the boards. This is accomplished by having the pile elevated above the ground and piling the boards up with cross pieces or " stickers " between each layer. It is important to have a rigid foundation to prevent sagging of the pile. The pile should also be given a slight pitch so as to drain off any rain.

The length of time for air seasoning will of course vary with such factors as the size and kind of wood, weather conditions, season of year cut, and manner piled. Because of these variables, the time may be anywhere from 3 months to 2 years. The process of air seasoning is said to be complete when the wood ceases to lose weight. However, it may still contain 10 to 15% moisture. If this amount is excessive, further reduction may be accomplished by inside storing or by kiln drying.

The objects of seasoning woods are several. One of the most important, from the standpoint of the lumberman, is to reduce the weight of the wood. The moisture content in many woods runs as high as 50% of the total weight. This fact obviously makes seasoning profitable from the standpoint of saving on freight charges. Seasoned wood is also less susceptile to decay or discoloration than green wood, in which defects develop rapidly, especially when the bark is left on the logs. Seasoning also makes the wood fit for commercial use by reducing the possibility of shrinkage or checking. If the moisture content is reduced to the point where there can be little further change in weight, warping of the finished product will be reduced to a minimum.

Kiln drying accomplishes the same results as air seasoning, only in a much shorter time. It is the process of drying out the lumber by the application of heat in a kiln. This method will accomplish in several days what it requires several months to do by natural seasoning. In addition to materially reducing the time, this process makes it possible to reduce the moisture content to a minimum, considerably below what could be attained by air seasoning. Because of this reduced time in the kiln, there is little chance for any deterioration of the wood. The process also drives off the volatile matter in the wood, hardens the resin, and to some extent makes the wood less susceptible to future absorption of moisture. A minimum moisture content in wood is an advantage in pattern work, because the wood is in better condition for taking glue and shellac and less likely to change in shape.

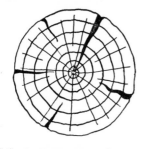 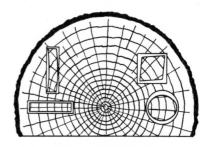

"A" SHRINKAGE CRACKS IN LOG

"B" EFFECT OF ANNUAL RINGS ON SHRINKAGE

 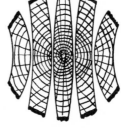

"C" DIRECTION OF WARP IN BOARDS

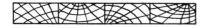

"D" METHOD OF GLUING BOARDS

FIG. 16.   Warping in Lumber and Method of Gluing.

Shrinkage and Warping of Wood

As previously stated, shrinkage in wood can be reduced to a minimum, but owing to wood's peculiar structure can never be entirely eliminated. With atmospheric changes the cells in the wood absorb or give off moisture with accompanying swelling or shrinkage. Figure 16*A*, showing the cross section of a log, displays the results of shrinkage. As may be noted, all the shrinkage takes place in the direction of the annular rings and not along the medullary rays, resulting in the checks around the circumference of the log. If this log were straight-sawed before seasoning, as shown at *C*, and allowed to dry, this shrinkage action along the annual rings would cause a warpage of the boards, as shown in the illustration at *D*. Note that the warping is away from the heart of the log and that the boards that are cut parallel to the medullary rays (" quarter-sawed ") have the least warp. At *B* is shown the effect of shrinkage on sections taken at various places in the section of a log. It is well to keep these facts in mind when patterns have to be built up by gluing boards together. It is advisable to alternate the grain as shown at *D* so as to eliminate the possibility of a serious warp in the finished assembly.

REVIEW QUESTIONS

1. What is a pattern? How does pattern work differ from cabinet work?

2. Name and sketch six types of patterns.

3. List and define the various allowances that must be considered in making a pattern.

4. What is a shrink rule? How does it differ from a standard rule?

5. What is meant by *draft* as applied to pattern work? Must all surfaces on a pattern have draft?

6. Under what circumstances would you recommend metal patterns?

7. What is shellac? Why is it used, and how is it applied?

8. Why should sharp corners on castings be avoided?

9. What are the principal woods used for patterns?

10. Name the essential wood-working machines for a pattern shop.

11. Describe the following tools and state what they are used for: back saw, smooth plane, bevel square, paring gouge, and fillet tool.

12. What is the difference between a jointer and a planer?

13. Why is wood seasoned, and how is it done?

14. What causes a board to warp? Show by sketch the direction that a board will warp.

15. What is the color scheme usually used in shellacking patterns?

16. Under what conditions are loose pieces used on patterns?

17. What is the proper cutting angle for wood tools?

18. Distinguish among a core, core print, and a core box.

19. How does a gated pattern differ from a match plate?

20. For what purpose is a " master pattern " used?

BIBLIOGRAPHY

" Cost of Castings Reduced by Skeleton Patterns," *The Iron Age*, August 20, 1936.

" Double End Tenoners and Their use," *ASME, Trans.*, 1932.

FAWCETT, L. H., " The Influence of Design on Brass and Bronze Castings," *Trans. AFA*, Vol. 40, pp. 360–74, 1932.

HANLEY, E. C., *Wood Pattern Making*, The Bruce Publishing Company, 1924.

HARBISON, C. B., " Designers Should Cooperate with Foundrymen," *Steel*, June 22, 1936, pp. 54, 56.

HOLLAND, KILEY, *Pattern Design*, 1st edition, International Text Book Company, 1939.

McCASLIN, H. J., *Wood Pattern Making*, 2d edition, McGraw-Hill Book Company, 1932.

MELMOUTH, F. A., " Design," *Product Engineering*, April 1935.

" Recommendations for Design of Non-Ferrous Castings," Report of AFA Non-Ferrous Division Committee, *Trans. AFA*, Vol. 40, pp. 518–26, 1932.

RICHARDS, W. H., *Principles of Pattern and Foundry Practice*, McGraw-Hill Book Company, 1930.

RITCHEY, MONROE, HALL, BEESE, *Pattern Making*, American Technical Society, 1933.

WHEELER, K. V., " Foundry Factors Affecting Steel Casting Design," *Trans. AFA*, Vol. 40, pp. 125–52, 1932.

CHAPTER 3

METAL CASTING

Cast Iron

Cast iron is a general term applied to a wide range of iron–carbon–silicon alloys in combination with smaller percentages of several other elements. It is an iron alloy containing so much carbon, or its equivalent, that it is not malleable. The principal difference between cast iron, steel, and wrought iron is the carbon content. The approximate carbon limits are:

Cast iron	$C > 2.0\%$
Steel	$C < 2.0\%$, but $> 0.1\%$
Wrought iron	$C < 0.1\%$

Quite obviously, cast iron may have a wide range of properties, since small percentage variations of its elements may cause considerable change in the physical properties. Cast iron should not be thought of as a metal containing a single element, but rather as one having in its composition at least six elements. All cast irons contain iron, carbon, silicon, manganese, phosphorus, and sulfur. If the iron is in the category of alloy cast iron, still other elements will be present. These elements which go to make up cast iron should not be thought of as impurities, for they all have important effects on the physical properties. Pure iron, known as ferrite, is very soft and has few uses in industrial work. All the desirable properties, such as strength, hardness, and machinability, are controlled by regulating the elements other than ferrite in the cast iron.

The principal raw material for iron castings is *pig iron,* the product of the *blast furnace.* Pig iron is obtained by smelting iron ore with coke and limestone, the final analysis depending primarily on the kind of ore used. Table 1 lists the principal iron ores used in the production of pig iron.

Figure 1 is a diagrammatic view of a blast furnace for producing pig iron. The average blast furnace is about 20 feet in diameter and around 100 feet in height. Daily capacities of such furnaces range from 600 to 1000 tons of pig iron per 24 hours. The raw materials — ore, coke, and limestone — are brought to the top of the furnace with

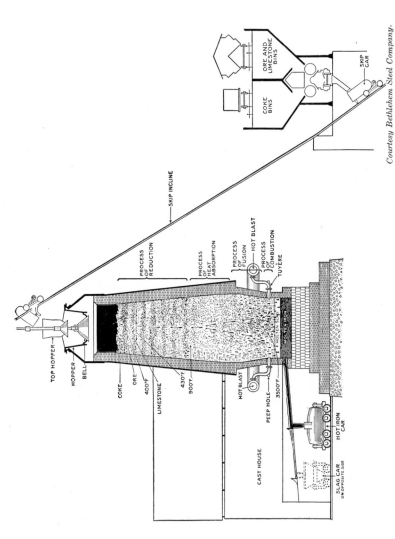

Courtesy Bethlehem Steel Company.

FIG. 1. Sectional View of a Blast Furnace.

a skip hoist and dumped into the double-valve hopper. The hot blast of air enters the furnace through *tuyères* placed around the furnace just above the hearth. As the coke burns, the ore is reduced by contact with the hot carbon monoxide gas. The limestone added with the charge combines with the gangue materials of the ore to

TABLE 1

IRON ORES

Name	Symbol	Color	% Metallic Iron	Location
Hematite	Fe_2O_3	Red	70	Lake Superior District
Magnetite	Fe_3O_4	Black	72.4	N. Y., Ala.; Sweden
Siderite	$FeCO_3$	Brown	48.3	N. Y., Ohio; Germany, England
Limonite	$Fe_2O_3 \cdot X(H_2O)$	Brown	60–65	Eastern U. S., Tex., Mo., Col., France

render it into a fluid slag. Slag floats over the molten iron and is withdrawn at frequent periods; the iron is tapped at intervals of 4 to 6 hours. In addition to the equipment shown in the figure, there are three or four *stoves* — large cylindrical towers for preheating the air blown into the furnace. These stoves are heated by blast-furnace gas taken from the top of the blast furnace and passed through suitable cleaners to remove ashes. The remainder of the gas is washed and used for generating power and as fuel in other furnaces about the plant.

By regulation of operating conditions and proper selection of ore mixtures, the composition of the pig iron can be controlled. Common grades of pig iron produced in the United States are shown in Table 2.

TABLE 2

CLASSIFICATION OF PIG IRON

Grade of Iron	Silicon	Sulfur	Phosphorus	Manganese
No. 1 Foundry	2.5 –3.0	Under 0.035	0.05–1.0	Under 1.0
No. 2 Foundry	2.0 –2.5	Under 0.045	0.05–1.0	Under 1.0
No. 3 Foundry	1.5 –2.0	Under 0.055	0.05–1.0	Under 1.0
Malleable	0.75–1.5	Under 0.050	Under 0.2	Under 1.0
Bessemer	1.0 –2.0	Under 0.050	Under 0.1	Under 1.0
Basic	Under 1.0	Under 0.050	Under 1.0	Under 1.0

Description of a Cupola

The production of iron castings is accomplished by remelting scrap along with pig iron in a furnace called a *cupola*. The construction of this furnace is simple, consisting of a vertical stack lined with a refractory material with provisions for introducing an air blast near the bottom. A cross section of a cupola is shown in Figure 2, with the principal parts labeled.

The entire cupola rests on a circular plate, which is supported above

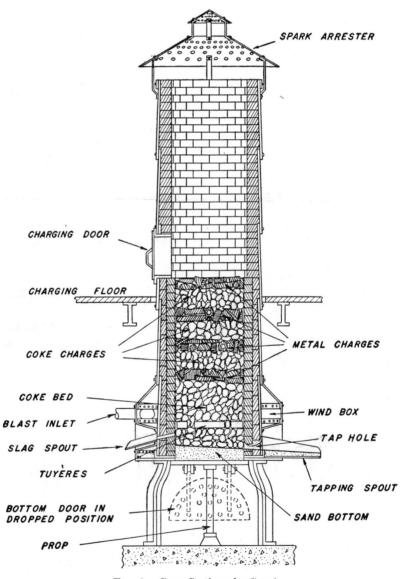

SPARK ARRESTER

CHARGING DOOR

CHARGING FLOOR

COKE CHARGES

METAL CHARGES

COKE BED

BLAST INLET

SLAG SPOUT

TUYÈRES

WIND BOX

TAP HOLE

TAPPING SPOUT

BOTTOM DOOR IN DROPPED POSITION

SAND BOTTOM

PROP

FIG. 2.　Cross Section of a Cupola.

the floor by four columns suitably spaced so that the hinged bottom doors can swing freely without hitting them. In operation, these doors are swung into horizontal position and held in place by a vertical prop. The charging door is located about halfway up the vertical shell, and the top of the cupola is open except for a metal shield or spark arrester.

The openings for introducing the air to the coke bed are known as *tuyères*. Usual practice is to have a single row around the circumference of the wall, although some large cupolas have two rows. The tuyères, flaring in shape with the large end on the inside to cause the air to spread evenly, are placed fairly close together to obtain as nearly uniform air distribution as possible. The number of tuyères varies with the cupola diameter, ranging from four on small cupolas to eight or more on large installations; and the combined area of the inlets is roughly one fourth of the cross-sectional area of the cupola. Normally, the bottom plate of the iron tuyère casting is about 20 inches above the bed of the cupola, although this height will vary according to whether the type of operation is intermittent or continuous. A shallow crucible is satisfactory for long heats, as less coke is required.

Surrounding the cupola at the tuyères is a wind box or jacket for the air supply. Small windows covered with mica are located opposite each tuyère so that conditions in the cupola can be inspected. The air blast, furnished by a positive displacement or centrifugal type of blower, enters the side of the wind jacket.

The opening through which the metal flows to the spout is called the tap hole. The breast surrounding the tap hole is made of fire clay and built up anew before each heat. Opposite the pouring spout at the rear of the cupola is located another spout for slag disposal. This opening is a few inches below the tuyères to prevent slag running into them and also to prevent possible chilling of the slag by the air blast.

Preparing and Charging Cupola

The first operation in preparing a cupola is to clean out the slag and refuse on the lining and around the tuyères from the previous run. Care must be exercised in doing this so as not to damage the refractory lining. Any bad spots or broken bricks are repaired with a daubing mixture of fire clay and silica sand or ganister. Brick and clay in the breast are removed preparatory to rebuilding it with new materials. After the lining is repaired, the bottom doors are swung into position and the prop placed under them. All cracks are closed with fire clay, and a layer of black molding sand is placed on the bottom. This sand is rammed down and given a slope towards the spout, the depth being not less than 4 inches at the lowest point. The breast opening at the spout is made up of a mixture of fire clay and sand, or a separate breast brick can be used. A small tap hole about 3/4 to 1 inch in diameter is provided.

The firing of a cupola is started $2\frac{1}{2}$ to 3 hours before the first metal is to be tapped. Kindling wood is thrown in the charging door after a few flat pieces are placed on the bottom to protect the packed sand. A sufficient amount of wood should be used to ignite a bed of coke. All the tuyères are open when the fire is started, and only a natural draft is used. Coke is added from time to time until the bed is built up to its proper height above the tuyères. The height of the bed coke is important, as it determines the height of the melting zone and affects both the temperature and oxidation of the metal. Another controlling factor is the pressure of the air blast, as with increased pressure a higher bed is necessary. Depending on the aforementioned conditions, the bed height may range from 20 to 50 inches above the top of the tuyères. The height of the bed should be gaged by dropping a gage bar from the charging door to coke level.

As soon as the coke bed is thoroughly ignited, the pig iron and scrap may be charged. The alternate charges of coke and iron are made in a ratio of 1 part of coke to 8 or 10 parts of iron, measured by weight. The ratio depends upon the heating value of the coke, the size of the iron pieces making up the charge, and the metal temperature desired. A ratio of 1 to 8 probably represents average practice. The size of the charge depends on the diameter of the cupola and also on the amount of coke that is necessary to provide sufficient fuel for the charge. In general, a uniform layer thickness of 6 to 8 inches is used between each charge. By calculating the weight of the coke necessary for this layer and knowing the coke–iron ratio, the weight of the iron charge may be obtained. If the coke layer in a small cupola weighs 60 pounds, the iron charge will be 480 pounds.

In addition to charging iron and coke, a fluxing material should also be used if the runs are to be long. The object of adding a flux is to remove impurities in the iron, protect the iron from oxidation, and render the slag more fluid for easy removal from the cupola. Limestone ($CaCO_3$) is the principal fluxing material, although fluorspar (CaF_2) and soda ash (Na_2CO_3) are also used. This material is applied over the coke charges in small lumps not exceeding 2 inches in diameter. Although limestone is the cheapest, fluorspar gives the slag more fluidity. Experience has proved that about 75 pounds of limestone should be used per ton of iron. Frequently fluorspar is added in a ratio of 1 to 3 or 4, replacing part of the limestone. The amount of fluxing materials is subject to considerable variation, depending on the amount of coke ash formed and the cleanness of the metal. Slag that is formed floats on the metal accumulated on the hearth and flows continuously from the slag hole at the rear of the cupola during the heat.

After the cupola is fully charged up to the charging door, it is desirable for the iron to soak in the heat about three fourths of an hour. No forced draft is used during this period, the only draft coming from the tuyère peep holes and the spout opening. Prior to turning on the blast, the tuyère openings should be closed. After the blast has been on a few minutes, molten metal starts accumulating in the hearth. The tap hole is then stopped up until a sufficient amount of molten metal is accumulated in the cupola to warrant pouring operations. During operation the cupola should be kept filled to the charging door by adding successive charges as rapidly as room is provided by the settling of the material. This is important, for the settling of the charges permits a rapid escape of the gases, and the iron that is charged loses the advantage of this heat. The length of a heat may be as long as 16 hours, although most runs are only a few hours in length.

Intermittently the tap hole of the cupola is opened, allowing the metal to flow into a large ladle. It is then closed again with a conical clay plug called a *bot*. This procedure is repeated until all the metal is melted and poured. At the end of the run, the blast is shut off and the prop under the bottom doors knocked down, allowing the remains in the cupola to drop to the floor. The bed of hot iron, slag, and coke is quenched with water as quickly as possible and removed from beneath the cupola. Any coke or iron in the remains is salvaged and taken into account in the cupola calculations for the heat.

Cupola Air Supply

The amount of air required to melt a ton of iron depends on the quality of coke and the coke–iron ratio. Theoretically, 113 cubic feet of air at 14.7 pounds per square inch and at 60F are required for one pound of carbon. For other operating conditions some correction should be made to get the correct volume of air. For coke the previous figure should be reduced slightly to compensate for the ash content. Assuming a 1 to 8 ratio, we find that 250 pounds of coke are required to melt one ton of iron. Multiplying 250 by 105 gives 26,250 cubic feet of air required to melt one ton. Actually in practice perfect combustion is not obtained, and a larger volume of air is required. A value of 30,000 cubic feet of air per ton of iron is frequently used in estimating the capacity of a blower.

The pressure of the blast to be maintained will depend on the size of the cupola, compactness of the charge, kind of iron being melted, and the temperature. Small cupolas may require a pressure of only 5 to 8 ounces; large cupolas may operate as high as 28 ounces. No definite rule for pressure can be laid down; the proper value can be obtained only by actual operating experience.

The best type of blower for cupola operation is the positive-displacement type, because it delivers a constant volume of air to the cupola, irrespective of changing furnace conditions. In making calculations a small air-slippage loss should be taken into account. This loss is a known value for a given blower. Variations in capacity are obtained by speed regulation. With centrifugal blowers the volume to the cupola varies according to the pressure in the wind jacket, which in turn is affected by the height of charge and other conditions in the cupola. Volume may be controlled by a gate in the blast line connected with suitable electric controls. The regulation is obtained by changing the power input of the motor driving the blower.

Brackelsberg Rotating Furnace

This furnace (shown in Figure 3), which burns pulverized coal as its fuel, is used in both American and European foundries for the production of malleable cast iron and high-test gray irons. It consists of a rotating cylindrical shell, open at both ends, which is mounted on a tilting mechanism so that either end may be raised for slagging, pouring, or charging. Close metallurgical control of all the elements is one of the important advantages of this furnace. In addition, the fuel cost is low, and oxidizing and metal losses are reduced to a minimum. If it starts up cold, the first heat requires about 3 hours; following heats require about 2 hours each. Capacities of these furnaces range from 1/2 to 10 tons.

Air Furnace

The air or reverberatory furnace shown in Figure 4 has been widely used for many years for the production of malleable-iron and high-test gray-iron castings. Early furnaces were hand-fired with bituminous coal, but most furnaces of today use pulverized coal or oil as

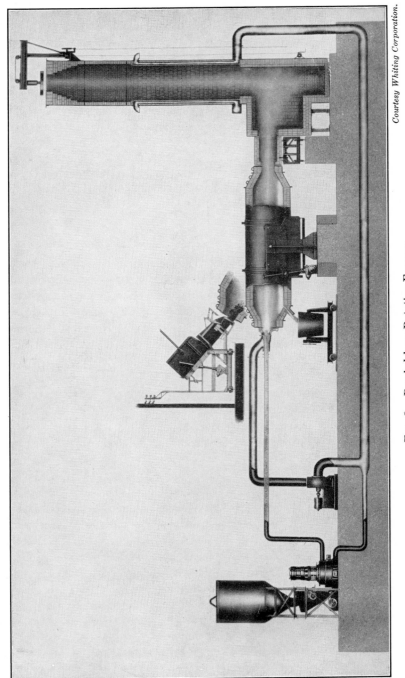

Fig. 3. Brackelsberg Rotating Furnace.

75

the fuel. Charging is done through the roof of the furnace by removing sections of the arch called *bungs*. This furnace lends itself
to close control, since the metal can be tested at intervals. Further-

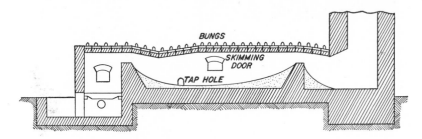

Fig. 4. Air Furnace.

more, the metal is not in contact with the fuel as it is in the cupola
furnace; and the analysis, particularly the carbon content, may be
held to close limits. Since the initial and operating costs of this
furnace are higher than those of a cupola, the air furnace does not
have wide application except for malleable iron castings. Capacities
of these furnaces range from 5 to 50 tons per heat.

Electric-Arc Furnaces

Although electric-arc furnaces are used principally for the production of steel and alloy-steel castings, they are also used to a limited
extent for high-test iron castings.
Furnaces of both the *direct-arc* and
indirect-arc types are used. In the
direct-arc furnace, shown in Figure
5, the current passes from the
electrodes to the metal, through the
metal, and back to the electrodes or
the hearth. Indirect-arc furnaces
have horizontal electrodes above
the metal which heat by radiation.
Heating costs for electric furnaces
are higher than for furnaces of other
types, but this increase may be
counteracted to some extent by using low-priced materials in the
furnace charges. Furthermore, electric furnaces lend themselves to
close temperature control, and the analysis of the metal may be held
to accurate limits.

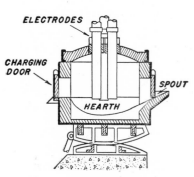

Fig. 5. Direct-Arc Furnace.

Kinds of Cast Iron

1. Direct-iron castings. This iron is a product of the blast furnace and is usually known as pig iron. It is not suitable for most commercial castings until it has been remelted in a cupola or furnace of other type. Furthermore, the output of a blast furnace is so large that it would be difficult to provide sufficient commercial molds to take care of the output.

2. Gray iron. Gray iron is the name given to the ordinary commercial iron, which is so called because of the grayish color of the fracture. This color is due to the carbon's being principally in the form of flake graphite. This iron is easily machined, has a high compression strength, a low tensile strength, and no ductility. The percentages of the several elements may vary considerably, but are usually within the following limits:

Carbon	2.75–3.50
Silicon	0.90–2.75
Manganese	0.40–1.00
Phosphorus	0.15–1.00
Sulfur	0.02–0.15
Iron	Remainder

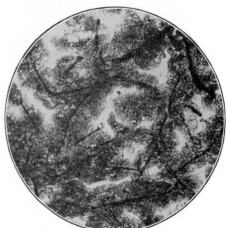

Both from University of Texas Shop Laboratories.

FIG. 6. Structure of Gray Iron (unetched). Magnification ×200.

FIG. 7. Structure of Gray Iron (etched). Magnification ×200.

Figures 6 and 7 are microphotographs showing the structure of gray cast iron. If a specimen is polished and examined under the microscope, the appearance is as shown in Figure 6. The dark lines are small flakes of graphite which greatly impair the strength of the iron. The strength of cast iron is increased if these flakes are small and uniformly distributed throughout the metal. Etching a specimen with a dilute solution of nitric acid results in the structure shown in Figure 7. The light-colored constituent is *ferrite* or pure iron; the other new constituent is known as *pearlite*. Pearlite (composed of alternate lamellae of ferrite and iron carbide) is found in most irons and is similar to the pearlite found in carbon steels. This constituent adds to the strength and wear resistance of the iron. The dark graphite flakes may also be seen in this microphotograph.

3. White iron. This iron shows a white fracture, which is due to the fact that the carbon is in the form of a carbide, Fe_3C. The

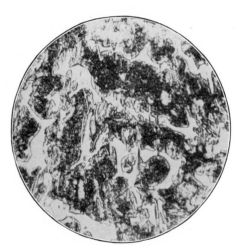

University of Texas Shop Laboratories.

FIG. 8. Structure of White Iron.
Magnification ×200.

carbide, known as *cementite*, is the hardest constituent of iron and makes it hard and brittle. White iron with a large percentage of carbide cannot be machined. The principal constituents shown in the micrograph (Figure 8) are cementite and pearlite. The dark area is the pearlite and the light area the cementite.

White cast iron may be produced by casting against metal chills or by regulating the analysis. Chills are used when a hard wear-resisting surface is wanted for such products as car wheels, rolls for crushing grain, and jaw-crusher plates. The first step in the production of malleable iron is to produce a white-iron casting by controlling the analysis of the metal. One specification* for the production of these castings is as follows:

Carbon	1.75–2.30
Silicon	0.85–1.20
Manganese	Less than 0.40
Phosphorus	Less than 0.20
Sulfur	Less than 0.12
Iron	Balance

* ASTM Specification A47-33, Grade 35018.

4. Mottled cast iron. This is a product intermediary between gray and white cast iron, the name being again derived from the appearance of the fracture. It is obtained in castings where certain wearing surfaces have been chilled. Improperly annealed malleable iron shows a similar mottled structure.

Malleable Cast Iron

Malleable iron castings when first cast require a white iron having all the carbon in the carbide form. Several types of furnaces, including the cupola, air furnace, and Brackelsberg furnace, are used for producing this type of iron. The castings obtained are packed in pots and placed in an annealing oven so arranged as to allow free circulation of heat around each unit. The annealing time lasts from 3 to 4 days at temperatures varying from 1500 to 1850 F. In this process the hard iron carbides are changed into nodules of *temper* or *graphitic carbon* in a matrix of comparatively pure iron, as shown in the micrograph in Figure 9. Malleable iron has a tensile strength of around 54,000 pounds per square inch and an elongation of 18%. Castings have considerable shock resistance

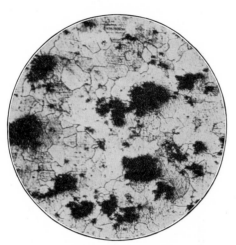

University of Texas Shop Laboratories.

Fig. 9.   Structure of Malleable Cast Iron.
Magnification ×200.

and good machinability. Malleable castings are used principally by the railroad, automotive, pipe-fitting, and agricultural-implement industries.

Effect of Chemical Elements on Cast Iron

Carbon. While any iron containing over 2.0% carbon is in the cast-iron range, gray cast iron has a carbon content of 3 to 4%. The amount depends on the carbon content in the pig iron and scrap used and that absorbed from the coke during the melting process. The final properties of the iron depend not only on the amount of carbon but also on the form in which it exists. The formation of graphitic carbon depends upon slow cooling and upon the silicon content. High silicon promotes the formation of graphitic carbon. Carbon in this state acts as a softener for the iron, reduces the shrinkage, and gives the iron machinability. The strength of the iron increases with the percentage of carbon in the combined form.

Silicon. Silicon up to 3.25%, is a softener in iron and is the predominating element in determining the amounts of combined and graphitic carbon. It combines with iron that otherwise would combine with carbon, thus allowing the carbon to change to the graphitic state. After an equilibrium is reached, additional silicon unites with the ferrite to form a hard compound. Hence silicon above 3.25% acts as a hardener. In melting, the average loss of silicon is about 10% of the total silicon charged into the cupola. High silicon content is recommended for small castings and low for large castings. When used in percentages from 13 to 17, an alloy having acid and corrosion resistance is formed.

Manganese. Manganese does not have an appreciable effect in small amounts, but in amounts over 0.5% it combines with sulfur to form a manganese sulfide. The mixture has a low specific gravity and is eliminated from the metal with the slag. In addition to its acting as a purifier for eliminating sulfur, it also has some action as a deoxidizer. Other effects are to increase fluidity, strength, and hardness of the iron. If the percentage is increased appreciably over the usual amounts, it will promote the formation of combined carbon and rapidly increase the hardness of the iron. Of the manganese originally charged, 10 to 20% is lost in the melting process.

Sulfur. There is nothing good to be said for sulfur in cast iron. It promotes the formation of combined carbon with accompanying hardness and causes the iron to lose fluidity, with resultant blow holes. Sulfur gets into the iron from the ore and also from the coke during the melting process. Each time the iron is remelted there is a slight pickup in sulphur, frequently as much as 0.03%. To counteract this increase, manganese should be added to the charge in the form of ferromanganese briquettes and spiegeleisen.

Phosphorus. The principal effects of phosphorus are to increase the fluidity of the molten metal and to lower the melting temperature. For this reason phosphorus up to 1% is used both in small castings and in those having thin sections. Large castings should have low phosphorus content, as additional fluidity in the iron is not required. There is a slight decrease in strength and shock resistance as the percentage increases. There is little change in the phosphorus content during the melting process, although in some calculations it is assumed that there is a pickup of about 0.02%. The action of any remelting process upon phosphorus is principally one of concentration, as this element does not oxidize readily except under special conditions. In order to control this element, care should be exercised in selecting the grade of scrap that is used.

Phosphorus also forms a constituent known as *steadite*, a mixture of iron and phosphide, which is hard, brittle, and of rather low melting point. It contains about 10% phosphorus, so that an iron with 0.50% phosphorus would have 5% steadite by volume. Steadite appears as a light structureless area under the microscope but may appear as a network if sufficient phosphorus is present.

Calculation of Cupola Charge

Careful consideration must be given to the materials charged into a cupola if a uniform product is desired from day to day. Raw materials vary in composition, and there are also changes that take place during the melting operation. For some elements (silicon and manganese) there is a definite percentage loss due to oxidation. Carbon also has a loss due to oxidation, but this is compensated by absorption from the coke. Sulfur does not suffer any loss from oxidation and actually picks up additional amounts from the coke. There is little change in the phosphorus content. Most cupolas operate on a fairly large percentage of return scrap, which is close to the desired analysis. In computing a charge, however, both return and new scrap must be included as well as the various grades of pig iron. The following problem illustrates the procedure involved in calculating a charge:

Problem: Given the following materials to work with, what would the final iron analysis be, using cupola melting? Assume a 3000-lb charge made up of metals of the following composition:

Iron	Carbon	Silicon	Manganese	Phosphorus	Sulfur
No. 1 pig iron	3.5	2.50	0.72	0.180	0.016
No. 2 pig iron	3.5	3.00	0.63	0.120	0.018
New cast-iron scrap	3.4	2.30	0.50	0.200	0.030
Returns from foundry	3.3	2.50	0.65	0.170	0.035

The material is to be used in the following proportions: No. 1 pig — 10%; No. 2 pig — 20%; returns — 40%; new scrap — 30%.

(a) Carbon content; oxidation loss = gain from coke

No. 1 pig	$3000 \times 0.10 \times 0.035 =$	10.5 lb
No. 2 pig	$3000 \times 0.20 \times 0.035 =$	21.0 lb
New scrap	$3000 \times 0.30 \times 0.034 =$	30.6 lb
Returns	$3000 \times 0.40 \times 0.033 =$	39.6 lb
		101.7 lb

$$\text{Per cent carbon} = \frac{101.7}{3000} \times 100 = 3.39$$

(b) Silicon content; oxidation loss = 10%

No. 1 pig	$3000 \times 0.10 \times 0.025 =$	7.5 lb
No. 2 pig	$3000 \times 0.20 \times 0.030 =$	18.0 lb
New scrap	$3000 \times 0.30 \times 0.023 =$	20.7 lb
Returns	$3000 \times 0.40 \times 0.025 =$	30.0 lb
		76.2 lb

$$\text{Per cent silicon} = \frac{76.2 - (76.2 \times 0.10)}{3000} \times 100 = 2.28$$

(c) Manganese content; oxidation loss = 20%

No. 1 pig	$3000 \times 0.10 \times 0.0072 =$	2.16 lb
No. 2 pig	$3000 \times 0.20 \times 0.0063 =$	3.78 lb
New scrap	$3000 \times 0.30 \times 0.0050 =$	4.50 lb
Returns	$3000 \times 0.40 \times 0.0065 =$	7.80 lb
		18.24 lb

$$\text{Per cent manganese} = \frac{18.24 - (18.24 \times 0.2)}{3000} \times 100 = 0.49$$

(d) Phosphorus; oxidation loss = 0

No. 1 pig	$3000 \times 0.10 \times 0.0018 =$	0.54 lb
No. 2 pig	$3000 \times 0.20 \times 0.0012 =$	0.72 lb
New scrap	$3000 \times 0.30 \times 0.0020 =$	1.80 lb
Returns	$3000 \times 0.40 \times 0.0017 =$	2.04 lb
		5.10 lb

$$\text{Per cent phosphorus} = \frac{5.10}{3000} \times 100 = 0.17$$

(e) Sulfur; oxidation loss = 0. Gain from coke is approximately 4% of sulfur in coke.

No. 1 pig	$3000 \times 0.10 \times 0.00016 =$	0.048 lb
No. 2 pig	$3000 \times 0.20 \times 0.00018 =$	0.108 lb
New scrap	$3000 \times 0.30 \times 0.00030 =$	0.270 lb
Returns	$3000 \times 0.40 \times 0.00035 =$	0.420 lb
		0.846 lb

Assuming a coke to iron melting ratio of 1 to 8 and a coke with a sulfur content of 0.50%, we have

$$375 \times 0.005 = 1.875 \text{ lb of sulfur}$$

$$\text{Pickup } 4\% = 0.075 \text{ lb}$$

$$\text{Then per cent sulfur} = \frac{0.846 + 0.075}{3000} \times 1000 = 0.0307$$

Steel Castings

Steel is a crystalline alloy of iron, carbon, and several other elements, which hardens when quenched above its critical temperature.

It contains no slag and may be cast, rolled, or forged. Carbon is the most important constituent because of its effect on the hardenability and strength of the steel. Steel castings may be classified as follows.*

STEEL CASTINGS

1. Carbon Steel
   - (*a*) Low carbon (less than 0.20%).
   - (*b*) Medium carbon (0.20 to 0.40%).
   - (*c*) High carbon (over 0.40%).

2. Alloy steel
   - (*a*) Low alloys (special alloying elements totaling less than 8.0%).
   - (*b*) High alloys (special alloying elements over 8.0%).

Medium-carbon–steel castings are the most frequently used in the carbon–steel range. They have ductility and good tensile strength in a normalized condition, ranging from 60,000 to 80,000 pounds per square inch. The range of chemical composition is given here:

MEDIUM-CARBON–STEEL CASTINGS

Carbon	0.20–0.40
Manganese	0.50–1.00
Silicon	0.20–0.75
Phosphorus	0.05–maximum
Sulfur	0.06–maximum
Ferrite	Remainder

Alloy castings may contain special elements in addition to those already listed or they may have more than the usual percentage of some normal element. Special elements frequently added to foundry steel are aluminum, nickel, chromium, cobalt, molybdenum, vanadium, and copper. Steels alloyed with silicon and manganese, both normal constituents, are also frequently used. A great variety of steels is possible in this range, differing widely in strength resistance to corrosion, high temperatures, and abrasion.

A typical microstructure of a medium-carbon cast steel is shown in Figure 10. The light areas are ferrite and the dark areas pearlite. The grain structure of most cast steels is large because of the high

* *Cast Metals Handbook*, American Foundrymen's Association.

casting temperature of the metal combined with relatively slow cooling. This defect can be remedied by subsequent heat treatment.

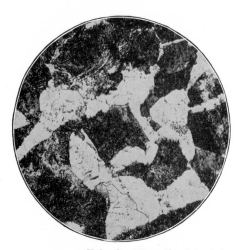

For the production of steel castings, four types of furnaces are used:

1. Open hearth (both acid and basic).
2. Electric (arc and induction).
3. Crucible.
4. Converter (acid).

The largest tonnage is produced in basic open-hearth furnaces. A furnace of this type is shown in Figure 11. Because of the large capacities of these furnaces, ranging from 25 to 100 tons, this process is used principally for large castings. The basic process is preferred to the acid process, because phosphorus can be controlled and sulfur can be partly eliminated.

University of Texas Shop Laboratories.

FIG. 10.   Structure of Medium-Carbon Cast Steel.   Magnification ×200.

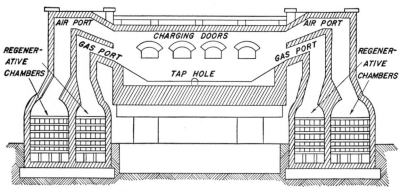

FIG. 11.   Open-Hearth Furnace.

The usual type of electric furnace is the direct-arc type as shown in Figure 5. This is primarily a remelting furnace using steel scrap as the raw material. The acid type of furnace is used when the raw material does not contain high phosphorus or sulfur. Foundry furnaces of this type range from 1/2 to 10 tons per heat. This furnace is the principal one used for small- and medium-steel castings. Electric induction furnaces are used primarily in the production of alloy-steel castings because of the accurate control of melting conditions and composition. These furnaces range in capacity from a few pounds to 4 tons.

The crucible process is the oldest process for making steel, but is little used today for steel castings. Wrought iron, washed metal, steel scrap, charcoal, and ferroalloys constitute the raw materials for this process. These materials are placed in crucibles having a capacity of around 100 pounds and melted in a regenerative furnace.

In making steel castings from a converter, liquid metal or cupola iron is poured into the converter, and the heat for the refining operation is produced by blowing air through the molten metal. The result is the oxidation of the silicon, manganese, and carbon. Sideblower converters, as shown in Figure 12, are used in foundry work and have a capacity of around 2 tons. Both the converter and crucible process have largely been replaced by electric-arc and induction furnaces.

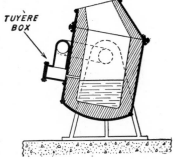

FIG. 12.   Side-Blower Converter.

Since the pouring temperature for steel castings is 2900 to 3200 F, it is necessary to use a highly refractory and permeable sand. Most molds for large and medium castings are either baked or skin-dried to eliminate gas troubles in the mold, but green sand may be used for light and intricate castings. Green sand has the advantage of offering less resistance to the normal contraction of the castings. Large risers must always be used on steel castings to compensate for the large amount of shrinkage.

Pouring and Cleaning Castings

After the molds are made, they are lined in rows on the foundry floor, and ladles of metal are carried to them. For jobbing and small-run production work, where there is considerable variation in mold size, this procedure is the most economical. In large production foundries, engaged in the mass production of castings, the problem of handling the molds and molten metal is solved by placing the molds on conveyors, as shown in Figure 13, and passing them slowly by a pouring station. The pouring station may be located permanently next to the furnace, or metal may be brought to certain points by overhead handling equipment, as shown in the figure. The job illustrated is especially adapted to conveyor handling, as all flasks are of the same size. The conveyor serves as a storage place for the molds while they are being transported to the cleaning room.

Courtesy Steel Founders' Society of America.

FIG. 13.   Filling Molds from Lip-Pour Ladle.

After a casting has solidified and cooled down to a suitable temperature for handling, it is shaken from the mold. Figure 14 shows an unloading station for a conveyor where the molds are shaken out mechanically. This is known as a ventilated mold shakeout, the dust being carried away by the cyclone dust collector, while the sand is collected underneath and transported to the conditioning station. Present-day safety codes frequently require the control of dust conditions. Dust collectors are used also in other casting cleaning operations, such as tumbling and sand blasting.

Nonferrous castings do not offer much of a cleaning problem, as they are poured at lower temperatures than iron or steel, and the sand has little tendency to adhere to the surface. Gates and sprues are cut off either in a sprue press or with a metal band saw. Hand or rotary machine brushing is usually sufficient to prepare the casting for machining operations.

Iron and steel castings, however, offer a great problem, for they are covered with a layer of sand and scale which is somewhat difficult to remove. The gates and risers on iron castings may be broken off, but to remove them from steel castings a cutting torch or a high-

Courtesy American Foundry Equipment Company.

FIG. 14. Cyclone-Dust-Collector Ventilating-Mold Shakeout at Locomotive Finished Materials Company.

speed cutting-off wheel is necessary. Figure 15 shows a high-speed cut-off machine built for removing gates and risers. The casting is held securely in a quick-acting clamping device, and the abrasive wheel used is 20 inches in diameter and 3/4 inch thick — capable of removing risers 6 inches in diameter. Gates 3 inches in diameter may be removed in 20 to 30 seconds, depending on the hardness of the material.

FIG. 15.   High-Speed Cut-off Machine Removing Risers from Casting.

To clean castings, several methods may be used, depending on the size, kind, and shape of the castings. The most common piece of equipment used is the rotating cylindrical tumbling mill. The cleaning is accomplished by the tumbling action of the castings upon one another as the mill rotates. A similar piece of equipment, known as Wheelabrator Tumblast, is shown in Figure 16. This is one of the smaller-sized machines and is recommended for small shops. It will clean 65 to 100 pounds of gray iron or malleable castings in 5 to 8 minutes. Larger machines of this type have capacities of over a ton per charge. The machine consists of a cleaning barrel formed by an endless apron conveyor. The work is tumbled beneath a blasting unit located just above the load, and metallic shot is blasted onto the castings. After striking the load, the shot falls through holes in the conveyor and is carried overhead to a separator and storage hopper. From there it is fed by gravity to the blasting unit. The unit is unloaded by reversing the apron conveyor. A dust collector is installed with the machine to eliminate dust hazards.

Sand-blasting units may be used separately for cleaning castings. Sharp sand is blown against castings inside a blasting cabinet. This

Courtesy American Foundry Equipment Company.

FIG. 16. Phantom View of Wheelabrator Tumblast.

removes all foreign matter completely and gives the casting a clean surface appearance. Castings which are to be plated or galvanized are frequently pickled in a weak acid solution and then rinsed in hot water. Large castings, which are difficult to handle, are often cleaned by hydraulic means. The casting is placed on a rotating table, and streams of water under considerable pressure wash away the sand.

In addition to these cleaning processes, many castings require a certain amount of chipping or grinding to remove surface and edge defects. Stand, portable, and swing-frame grinders are all used for this work. In Figure 17 several swing-frame grinders are shown in operation with dust-exhaust booths provided to prevent dust hazards. Fast free-cutting abrasive wheels, operating at a cutting speed of around 9500 feet per minute, are recommended for this type of grinding. Swinging-frame grinders are used in steel mills for removing defects on ingots and are widely used as well in steel and iron foundries.

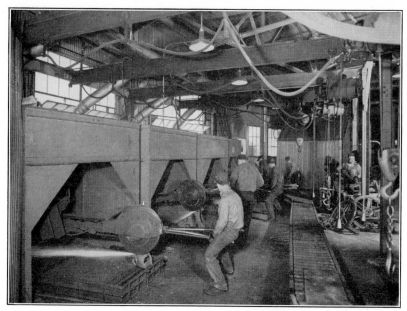

Courtesy Steel Founders' Society of America.

FIG. 17. Cleaning Castings with Swing-Frame Grinders.

Nonferrous Casting

The foundry practice for making nonferrous castings differs little from that used for iron castings. Molds are made in the same way and, in general, by the same tools and equipment, except for the kind of sand and the type of melting furnace used. The molding sand is usually of finer grain size, since most castings are fairly small and a smooth surface is desired. The sand need not be so refractory as sand for iron and steel castings, because the melting temperature for nonferrous alloys is lower.

The crucible furnace shown in Figure 18 is frequently used for this work. The furnace may be either the stationary or the tilting type. Coke is commonly used as the fuel for the stationary-pit furnaces, although oil or gas may be used equally well if available. The latter fuels have the advantage of melting more quickly than coke. Electrical-resistance, indirect-arc, and induction furnaces may be used under certain conditions. Such furnaces possess the advantages of accurate temperature control and low melting losses. Electric furnaces are widely used for laboratory and research work as well as for installations requiring large production.

Crucibles used in nonferrous melting are made of a mixture of graphite and clay. Although these crucibles are quite fragile when cold and must be handled with care, they possess considerable strength when heated. New crucibles contain a small percentage of moisture and should be dried out slowly and uniformly before use. When heated, a crucible becomes somewhat plastic, and serious strains are imposed upon it if the tongs do not fit properly.

In preparing a heat in a coke-fired furnace, as shown in Figure 18, a wood fire is first built in the furnace. From this fire a bed of coke is started and built up to a depth of about 12 inches. The crucible charged with metal is placed on this bed, and small pieces of coke are packed around the crucible up to the level of the top. The cover is then placed on the furnace, and the metal is heated with a natural draft through the bed of coke. As the metal in the crucible is melted, additional metal must be added from time to time. To remove the crucible from the furnace, special tongs are required which conform to the outside of the crucible. Before pouring, any coke, oxides, or other foreign materials should be skimmed from the surface.

Nonferrous Metals and Alloys

The common elements used in nonferrous castings are copper, aluminum, zinc, tin, and lead. Many alloys, however, have small amounts of other elements, such as antimony, phosphorus, manganese, nickel, and silicon.

Two of the most common alloys using copper are brass and bronze. *Brass* is essentially an alloy of copper and zinc. The percentages of each element may vary considerably, but in most cases the zinc percentage ranges from 10 to 40. The strength and hardness of the alloy are increased as the percentage of zinc is raised up to 40. Large percentages of zinc are not desirable, owing to a rapid decrease in strength and the tendency for the zinc to volatilize in melting. An addition of a small percentage of lead increases machinability. Brass has a wide application in industry because of its strength, appearance, resistance to corrosion, and ability to be either rolled, cast, or extruded. Typical brass compositions are shown in Table 3.

Bronze is an alloy having copper and tin as the principle alloying elements. Many alloys classified as bronzes contain large percentages

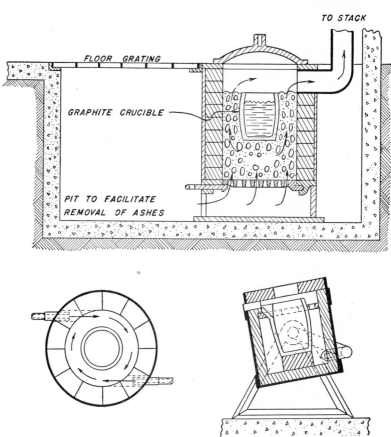

FIG. 18. Crucible Furnace for Nonferrous Metals.

of other elements. Bronze is widely used for bearings because of the hardness imparted to the alloy by the tin. The percentage of tin seldom exceeds 10, since above this amount brittleness increases rapidly with a corresponding decrease in ductility. Table 3 gives several analyses of bronze alloys.

Because of their light weight and ability to resist many forms of corrosion, aluminum alloys have a wide application in industry today. Many of them respond to heat treatment and are suitable where high strength is needed. The usual casting aluminum of 92% aluminum and of 8% copper is widely used for miscellaneous castings. The copper adds to the hardness and strength of the alloy. An aluminum–silicon alloy containing 93–95% aluminum, 9–6% silicon, 0.4% iron is used for gears, propellor blades, and parts requiring resistance to salt water. An important alloy used in airplane work is Duralumin. It contains 95% aluminum, 4% copper, 0.5% manganese, and 0.5% magnesium. This alloy responds to heat treatment and may have a tensile strength as high as 55,000 pounds per square inch.

TABLE 3

COPPER – ZINC – TIN ALLOYS*

Name	Cu	Sn	Zn	Pb	Ni	Si	Mn	Al	Fe	Use
Red brass	90		10							Hardware
Yellow brass	70		30							Cartridges, tubes
Leaded red brass	85	5	5	5						Castings, machinery
Leaded yellow brass	72	1	24	3						Plumbing fixtures
Tin bronze	88	10	2							Bearings, ship hardware
Bell metal	80	20								Bells
Bearing bronze	85	10		5						Machine bearings
Silicon bronze	95					4	1			Castings
Manganese bronze	60	1.5	23	1			1.5	1.5	1.5	High-strength parts
Aluminum bronze	78				5		3	10	4	Corrosion-resisting parts
Nickel silver	65	4	6	5	20					Dairy and laundry equipment

*Cast Metals Handbook, 1943, American Foundrymen's Association.

Several alloys utilizing magnesium have been developed by the Dow Chemical Company for sand-casting work. The casting of Dowmetal requires some special equipment, and slightly different molding and pouring methods are used. Dowmetal H is used for most commercial and aircraft sand castings, as it offers excellent resistance to salt water and alkalies. Its composition is approximately 6% aluminum, 0.2% manganese, 3% zinc, 0.5% silicon, and 90% magnesium. Alloy P, also used for sand casting, is similar in composition except that the aluminum content is 10% and the zinc content is reduced to 1%. This alloy has increased hardness and tensile strength but is not recommended for shock conditions. Magnesium alloys are receiving much attention where light weight is essential. These alloys weigh about two thirds as much as ordinary aluminum alloys.

Many other nonferrous alloys of copper, aluminum, and magnesium are available where a combination of lightness, good strength, and machinability, is desired. It is not within the scope of this book to list and discuss these various alloys, but complete information as to analysis and physical properties may be found in the *Cast Metals Handbook*, published by the American Foundrymen's Association.

REVIEW QUESTIONS

1. Sketch a vertical section of a cupola and label all essential parts.

2. List the raw materials and products of a blast furnace; a cupola.

3. How is a cupola prepared for a heat and charged?

4. How is pig iron made, and for what purpose is it used?

5. How much iron will 50 pounds of coke melt?

6. Give a typical analysis of cast iron.

7. Define cast iron, brass, and bronze.

8. In what forms does carbon exist in cast iron? State the influence of each form on cast iron.

9. What different kinds of cast iron are there?

10. List the elements in cast iron and state the influence of each.

11. List four methods of cleaning castings.

12. List the principal iron ores and give symbol and color of each.

13. List the important elements used in nonferrous castings.

14. What is a Brackelsberg furnace, and how does it operate?

15. How are malleable iron castings made?

16. What is steel? How are steel castings classified?

17. What furnaces are used for making steel castings?

18. Sketch a crucible furnace for making nonferrous castings.

19. What is the difference between red brass and yellow brass?

20. What is Dowmetal, and for what purposes is it used?

BIBLIOGRAPHY

Alloy Cast Irons, American Foundrymen's Association.

BENNETT, J. S., " Essentials in the Production of Sound Steel Castings," *Foundry Trade Journal*, April 11, 1935, pp. 253, 256.

BRIGGS, C. W., and GEZELIUS, R. A., " Studies on Solidification and Contraction in Steel Castings," *Trans. AFA*, Vol. 43, pp. 274–302, 1935.

CAMPBELL, H. L., *Metal Castings*, John Wiley & Sons, 1936.

Cast Metals Handbook, 3d edition, American Foundrymen's Association, 1944.

CHARNOCK, G. F., and PARTINGTON, F. W., *Mechanical Technology*, Constable & Company, Ltd., London, 1934.

MASSARI, S. C., " The Properties and Uses of Chilled Iron," *ASTM*, Vol. 38, part 2, pp. 217–34, 1938.

Metals Handbook, American Society for Metals, 1939.

Steel Castings Handbook, 1st edition, Steel Founders' Society of America.

WENDT, R. E., *Foundry Work*, 3d edition, McGraw-Hill Book Company, 1928.

CHAPTER 4

SPECIAL CASTING METHODS

Castings from various types of sand molds probably have fewer limitations than those produced by any other casting processes. All metals may be cast in sand molds, and there is no limitation as to size. However, sand molds are single-purpose molds, being completely destroyed after the metal has solidified. Quite obviously, the use of a *permanent mold* would effect considerable saving in labor cost. Great strides have been made in this field, particularly in the die casting of nonferrous alloys. *Centrifugal casting* is another promising method, in which the molds may be either of single-purpose or permanent materials. Of the nonmetallic molds, the *"lost-wax"* process and the *plaster-of-Paris* molds for precision castings have gained a new importance through wartime research. A summary of the various special casting methods which will be discussed in this chapter is as follows:

    1. Casting in metallic molds
       (*a*) Gravity casting.
       (*b*) Slush casting.
       (*c*) Pressed or Corthias casting.
       (*d*) Die casting.
          (1) Hot-chamber machines.
          (2) Cold-chamber machines.
       (*e*) Centrifugal casting.
          (1) True centrifugal.
          (2) Semicentrifugal.
          (3) Centrifuge type.
    2. Casting in nonmetallic molds
       (*a*) "Lost-Wax" method.
       (*b*) Plaster-of-Paris molds.
       (*c*) Molds of wood, paper, rubber, and the like.

Methods of Casting in Metallic Molds

Permanent molds must be made of metals capable of withstanding high temperatures. Because of their high cost they are recommended only when many castings are to be produced. While permanent molds for large castings and alloys of high melting temperatures

FIG. 1. Gravity Pouring in
Permanent Mold.

Courtesy Aluminum Company of America.

FIG. 2. Permanent Mold Open Showing
Mold Construction and Aluminum Casting
Just Poured.

would be impractical, they can be used advantageously for small and medium nonferrous castings that are manufactured in large quantities.

Gravity casting. This method consists of filling a mold as in sand casting. Figure 1 shows the pouring of aluminum into a metal flask. No pressure is used except that obtained from the head of the metal in the mold. As soon as the metal solidifies sufficiently to hold its shape, the mold is opened as shown in Figure 2. The aluminum casting with accompanying sprue is shown in the foreground of the casting table. Both metal and dry-sand cores may be used in molds of this type. If metal cores are used, they are usually withdrawn as soon as the metal starts to solidify.

To prevent excessive chilling, the molds should be heated up to the proper temperature by filling the molds with hot metal several times. As the proper temperature is obtained, the surface defects disappear, and there is no longer any evidence of excessive chilling. Further control is obtained by regulating the rate at which the castings are produced. Castings should be removed before the metal solidifies to eliminate cracks from developing as a result of shrinkage strains. Both ferrous and nonferrous castings are produced successfully by this method; however, extremely thin section castings are not recommended because of the chilling effect from the metal mold. Products of this type of molding include aluminum pistons for internal-combustion engines, cooking utensils, and various small castings of uniform section.

Slush casting. Slush casting is a method of producing hollow castings in metal molds without the use of cores. Molten metal is poured into the mold, which is turned over immediately so that the metal, remaining liquid, can run out. A thin-walled casting results, the thickness depending on the chilling effect from the mold and the time of the operation. The casting is removed by opening the halves of the mold. This method of casting is used only for ornamental objects, statuettes, toys, and other novelties. The metals used for these objects are lead, zinc, and various low-melting alloys. Parts cast in this fashion are either painted or finished in a way to represent bronze, silver, or other more expensive metals.

Pressed or Corthias casting. This method of casting resembles both the gravity and slush processes but differs somewhat in the manner in which the operation is performed. A definite amount of metal is poured into an open-ended mold, and a close-fitting core is forced into the cavity. This causes the metal to be forced into the mold cavities with some pressure. The core is removed as soon as the metal sets, leaving a hollow thin-walled casting. This process, developed in France by Corthias, is limited in use mainly to ornamental casting of open design.

Die casting. Die casting, as practiced in the United States, refers to the forcing of molten metal under pressure into a metal die. The term *die* used in this process implies a metallic mold which is filled under pressure. Pressures vary according to the kind of metal being cast and numerous other factors, ranging from 5 to 30,000 pounds per square inch. The usual pressures for small-parts cast of low-melting metals are 80 to 125 pounds per square inch. Regardless of the amount, the pressure is maintained until solidification is completed.

Die casting, the most widely used of any of the permanent-mold processes, is done by two different methods: the *hot-chamber method* and the *cold-chamber method*. In the former, a melting pot is included with the machine, and the injection cylinder is immersed in the molten metal at all times, the injection cylinder being actuated by either air or hydraulic pressure which forces the metal into the dies to complete the casting. Machines using the cold-chamber process have a separate melting furnace, and metal is introduced into the injection cylinder by hand or mechanical means. Hydraulic pressure then forces the metal into the die.

HOT CHAMBER MACHINES. Die-casting machines have been in gradual process of development since 1847, when the first machine was patented. This machine, for casting printer's type in lead, employed a specialized form of die casting. Modern machines, operated from a keyboard, are now used daily for producing type from lead-base alloys.

The first machines developed for the production of miscellaneous castings were limited to low-melting alloys and were hand-operated. Some machines of this type are still in operation but are confined to small-production runs and to castings of rather simple design. Most production machines of today are either semiautomatic or completely automatic. The essential parts of a die-casting machine using the hot-chamber method are the container for molten metal, a heating chamber, means for forcing the metal into the die, stationary and movable dies, a mechanism for opening and closing dies, an ejector mechanism for removing the casting, and the necessary framework for the machine. Various modifications will be discussed in connection with typical designs.

Metal is forced into the mold and pressure maintained during solidification by either of two common methods: (1) A *plunger,* or (2) *compressed air.* Both methods are shown diagrammatically in Figure 3. The plunger-type machine, shown in the upper part of the figure, is hydraulically operated for both the metal plunger and the mechanism for opening and closing the die. In this machine the plunger operates in one end of a gooseneck casting which is submerged in the molten metal. With the plunger in the upper position, metal flows by gravity into this casting through several holes just below the plunger. On the down stroke these holes are closed by the plunger, and pressure is applied on the entrapped metal, causing it to be forced into the die cavity under pressure. Pressures over 5000 pounds per square inch are used in some machines of this type, resulting in castings of dense structure. As soon as the casting is solidified, the pressure is relieved, the dies are forced open, and the casting is ejected by means of knockout pins. The sprue is removed with the runner and the castings.

Plunger-type machines are used with low-melting alloys because of the difficulties encountered in plunger fits at higher temperatures

and increased corrosion of the parts. Alloys of zinc, tin, and lead
are particularly adapted to these machines.

Air-operated machines, such as the one shown in the lower part of
Figure 3, have a gooseneck casting that is operated by a lifting
mechanism. In the starting position the casting is submerged in the

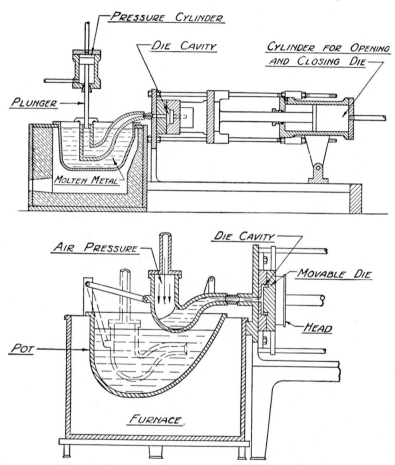

FIG. 3.   Diagrammatic View of Plunger and Pneumatic Die-Casting Machine.

molten metal and is filled by gravity. It is then raised so that the
nozzle is in contact with the die opening, and locked in position.
Compressed air, at pressures ranging from 80 to 500 pounds per
square inch, is applied directly on the metal, thus forcing it into the
die. When solidification is about complete, the air pressure is turned
off and the gooseneck lowered into position to receive more metal.

The operation of opening the dies, withdrawing cores, and ejecting the castings is the same as for the plunger-type machine. Frequently this operation is controlled by a hydraulic cylinder which produces smoother action than compressed air.

A small fully pneumatic plunger machine, known as a Kippcaster 215B, is shown in Figure 4. This machine is used primarily for zinc,

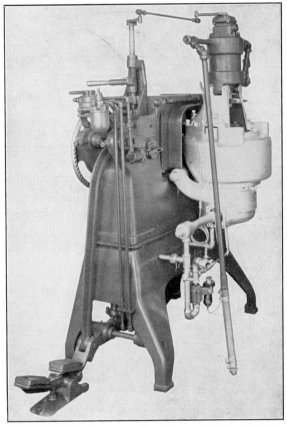

Courtesy Madison-Kipp Corporation.

Fig. 4. Kippcaster No. 215B, Plunger-Loaded Gooseneck; Operated by Air Pressure; Used for Zinc, Lead, and Tin Alloys.

lead, and tin alloys and has a capacity per shot of $1\frac{1}{2}$ pounds of zinc with air pressures ranging from 80 to 125 pounds. Standard die sizes are 6 inches by 6 inches, with an over-all thickness of 4 inches. The volume of air required depends upon the speed of operation, but, for small castings at five shots per minute, a compressor having a capacity of 12 cubic feet of free air per minute is sufficient to operate two machines.

In operating this machine, the dies are closed by means of an air valve at the rear of the machine and locked in place. Next the air valve controlling the plunger air cylinder is opened, and air pressure is applied on the displacement cylinder, forcing the metal into the dies. The lever is then returned so that the plunger rises to its original position. As the dies are opened, the castings are automatically ejected.

In the hot-chamber process there are limitations as to the kinds of metal which can be used and the maximum pressure that can be exerted on the molten metal. Since many metals have an affinity for iron, only those which do not attack the immersed metal parts can be used. The alloys of zinc, tin, and lead are recommended for the process. Metals which require extremely high pressure to obtain desired densities are cast in cold-chamber machines.

COLD-CHAMBER MACHINES. The die casting of brass, aluminum and magnesium requires higher pressures and melting temperatures and necessitates a change in the melting procedure from that previously described. These metals are not melted in a self-contained pot, as the life of the pot would be very short. The usual procedure is to heat the metal in an auxiliary furnace and ladle it to the plunger cavity next to the dies. It is then forced into the dies under hydraulic pressure. Machines operating by this method are built very strong and rigid to withstand the heavy pressures exerted on the metal as it is forced into the dies. Two types of machines are in general use: in one, the plunger is in a vertical position; in the other, horizontal.

A diagrammatic sketch illustrating the operation of horizontal plunger cold-chamber machines is shown in Figure 5. In the first figure the dies are shown closed, with cores in position, and the molten metal ready to be ladled in. As soon as the ladle is emptied, the plunger moves to the left and forces the metal into the two cup-shaped molds. After the metal solidifies, the cores are first withdrawn, and then the dies opened. In the third figure the dies are opening, and the casting is shown as ejected from the stationary half. To complete the process of opening, an ejector rod comes into operation and ejects the casting from the movable half of the die. This operating cycle is used in a variety of machines (made by the Reed-Prentice Corporation) which operate at pressures ranging from

5600 to 22,000 pounds per square inch. These machines are fully hydraulic and semiautomatic. After the metal is ladled in, the rest of the operations are automatic.

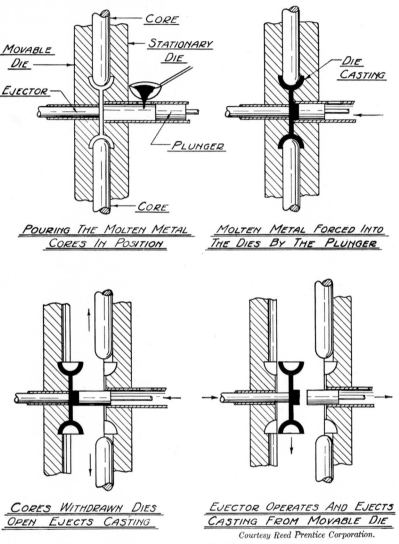

POURING THE MOLTEN METAL CORES IN POSITION

MOLTEN METAL FORCED INTO THE DIES BY THE PLUNGER

CORES WITHDRAWN DIES OPEN EJECTS CASTING

EJECTOR OPERATES AND EJECTS CASTING FROM MOVABLE DIE

Courtesy Reed Prentice Corporation.

Fig. 5. Die Casting of Brass, Aluminum, and Magnesium.

The hydraulically operated machine shown in Figure 6 is equipped with a hand-ladling injection unit for brass, aluminum, and magnesium die casting. When casting, metal from a near-by furnace is ladled to the well opening on the plunger injection unit located to the rear of the stationary die plate. Aside from the ladling procedure, the operation of the machine is the same for hot-chamber machines. The hand-ladling injection unit on this machine can be readily replaced by a plunger-gooseneck unit for casting low-melting alloys.

The manufacture of brass die castings is a comparatively recent achievement. The difficulties of the high temperatures involved and

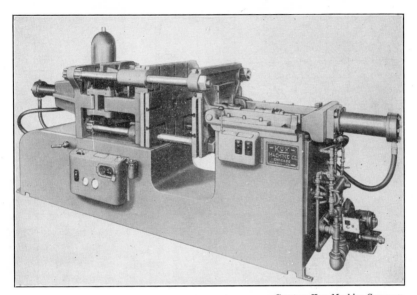

Courtesy Kux Machine Company.

FIG. 6.   Hydraulic Die-Casting Machine with Hand-Ladling Injection Unit for Aluminum, Brass, and Magnesium.

the resulting rapid oxidation of the steel dies have been largely overcome by improvements in die metals and by casting at as low a temperature as possible. A machine developed in Czechoslovakia, known as the Polak machine, is used successfully for the production of these castings by the Titan Metal Manufacturing Company of the United States. This machine is designed to use metal in a semiliquid or plastic state to permit operation at lower temperatures than those used for liquid metal. To protect the dies further from overheating, water is circulated through plates adjacent to the dies. Metal is maintained under close temperature control and is ladled by hand to the compression chamber. The pressure used in this machine is **9800**

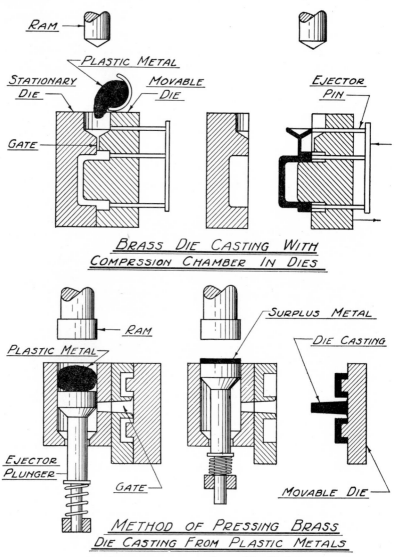

FIG. 7. Construction of Dies for Pressing Brass Die Castings.

pounds per square inch; and 100 to 200 shots per hour can be made, depending on the size of the machine.

Two variations of this process, each with the injection plunger in a vertical position, are diagrammatically illustrated in Figure 7. In the lower figure the compression chamber, into which the plastic metal is ladled, is separate from the dies. The metal is poured into this cavity onto a spring-backed plunger. As the ram descends, this plunger is forced down until the gate opening is exposed, permitting the metal to be forced into the die cavity. As the ram returns to its upper position, the ejector plunger also moves upward, carrying with it any surplus metal. As the die opens up, the casting is ejected.

A variation of this machine, with the compression chamber a part of the die, is shown in the upper part of the figure. Metal is poured

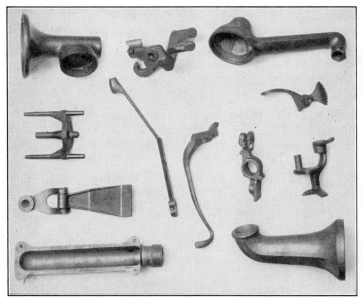

FIG. 8.   Variety of Brass Die Castings.

into this chamber at the upper part of the die and forced by pressure into the die cavity as the ram descends. As soon as the ram moves up, the dies open, and the casting is ejected by means of the ejector pins. The sprue and excess metal are trimmed off in the finishing operation.

A wide variety of parts can be cast by this method, as illustrated in Figure 8. In addition to rather intricate irregular shapes, both gear teeth and threads may be successfully formed by this process. Those shown in the figure are of brass, but alloys of zinc, aluminum, and lead can also be cast in the plastic state.

Die-casting dies.   Dies for both the hot- and cold-chamber methods are similar and in general have the appearance of the one shown in Figure 9. This die is designed to produce 75-mm windshields for armor-piercing shot from an aluminum alloy. Dies are made in two sections to provide means for removing the castings and are usually provided with heavy dowel pins to keep the halves in proper alignment. Metal enters the stationary side when the die is locked in closed position. As the die opens, the ejector plate in the movable

FIG. 9.   Die For Producing 75-Mm Windshields for Armor-Piercing Shot.

half of the die is advanced so that pins or sleeves project through the die half and force the casting from the cavity or fixed core. The dies are provided with a separate mechanism for moving the ejector plate or movable cores. It is desirable to provide vents and small overflow wells (see Figures 2 and 9) on one side of the die to facilitate the escape of air and to catch any metal which has passed through the die cavity. Most dies are provided with channels for water cooling to keep the die at correct temperature for rapid production.

Advantages and disadvantages of die casting.   The design engineer has many problems to solve besides those involving either kinematics or the forces acting on the various machine elements. Among these problems are the selection of proper materials and methods of manufacture. In each case a decision must be made as to whether a part should be cast, forged, stamped, welded, or fabricated in some other manner. If the part is to be cast, several methods are available, and the design engineer must know their possibilities and limitations.

One of the main advantages of die castings over sand castings is the rapidity of the process, since both molds and cores are permanent. This feature alone is sufficient to warrant its consideration in the mass production of castings; however, it has other advantages. For one thing, the metal molds give die castings a smooth surface which not only greatly improves their appearance but also minimizes the work required to prepare them for plating or other finishing operations. By this method, size is so accurately controlled that little or no machining is necessary. Because of the uniformity in wall thickness, less material is required in die castings than in sand castings. Furthermore, there are no possibilities for sand inclusions, and a strong dense metal structure is obtained. Finally, because of the accurate tolerance that can be maintained, the process eliminates such machining operations as drilling and certain types of threading.

One of the limitations of die casting is the high cost of the equipment and dies. This is not an important factor in mass production, but it does eliminate its use in short-run jobs. There is also a rapid decrease in the life of the dies as the metal temperature increases. In some cases there is an undesirable chilling effect on the metal unless high temperatures are maintained. Metals having a high coefficient of contraction must be removed from the mold as soon as possible because of the inability of the mold to contract with the casting. There are certain limitations as to the shape of die castings, and the process is not adapted to the production of large castings. For these reasons, die casting has to a large extent been limited to low-melting alloys, but with a gradual improvement of heat-resisting metals for dies this process can now be used for numerous alloys.

Die-Casting Alloys

Although a large variety of alloys can be successfully die-cast, a proper selection is somewhat difficult, because these metals have not been completely listed according to any one standard. The present ASTM designation is probably the most complete, but there are numerous other commercial alloys used that have other designations.

The principal base metals used are zinc, lead, tin, aluminum, and magnesium. Other alloys not so widely used are brass, bronze, cast iron, and steel. These may be further classified as low-temperature alloys and high-temperature alloys; those having a melting temperature below 1000 F, such as zinc, tin, and lead, are in the low-temperature class. The latter have the advantages of lower cost of production and lower die-maintenance costs. As the casting temperature increases, special steels are required to resist oxidation and checking of the die surfaces. The destructive effect of high temperature on the dies has been the principal factor in retarding the development of high-temperature die castings.

Another factor governing the choice of alloy is the corrosive or chemical action of the molten metal on the respective machine parts and dies. This action increases with temperature, although it is worse with some alloys than with others. Aluminum, in particular, has a destructive action on ferrous metals and for this reason is seldom melted in the machine.

Other considerations which influence selection of alloys are the physical properties required, weight, machinability, resistance to corrosion, surface finish, and cost. Obviously, the lowest-cost alloy that will give satisfactory service should be selected.

Zinc-base alloys. By far the largest number of die castings are made of zinc-base alloys. This alloy casts well with a good finish, has considerable strength, is low in cost, and can be melted at fairly low temperatures. Pure zinc should be used, since such elements as iron, lead, and cadmium are impurities that cause serious casting defects. The usual elements alloyed with zinc are aluminum, copper, and magnesium; all are held to low percentages. Aluminum in amounts around 4% greatly reduces the tendency of the metal to dissolve iron. Copper increases the ductility, impact strength, and resistance to corrosion. Magnesium, which is usually held to 0.03%, adds to the stability of the alloy.

One of the most widely used zinc alloys is ASTM XXI, which corresponds to SAE 921 and to the New Jersey Zinc Company's alloy Zamak 2.* This alloy has 4.10% aluminum, 2.70 copper, 0.03 magnesium, and the remainder pure zinc. The melting point of the alloy is 733.6 F, and the shrinkage is 0.149 inch per foot. Average tensile strengths of 47,300 pounds per square inch are obtained from cast specimens 6 months after the casting date. Another important alloy, ASTM XXIII (SAE 903 or Zamak 3), contains 4.1% aluminum,

* Properties of Die-Cast Zamak Alloy, The New Jersey Zinc Company.

and 0.04% magnesium, with the remainder pure zinc. Other elements totaling 0.2% may be present in this alloy but are not desired. It has slightly lower tensile strength and hardness, but is distinguished by its excellent retention of impact strength and dimensions.

Zinc alloys are widely used in the automotive industry and for such numerous items as small pulleys, business-machine parts, bearing retainers, spray guns, small machine tools, toys, and household appliances.

Lead-base alloys. Pure lead, which melts at 621 F, will melt at around 570 F when alloyed with 17% antimony. This element is the principal one used with lead, and its percentage ranges from 9 to 17. Antimony hardens lead and reduces its shrinkage value. Lead alloys have low physical properties, but are cheap and easily cast. Their use is principally for light-duty bearings, weights, battery parts, X-ray shields, and for applications requiring a noncorrosive metal. ASTM Standard Specification B 23–26, Grades 12 and 11, have respectively 10 and 15% antimony content, with copper held to 0.50% maximum and arsenic to 0.25% maximum.

Tin is also used as an alloying element with lead and antimony. Tin increases the fluidity, hardness, and strength of the alloys, thus improving their use for bearing purposes. The ASTM Standard Specifications B 23–26 have several lead–tin–antimony alloys, with tin content ranging from 5 to 10%.

Tin-base alloys. Tin-base alloys, known as " babbitt metal," are used principally for bearings. The main alloying elements for this group are copper, antimony, and lead, although very small percentages of iron, arsenic, and bismuth frequently exist and constitute impurities. Although the melting point of tin is 450 F, the complete liquefaction point when alloyed is raised, necessitating casting temperatures for die-casting alloys of 700 to 800 F. The effect of the antimony is to harden the alloy and increase its antifriction properties.

ASTM Standard Specification B 23–26, Grade 1 (SAE 10), contains 91% tin, 4.50 copper, and 4.50 antimony. Grade 2 contains 89% tin, 3.50 copper, and 7.50 antimony. Both may contain small fractional percentages of several other alloys. These two alloys are extensively used in automotive and aircraft bearings, the latter being slightly harder and stronger. Another alloy (SAE 12) having 60% tin, 3.5 copper, 10.5 antimony, and 26.0 lead is somewhat cheaper, because of the high lead content, and has considerable use for light-duty bearings. The alloys previously mentioned are corrosion-resisting, and some of them are used in the handling of food products, for soda-fountain hardware, and similar applications. The shrinkage of tin- and lead-base alloys is exceptionally low, ranging from 0.002 to 0.003 inch per inch. This characteristic facilitates the casting of parts to accurate size and also lessens the danger of shrinkage cracks in casting.

Magnesium-base alloys. Magnesium is alloyed principally with aluminum, but may contain small amounts of silicon, manganese, zinc, and copper. Its alloys are the lightest in weight of all die-cast metals, being about two-thirds the weight of alloys of aluminum. Although the price per pound is slightly higher than for aluminum, the extra cost is compensated for by light weight and improved machinability.

ASTM Specification B 94–39T, Alloy 12 (Dowmetal K), is one of the principal die-casting alloys, having good casting characteristics and fairly high physical properties. This alloy contains about 10% aluminum, 0.30 zinc, 0.10 manganese, 0.70 silicon, and the remainder magnesium. By decreasing aluminum, greater ductility is obtained with some loss in strength. Manganese is sometimes added in percentages up to 1.5 to increase the resistance to salt water. Silicon in small amounts adds to the casting properties.

Magnesium alloys are cast in much the same manner as aluminum alloys and require a casting temperature around 1200 to 1300 F. Best results are obtained in ladle plunger machines at high pressures. The lightness of these alloys, combined with good physical properties, fits them admirably for aircraft motor and instrument parts, portable tools, household appliances, and many other similar applications.

Copper-base alloys. Die castings of brass and bronze have presented more of a problem in pressure casting because of their high melting temperatures. These temperatures range from 1600 to 1900 F and make it necessary to use heat-resisting alloy steel for the dies to reduce their rapid deterioration. Because of these high temperatures copper-base alloys are melted in an auxiliary furnace and ladled to the machine in either a liquid or plastic state. The latter method is used a great deal, as it permits operating at temperatures considerably lower than the melting temperatures of the molten metal. A plunger-type machine is used in this work.

Most of the alloys have a copper content around 60, with the remainder zinc, since this mixture is economical and has good casting qualities. The aforementioned alloy* is known as yellow brass and contains in addition to copper and zinc, about 1% tin, 0.75 lead, 0.25

---

* Corresponds to SAE 43 copper-base alloy.

manganese, and 0.10 aluminum. The tensile strength of this alloy is around 65,000 pounds per square inch. Tinicosil No. 20* contains 42% copper, 41 zinc, 1 lead, 16 nickel, and has a tensile strength of about 90,000 pounds per square inch. This alloy is used extensively in the Polak machines previously described and is cast in the plastic state. The American Brass Company has two patented alloys, Anaconda Brass Alloy No. 624 and Anaconda Bronze Alloy No. 1026.† The first mentioned contains 60% copper, 37.75 zinc, 1 tin, 1 lead, 0.10 aluminum, 0.15 silicon, and has a tensile strength of 55,000 pounds per square inch. The bronze alloy contains 81.5% copper, 14.1 zinc, 4.25 silicon, 0.15 manganese, and has a minimum tensile strength of 85,000 pounds per square inch. German silver or white brass may also be used in pressure casting.

Copper-base alloys have extensive use in miscellaneous hardware, electric machinery parts, small gears, marine fittings, chemical apparatus, automotive fittings, and numerous other small parts. These alloys are used principally where high strength or resistance to corrosion is required.

Aluminum alloys. A wide variety of die castings are made of aluminum alloys because of their lightness in weight and resistance to corrosion. However, compared to zinc alloys, they are lower in physical properties and more difficult to die-cast. Typical die castings made from aluminum alloys are shown in Figure 10.

Since molten alloys of aluminum will attack steel if kept in continuous contact with it, the cold-chamber process is used in casting. Although more expensive to operate than the air-injection type of machine, it has the advantage of producing sounder castings. The melting temperature of aluminum alloys is around 1185 F.

The principal elements used as alloys with aluminum are silicon, copper, nickel, and magnesium. Silicon adds to the hardness and corrosion-resisting properties; copper improves the physical properties slightly; nickel improves surface appearance; and magnesium increases the lightness and resistance to impact. ASTM V (SAE 305 or Alcoa 13) contains 0.6% copper, 12.0% silicon, 0.5% nickel, and the remainder aluminum except for small percentages of iron and several other elements. This alloy can be used for either large or intricate castings and has excellent corrosion resistance. ASTM XII (SAE 312 or Alcoa 81) contains 7.0% copper, 3.5% silicon, 0.5% nickel, traces of several other elements, and the remainder aluminum. This alloy

---

* Registered alloy by Titan Metal Manufacturing Company.

† Anaconda Copper and Copper Alloys, The American Brass Company.

enjoys wide use for average general-purpose casting. Alcoa 218 contains 8.0% magnesium with the remainder aluminum, which pro-

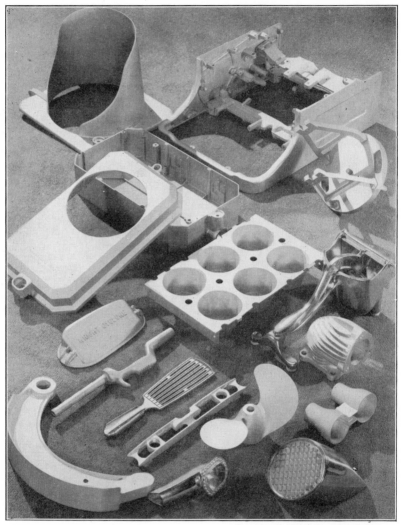

Courtesy Aluminum Company of America.

FIG. 10.　Die Castings From Aluminum Alloys.

vides the best combination of strength and ductility of any of the aluminum alloys. In addition to the aluminum alloys just described, there are six other ASTM alloys available for die-cast processing.

Centrifugal Casting*

Centrifugal casting is the process of rotating a mold while the metal solidifies, so as to utilize centrifugal force to position the metal in the mold. The metal is forced against the walls of the mold with much greater pressure than that obtained by static pressure in ordinary sand casting. Greater detail on the surface of the casting is obtained, and the dense metal structure has superior physical properties. Castings of symmetrical shape lend themselves particularly to this method, although many other types of castings can be produced. The methods of centrifugal casting may be classified as follows:†

1. True centrifugal casting.
2. Semicentrifugal casting.
3. Centrifuging.

True centrifugal casting is used for pipe, liners, and symmetrical objects which are cast by rotating the mold about its horizontal or

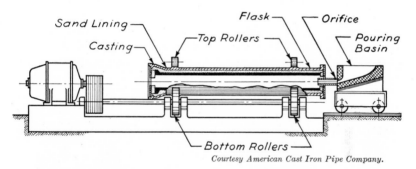

Courtesy American Cast Iron Pipe Company.

FIG. 11. Centrifugal Casting Machine for Casting Steel or Cast-Iron Pipe.

vertical axis. The metal is held against the wall of the mold by centrifugal force, and no core is required to form a true cylindrical cavity on the inside. This method is illustrated by the casting machine shown in Figure 11 designed for the production of steel or cast iron pipe. The wall thickness of the pipe produced is controlled by the amount of metal poured into the mold.

Another example of true centrifugal casting is shown in Figure 12, which illustrates two methods that may be used for casting radial-

* Acknowledgment is given to the Centrifugal Casting Company for supplying certain illustrations used in this discussion.

† S. D. Moxley, "Centrifugal Casting of Steel," *Mechanical Engineering*, October 1944.

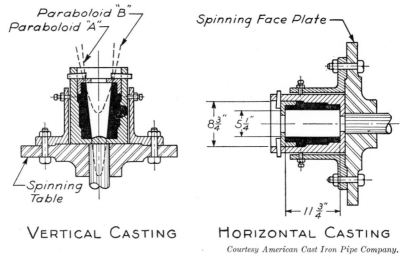

VERTICAL CASTING   HORIZONTAL CASTING

Courtesy American Cast Iron Pipe Company.

FIG. 12. True Centrifugal Method of Casting Radial-Engine Cylinder Barrels.

engine cylinder barrels. The horizontal method of casting is similar to the process followed in casting pipe lengths, and the inside diameter is a true cylinder requiring a minimum amount of machining. In

FIG. 13. Radial-Engine Cylinder Barrels Shown Before and After Machining.

vertical castings the inside cavity takes the form of a paraboloid as illustrated by the figure. The slope of the sides of the paraboloid depends on the speed of rotation, the dotted lines at *A* representing a higher rotational speed than shown by the paraboloid B. In order to reduce the inside-diameter differences between the top and bottom of the cylinder, spinning speeds are higher for vertical casting than for horizontal casting. Radial-engine cylinder castings produced by this process are shown in Figure 13 and illustrate the rough casting, the cylinder after rough machining, and the finished cylinder with its cooling fins.

As the name implies, *semicentrifugal* casting is the rotating of the mold about its vertical axis. In this method the center of the casting is usually solid, and the center cavity is machined out later or is formed by the metal passing down around the outside of a core. This method, often used in stack-molding, is illustrated in Figure 14, where five track wheels are cast solid in one mold. The number of castings made in a mold depends on the size of the casting and the convenience in handling and assembling the molds. Rotational speeds for this form of centrifugal casting are not so great as for the true centrifugal process. The process produces a dense structure at the outer circumference where it is needed, while the center metal is machined out.

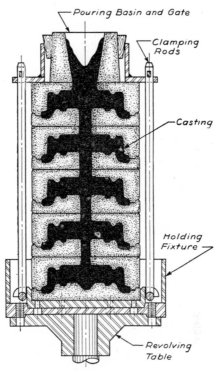

Courtesy American Cast Iron Pipe Company.

Fig. 14. Semi-centrifugal Stack Molding of Track Wheels.

In the *centrifuge* method several casting cavities are located around the outer portion of a mold, and metal is fed to these cavities by radial sprues or gates from the center of the mold. Either single or stack molds can be used. The mold cavities are filled under pressure from the centrifugal force of the metal as the mold is rotated. In Figure 15 are shown five castings made in one mold by this process. The internal cavities of these castings are irregular in shape and are formed by dry-sand cores. The centrifuge method, not limited to symmetrical objects, can produce castings of irregular shape such as bearing caps or small brackets. For many years the dental profession has used this process for the casting of gold inlays.

The average rotational speed* for miscellaneous centrifugal castings approximates 600 surface feet per minute, although for cast-iron pipe the speeds may be as high as 1250 feet per minute. Too great a speed may result in surface cracks caused by high stresses set up in the mold. The rotational speed depends on the kind of mold used (whether sand or metal), the manner of rotation (whether horizontal or vertical), the size of casting, and the kind of metal being cast.

FIG. 15.   Centrifuged Castings with Internal Cavities of Irregular Shape.

Centrifugal casting reduces cost by its advantages over other methods. Cores in cylindrical shapes and risers or feedheads are both eliminated. The castings have a dense metal structure with all impurities forced back to the center where frequently they can be machined out. Because of the pressure exerted on the metal, thinner sections can be cast than would be possible in static casting. Finally, any metal can be cast by this process. It must be remembered, however, that all castings cannot be made centrifugally, since there are definite size and shape limitations. Very often, sand castings can be produced much cheaper; but where finish and dimensional accuracy are demanded, die castings will continue to be used.

Methods of Casting in Nonmetallic Molds

Nonmetallic molds are not restricted to either high or low temperatures. Each mold material has its own temperature limitations as to the kind of metal for which it is suitable. In many cases nonmetallic molds are used in precision casting, as the dimensional accuracy obtained with accompanying smooth-surface finish tends to

* G. E. Stedman, "Unique Centrifugal Steel Casting Method," *Metals and Alloys*, August 1944.

offset the higher cost of these castings. Many unique castings made in nonmetallic molds are statuary, dental inlays, printing type, and intricate airplane castings.

" Lost-wax " casting process. This process* derives its name from the fact that the wax pattern used in the process is subsequently melted from the mold, leaving a cavity having all the details of the original pattern. The process as originally practiced by artisans in the 16th century consisted of forming the object in wax by hand. The wax object or pattern was then covered by a plaster investment. When this plaster became hard, the mold was heated in an oven, melting the wax and at the same time further drying and hardening the mold. The remaining cavity, having all the intricate details of the original wax form, was then filled with metal. Upon cooling, the plaster investment was broken away leaving the desired casting. One advantage of this process was that intricate forms having under-cuts could be reproduced, since the mold did not have to be opened for the pattern to be removed. In large castings, such as statuary,† plaster cores were used to provide relatively thin walls in the casting.

The need for precision-cast parts requiring little machining has been emphasized by the war effort, and steps have been taken to improve the process by developing suitable refractory investments and means for producing accurate wax patterns on a quantity basis. The procedure,‡ worked out by The General Electric Company, consists of first preparing a master pattern of steel from which a split lead-alloy mold is cast, as shown in Figure 16. This mold is then gated and is used in casting the wax patterns. These are usually cast from a wax-injection machine by placing the lead-alloy mold next to the nozzle of the machine. Upon solidification, the wax patterns are removed and several gated together.

To prepare the mold, the patterns are first dipped into a fine silica-flour mixture (shown in Figure 17) which forms a heavy coating around the wax patterns. They are then placed in a container, and plaster is poured around them to complete the mold. When dry, the mold is placed upside down and heated in an oven to melt out the

* J. D. Wolfe, " Precision Castings for Ordnance and Aircraft," *Metal and Alloys*, Vol. 18, October 1943.

† Benvenuto Cellini's famous chapter on the casting of his bronze statue, Perseus, contains much interesting information on methods of molding used during the Renaissance. Cellini used a form of " lost-wax " process.

‡ W. E. Ruder, " War Producer Selects ' Lost Wax Process ' for Complicated Parts," *American Machinist*, March 2, 1944.

wax and to dry and bake the mold thoroughly. The final casting is produced by gravity pouring, by forcing the metal in under pressure, or by centrifugal casting.

The latter process lends itself to the production of small parts of intricate design from metals which are difficult to machine. Close

Courtesy General Electric Company.

FIG. 16. Split Lead-Alloy Mold Shown with a Wax Pattern Being Lifted from It.

FIG. 17. Wax-Pattern Assembly Is Shown Precoated with Silica Flour Suspended in Suitable Binder.

tolerance can be secured and very little finish is required on the castings. Brass, bronze, beryllium, stellite, stainless steel, and aluminum alloys are now made by centrifugal castings, formerly used principally for precious metals. Castings produced in this fashion include small gears, cams, gage and instrument parts, turbine blades, small parts for guns, and numerous small parts on aircraft.

Plaster-mold casting. The use of plaster as a casting investment has had limited use for many years, but recent improvement in its ability to dry quickly with sufficient porosity has greatly accelerated its use as a modern casting material. Compared with sand molds, it has a higher molding cost, but the advantages gained by close tolerance, fine detail, and good surface finish enable it to be economically used in short and medium production runs.* The molds are not permanent, being destroyed in the process of removing the castings.

A gypsum plaster, dry-mixed with added strengtheners and setting agents, is mixed with water.† This mixture is then poured around the patterns. Before the plaster is completely set, the patterns are removed from the mold, and shortly thereafter the flask is also removed. Metal patterns mounted loose on a match plate are used

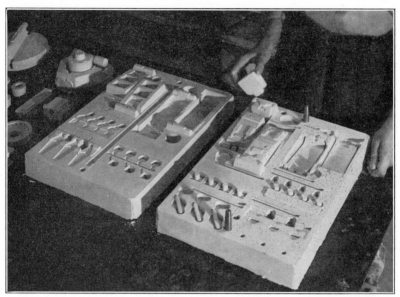

Courtesy Briggs Manufacturing Company.

FIG. 18.   Assembling Cores in Plaster Mold Prior to Closing.

to facilitate the withdrawing of the patterns and to permit the cope and drag to be made separately. The halves of the mold are then separately baked in an oven conveyor which drives off all extra moisture. Coming from the oven, they are assembled with necessary cores, and the mold is closed ready for pouring as shown in Figure 18. Castings are removed by breaking up the mold, any surplus plaster being removed by a washing operation.

Mold porosity for the removal of any gases developed in the mold

* Herbert Chase, " Which Form of Non-ferrous Casting," *Metals & Alloys*, Vol. 18, September 1943.

† *Plaster Mold Castings*, Briggs Manufacturing Company, 1944.

is controlled by the water content of the plaster. When the mold is dried, the water driven out leaves numerous fine passageways which act as vents. The amount of water added originally is in excess of what is needed for setting of the plaster and provides the excess of voids needed for venting. In addition to having adequate porosity, plaster molds have the necessary structural strength for casting plus enough elasticity to allow some contraction of the metal during its cooling.

Plaster molds are suitable only for nonferrous alloys having casting temperatures not much over 2100 F. While plaster has proved to be an excellent mold material for yellow brass, certain bronzes and aluminum alloys may also be used. The wide variety of small-size castings made by this process include miscellaneous airplane parts, small gears, cams, handles, pump parts, small housings, and numerous other intricate castings.

One of the principal advantages of plaster-mold casting is the resulting high degree of dimensional accuracy. This, coupled with the smooth surface obtained, enables the process to compete favorably with sand casting in producing parts requiring a considerable amount of machining. Because of the low thermal conductivity of plaster, the metal does not chill rapidly, and very thin sections may be cast. There is little tendency towards internal porosity in plaster-mold castings, and no difficulty is experienced with sand or other inclusions. In general, the process competes more successfully with die casting using the high-temperature alloys such as brass rather than metals such as zinc and aluminum. At high temperatures, metal molds have a relatively short life; with plaster molds, which are used only once, the temperature is no problem.

Molds of other materials. Various materials such as rubber, paper, and wood can be used for molds for casting of low-melting-temperature metals. Costume jewelry and similar small items are successfully cast in rubber molds.* A two-piece rubber casing is vulcanized over a mold or pattern which, when complete, serves as the mold. Casting is by centrifugal means, and metal temperatures up to 600 F can be used. The flexibility of the mold permits intricate designs with undercuts, but close dimensional accuracy cannot be maintained. An alloy of 98% tin, 1% copper, and 1% antimony is frequently used in this work.

Full-page newspaper type is cast in a mold (called a " mat ") upon which the type and illustration impressions have been made on

---

*Lupke, Paul, Jr., " Making Cast Models in Rubber Molds," *Mechanical Engineering*, June 1945.

damp paper. The type metal is poured into the mold after the paper is dry. End-grain wood may also be used as a mold material for low-melting-alloys where only a limited number of simple castings are desired.

REVIEW QUESTIONS

1. List the various methods of permanent-mold casting.
2. Discuss the process of slush casting, and state for what purpose it is used.
3. Define " die casting."
4. What are the advantages of the die-casting process over sand casting?
5. Distinguish between hot- and cold-chamber methods of die casting.
6. Show by sketch how metal is forced into the die in the plunger-type hot-chamber machine.
7. What are the limitations of the hot-chamber method of die-casting?
8. What metals are usually die-cast by the cold-chamber process?
9. List the principal metals or alloys used in die casting.
10. What group of alloys are most widely used in die casting?
11. What is " babbitt?" Give its approximate composition.
12. What are the limiting factors which prevent the die casting of all metals and alloys?
13. State the differences between true centrifugal casting, semicentrifugal casting, and centrifuging.
14. How is cast-iron pipe made? Illustrate by sketch.
15. What is the average rotational speed for centrifugal casting and upon what does the speed depend?
16. State the advantages of centrifugal casting over other methods.
17. Describe the " lost-wax " casting process.
18. What alloys may be cast in plaster molds?
19. What are the principal advantages claimed for plaster-mold castings?

BIBLIOGRAPHY

Anaconda Copper and Copper Alloys, American Brass Company.

ANDERSON, E. A., and WERKY, G. L., *Zamak Alloys for Zinc Alloy Die Casting*, New Jersey Zinc Company, 1944.

CHARNOCK, G. F., and PARTINGTON, F. W., *Mechanical Technology*, Constable & Company, London, 1934.

CHASE, HERBERT, (a) " Die-Cast or Sand-Cast, *Product Engineering*, February–April 1940; (b) " Die-Cast or Stamped," *Product Engineering*, August–September 1938; (c) *Die Casting*, John Wiley & Sons, 1934; (d) " Which Form of Non-ferrous Casting," *Metals & Alloys*, September 1944.

Die Casting for Engineers, New Jersey Zinc Company, 1942.

Dowmetal, Magnesium Alloy, Dow Chemical Company.

HARVILL, H. L., *High Pressure Die Casting*, H. L. Harvill Manufacturing Company, 1945.

LUPKE, PAUL, JR., " Making Standard Cast Models in Rubber Molds," *Mechanical Engineering*, June 1945.

MOXLEY, S. D., " Centrifugal Casting of Steel," *Mechanical Engineering*, 1944.

Plaster Mold Castings, Briggs Manufacturing Company, 1944.

SAGER, ALFRED, " Permanent Mold Castings," *Metals & Alloys*, April 1945.

STERN, MARC, *Die Casting Practice*, McGraw-Hill Book Company, 1930.

CHAPTER 5

PLASTIC MOLDING

In general, the term " plastic " is applied to all materials capable of being molded or modeled. Modern usage of this word has changed its meaning to include a large group of synthetic organic materials that become plastic by the use of heat and are capable of being formed to shape under pressure. Cellulose derivatives, natural and synthetic resins, and protein matter constitute the principal materials from which the plastics are made. These plastic materials lend themselves to a variety of methods of manufacture including molding, casting, extruding, and the production of various coatings and laminates.

Plastic molding as a large-scale manufacturing process is of comparatively recent date. The discovery of ebonite or hard rubber by Charles Goodyear in 1839 and the developement of celluloid by Hyatt about 1869 marked the beginning of plastic products. It was not until 1909, however, that one of the most important materials, phenol formaldehyde resin, was developed by Dr. L. H. Baekeland and his associates. Since then numerous other synthetic materials which vary widely in physical properties have been developed.

Products made from plastic materials can be produced rapidly with close-dimensional tolerance and excellent surface finish. In many cases they have replaced metals where lightness in weight, corrosion resistance, and dielectric strength are to be considered. Another important characteristic of these products is that they may be made either transparent or in colors. Plastics are physically adequate for many applications, but they do lack the strength, hardness, and ability to resist high temperature found in most of the metals. The high cost of dies and equipment tends to limit production, but even this ceases to be a limitation if the demand for the article warrants mass production.

Plastic Materials

Plastic materials may be broadly classified in two groups — *thermosetting* and *thermoplastic*. Thermosetting compounds are formed to shape under heat and pressure, resulting in a product that

122

is permanently hard. The heat first softens the material, but, as additional heat and pressure are applied, the plastic is hardened by a chemical change known as *polymerization*.* A group of parts molded from thermosetting resins is shown in Figure 1. Thermoplastic materials undergo no chemical change in molding and do **not**

Courtesy The Hydraulic Press Manufacturing Company.

FIG. 1. A Group of Parts Molded from Thermosetting Resins. A Majority of These Parts Were Compression-Molded.

become hard with the application of pressure and heat. They remain soft at elevated temperatures until they are hardened by cooling, and may be softened repeatedly by the application of heat.

In actual production these molding compounds or resins are frequently mixed with other materials to provide special properties or to decrease the cost. Fillers of wood, flour, cotton, rag fibers, asbes-

* Polymerization is a chemical process resulting in the formation of a new compound whose molecular weight is a multiple of that of the original substance.

tos, powdered metals, graphite, and other materials are used for this purpose. The selection of the filler is based upon the properties wanted in the final products. In addition to fillers, dyes may be added to give the product the color desired. Plasticizers or solvents are used with some compounds to soften them or to improve their flowability in the mold. Finally, lubricants may be added to improve the molding characteristics of the compound. These materials are mixed with the granulated resins prior to molding.

Some of the common plastic compounds are as follows:

Phenol formaldehyde. This compound, originally developed by Dr. Baekeland, is one of the principal thermosetting plastics used in industry today. The synthetic resin made by the reaction of phenols with formaldehyde forms a hard high-strength durable material which is capable of being molded under a wide variety of conditions. This material has high heat and water resistance, and may be produced in a wide range of colors. It is known under such trade names as Bakelite, Durez, Textolite, Resinox, and is used for products such as molded cases, buttons, bottle caps, switch plates, knobs, hand wheels, and for laminated products.

Urea formaldehyde. This plastic component, also thermosetting, is generally known under the trade names of Bakelite-Urea, Beetle, or Plaskon. In its manufacture, the chemical combination of urea with formaldehyde forms a colorless resin especially adapted to light-colored articles which resist ultraviolet light and remain colorless for a long period of time. These products also have a hard surface, high dielectric strength, and are light in weight. Typical applications are tableware, light fixtures, buttons, instrument dials, veneer bonds, and clock cases. An interesting use of urea resins is that of a binder for sand cores in the casting of light metals.

Phenol furfural. Because of the scarcity of formaldehyde during the war, furfural, made from waste farm products, assisted in relieving a serious situation in raw materials. This resin flows readily at low molding temperatures and cures rapidly when the proper temperature is reached. Products of furfural are dark in color and water-resistant and can be fabricated by most processes. Commercial products of this material, known as Durite, are used as liners for Army helmets, cafeteria trays, binder for grinding wheels, munition parts, and numerous electrical articles.

Melamine. Melamine resins, principally melamine formaldehyde, are a comparatively recent addition to the plastic family. The compound melamine, made from carbon, nitrogen, and hydrogen, produces an excellent shock- and heat-resisting product. It is a thermosetting plastic and is adapted to processing by either compression or transfer molding. Being arc-resistant and having high dielectric strength, it is highly useful for electrical parts such as telephone sets, circuit breakers, and terminal blocks. Other uses include laminated products, tableware, buttons, and enamel. Trade names for this product are Melmac, Monsanto, Melamine, Catalin Melamine, and Plaskon Melamine.

Cellulose derivatives. Cellulose derivatives are thermoplastic and are widely used in the United States because of the availability of cellulose material in cotton and wood. Cellulose nitrate, the first to be used, is highly inflammable, but has the advantage of being extremely tough, water-resistant, and clear in color. Cellulose acetate is a more stable compound, having considerable mechanical strength and the ability to be fabricated into sheets or molded by injection, compression, and extrusion. Military insignia, pens, knobs, flashlight cases, bristle coating for paint brushes, radio panels, and extruded strips are successfully made of this compound. Cellulose acetate–butyrate molding compound is similar to cellulose acetate, and both are produced in all colors and by the same processes. In general, cellulose acetate butyrate is recognized for its low moisture absorption, dimensional stability under various atmospheric conditions, and its ability to be continuously extruded. It is used in wartime products such as kit cases, blackout lenses, whistles, foot tubs, radio dials, control knobs, and gas-mask parts. Yarns made from extruded fibers have proved successful for certain fabrics and woven materials. Ethyl cellulose, one of the newer cellulose plastics, is the lightest of the cellulose derivatives. In addition to its use as a base for coating materials, it is also used extensively in the various molding processes because of its stability and resistance to alkalies. Other outstanding properties are its low temperature, flexibility, and high impact strength. It may also be produced in thin sheets and in various extruded shapes. Cellophane (regenerated cellulose) is produced in thin sheets by an extruding process and is useful for packaging materials since it provides a protective coating against moisture and other contaminating influences.

Polystyrene. Polystyrene is a thermoplastic material especially adapted for injection molding. Some of its outstanding characteristics are low specific gravity (1.07), availability in a color range from clear to opaque, resistance to water and chemicals, dimensional stability, and insulating ability. Styrene resins are used in the production of synthetic rubber and in addition are molded into such products as battery boxes, dishes, radio parts, lenses, and insulators. Commercial products are made under the trade names Styron, Loalin, Lustron, and others.

Vinyl resins. A number of vinyl resins commercially available at the present time include copolymers of vinyl chloride and vinyl acetate, polyvinyl butyrals, polyvinyl chloride, vinylidene chloride, and polyvinyl acetate. All are thermoplastic materials capable of being processed by compression, injection, or extrusion, into a wide variety of products. The copolymers* of vinyl chloride and vinyl acetate (Vinylite and Geon) are obtained with a wide range of properties by varying the ratio of the two resins. They are especially suitable for surface coatings and for both flexible and rigid sheeting. In addition they are extruded and molded into many products, a special grade being used in the manufacture of Vinyon fibers which have chemical resistance and considerable strength. Polyvinyl butyral, used for interlayers in safety glass, raincoats, sealing fuel tanks, and flexible molded products, is a clear tough resin. It has resistance to moisture, great adhesiveness, and stability toward light and heat. Polyvinyl chloride resin has a high degree of resistance to many solvents and will not support combustion. It has found wide industrial use in resilient rubberlike products. Polyvinylidene chloride, known as Saran, Vec, or Velon, is another material capable of being processed by most commercial methods into a wide variety of products ranging from fabrics to flexible tubes. It has excellent physical properties and good colorability, is resistant to liquids, and will not support combustion. Polyvinyl acetate is used as an adhesive for bonding many materials and as a base for various coatings, lacquers, ink, and plastic wood.

In addition to the materials previously listed, many other synthetic and natural resins, protein substances, and other materials find application in the manufacture of plastics. Nylon monofilaments are used for hosiery, parachute shroud line, glider tow ropes, and brush bristles. The shellac resins are used as coating material, binder for abrasive wheels, phonograph records, and insulators. A variety of cold-molding compounds consisting of materials such as asbestos fibers with bituminous, cement, or shellac binders are made into such products as knobs, handles, connector plugs, and arc shields. Casein

---

* A copolymer is a complex product formed by simultaneous polymerization of two substances, the properties of which are different from either of the original polymers.

and protein plastics, nonflammable and in all colors, are made into buttons, novelties, trimming accessories, sheets, and tubing.

Methods of Processing

Plastic materials differ greatly from each other and lend themselves to many processing methods: Compression molding, transfer molding, injection molding, jet molding, casting, extrusion, blowing, and laminating. Each material is best adapted to some one of the methods, although many can be fabricated by several of the processes listed.

Courtesy General Electric Company.

FIG. 2.   Close-up View of Rotary Performing Press Used in Making Disk Pellets of Various Molding Compounds.

In most processes the molding material is in powder or granular form, although for some there is a preliminary operation of preforming the material before use.

Preforming. This operation consists of compressing a powder into small pellets of a size and shape which conforms to a known mold cavity. All preforms are of the same density and weight, and the operation avoids waste of material in loading molds and, in general, speeds up production by rapid-mold loadings. Also, there is no possibility of overloading the molds. In the performing operation the thermosetting powder is cold-molded, and no curing takes place. Preforms are used only in compression and transfer-molding processes.

A rotary preforming press used in making disk pellets of various molding compounds is shown in Figure 2. The powder is fed by gravity from the hopper into the mold cells, and any excess powder is scraped off. The amount of material fed into each cell is controlled by regulating the lower punch. As the table revolves, pressure is applied uniformly on both sides, compressing the powder charge, and at the end of the cycle the table is ejected. In some cases tablets of more than one size are made at the same time, the only objection to this procedure being that there is some difficulty encountered in sorting the tablets.

Reciprocating machines, having only a single set of dies, are also used for a wide variety of preforming operations. The dies can be changed quickly, but the output is much lower than that of the rotary machines. When preforms are used in multiple-cavity molds, they are first transferred to a loading tray. The tray locates them accurately with reference to the mold cavities in the machine, and all preforms are molded simultaneously.

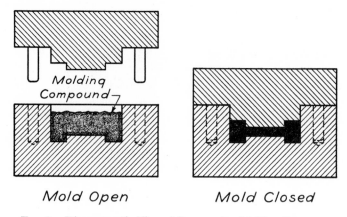

Mold Open   *Mold Closed*

FIG. 3.   Diagrammatic View of Compression-Molding Process.

Compression molding. Compression molding is illustrated in Figure 3. A given amount of material is placed into a heated metallic mold, and, as the mold closes, pressure is applied, causing the softened material to flow and conform to the shape of the mold.

The material can be used either in a granulated state or preformed into a tablet. Pressures used in compression molding vary from 1000 to 10,000 pounds per square inch, depending on the material used and size of the product. The temperature range is from 250 F to 400 F. Heat is very important for thermosetting resins, as it is required first to plasticize and then to polymerize or make them hard. Uniform heating of the powder is desirable, but not always easy to attain, because of the poor heat conductivity of the material. An important detail of the entire operation cycle is the unloading of the mold. The designer must make sure that pieces can be removed quickly, without damage or distortion.

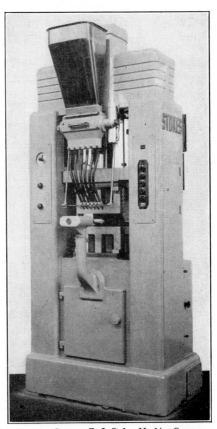

Some thermoplastic materials are processed by compression, but the cycle of rapid heating and cooling of the mold adds to the difficulty in using such material. Unless the mold is sufficiently cooled before ejection, distortion of the piece is apt to result.

A large variety of presses, ranging from hand-operated to completely automatic, are available for compression molding. The function of the press is to apply the necessary pressure and at the same time sufficient heat to plasticize properly and cure the plastic materials. Heat may be transferred from heated platens or applied directly to the

Courtesy F. J. Stokes Machine Company.

Fig. 4. Automatic Compression-Molding Machine.

metal mold. It is supplied by steam, heated liquids, electrical resistance, or ultrahigh-frequency electric currents.

Several automatic compression machines have been developed to take care of the mass production of small parts, such as bottle caps, electrical parts, buttons, and knobs. Figure 4 shows a 50-ton compression press equipped with an automatically controlled powder-measuring and feeding mechanism. The entire operation is automatic and accurately timed so as to maintain constant molding conditions. Automatic presses require a minimum of attention, are rapid in operation, and produce parts with uniform properties. They also can be equipped with attachments for automatic unscrewing and ejecting threaded parts, closures, and the like.

Transfer molding. In the process of transfer molding, the plastic powder or preforms are placed, not into the mold cavities, but into a

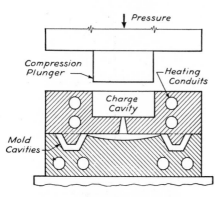

FIG. 5.   Arrangement for Transfer-
Molding.

pressure chamber above them, as illustrated in Figure 5. They are then plasticized by heat and pressure and injected into the mold cavities as a hot liquid where the material is cured and becomes hard. This process is especially desirable for producing parts requiring small metal inserts, since the hot plastic material enters the mold gradually and without great pressure. Intricate parts and those having large variation in section thickness can also be produced to advantage by this method. The process differs from the injection molding of thermoplastic materials in that the mold is kept heated at all times and parts are ejected without cooling. A recent development in the process of molding is the use of high-frequency electric current as a means of heating the mold.

A self-contained transfer-molding press for plastics is shown in Figure 6. The mold is closed and clamped by the upward-acting press platen. A center opening in the upper grid or bolster permits the operator to drop the molding material into the die sleeve or well, where the material is preheated. A downward-acting plunger forces the molding material into the mold where it is cured under heat and pressure. This press can be used universally for regular transfer molding, conventional compression molding, or electronic molding, which requires the addition of a high-frequency unit for plasticizing the molding material.

Injection molding. The principle employed for injection molding is shown in the diagrammatic sketch in Figure 7. The operations of this process are very similar to that of plunger-type die-casting machines. Molding material is fed by gravity from a hopper to a circular heating chamber, where it is compressed, softened, and finally injected into the closed mold under considerable pressure. The finished product is hardened in the mold by the cooling effect of water circulated through conduits in the mold. As the injection plunger retracts, the mold is opened and the product ejected.

An automatic-injection-molding machine of 16-ounce capacity is shown in Figure 8. A measured quantity of solid granular material is fed into the cylinder, where it is plasticized by heat and then forced into the mold by hydraulic pressure. The injection heating chamber for most machines is kept at a temperature between 350 F to 420 F, depending on the kind of material being charged and the size of the mold. Usually heat is furnished by a series of electrical-resistance coils thermostatically controlled, but in some machines heated oil is used. Another automatic injection machine of different design in illustrated

Courtesy The Hydraulic Press Manufacturing Company

Fig. 6. A Self-Contained Transfer or Compression-Molding Press.

in Figure 9. In this machine the injection cylinder unit is in a vertical position rather than horizontal; otherwise, the general features of the two machines are similar. The vertical arrangement lends itself to simple construction and ease in feeding the material into and through the heating chamber. The entire heating cylinder

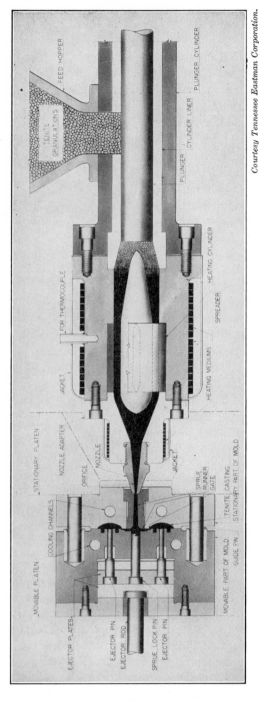

Courtesy Tennessee Eastman Corporation.

Fig. 7. Sectional View Illustrating General Method of Injection Molding.

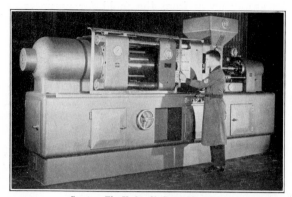

Courtesy The Hydraulic Press Manufacturing Company.

FIG. 8.   A Fully Automatic 16–Ounce Capacity Plastics
Injection-Molding Machine.

Courtesy Lester Engineering Company.

FIG. 9.   An Automatic Injection-Molding Machine, 22-Ounce Capacity, Having
the Heating-Cylinder Assembly in a Vertical Position.

unit can be swung away from the mold for purposes of cleaning, repair, and change of nozzles. This 22-ounce machine with a maximum output of around four shots per minute is especially adapted for the manufacture of steering wheels and other large parts.

FIG. 10. Close-up View of 10D 8-Ounce Plastic Injection-Molding Machine Showing Mold in Open Position.

Although the capacities of injection machines vary from 2 to 32 ounces, the custom production of small-parts machines of from 8- to 16-ounce capacity is most popular. A close-up view of an 8-ounce machine with the mold in open position is shown in Figure 10. In this set-up a 24 cavity mold is used in the production of a small part.

Thermoplastic materials are generally used in injection molding, as they are especially suited for rapid production. Compared to compression molding, this process is much faster, since the mold does not have to be alternately heated and cooled. The mold is maintained at a constant temperature, usually 165 F to 200 F, by circulating water; and a production cycle of from two to six shots per minute is possible. Mold costs are lower, as fewer cavities are necessary to maintain equivalent production by compression molding.

Courtesy Tennessee Eastman Corporation.

Fig. 11. Automobile-Heater Switch-Knob Castings Made in an Injection-Molding Machine.

Articles of difficult shapes and thin walls are successfully produced, as illustrated in Figure 11. Metal inserts, such as bearings, contacts, or screws, can be applied in the mold and cast integrally with the product. Material loss in the process is low, as sprues and gates can be reused.

Thermosetting materials can be injection-molded by a process known as *jet molding*. With a few minor changes, nearly any standard thermoplastic injection-molding machine can be converted to a jet-molding machine. The torpedo spreader in the heater is removed, and the main heating of the resin materials is concentrated at the nozzle passage to the sprue. As soon as the mold is filled, the nozzle area is cooled by circulating water, and the pressure in the chamber is released. No further chemical changes of the material take place until the cycle is repeated.

Extruding. Thermoplastic materials such as the cellulose derivatives, vinyl resins, polystyrene, and others, may be extruded through dies into simple shapes of any desired length. Thermosetting compounds are not adapted to this process because of the rapidity with which they harden, but they may be used to a limited extent.* A schematic diagram of a typical extruding press is shown in Figure 12. Granulated or powdered material is fed into the hopper and then forced through a heated chamber by a spiral screw. In the chamber the material becomes a thick viscous mass, in which form it is forced through the die. As it leaves the die, it is cooled by air or water

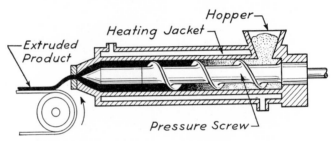

FIG. 12. Schematic Diagram of Typical Plastic Extrusion Press.

and gradually hardens as it rests on the conveyor. Long tubes, rods, molding sections, and many special sections are readily produced in this manner. Because they can be bent or curved to various shapes after extrusion by immersing in hot water, such products as conduits for electric conductors and for handling chemicals are made by this process.

Casting. Most casting is done with the phenolic and acrylic resins. Both of these resins can be produced in numerous attractive color combinations. Polystyrene and methyl methacrylate also can be cast where transparent articles are desired. Casting of plastics is done when there is not a sufficient number of parts desired to justify the making of expensive dies. Frequently open molds of lead are formed by dipping a steel mandrel of desired shape into molten lead and stripping the shell from the sides of the mandrel after it solidifies. Cores of lead, plaster, or rubber may be introduced if desired. Hollow castings are also produced by the slush-casting method. Solid objects may be made from molds of plaster, glass, wood, or

* A process for the continuous extrusion of thermosetting materials has been developed by Chrysler Corporation.

metal. In cases where parts have numerous undercuts, the molds are made of synthetic rubber. Thermoplastics have only a limited use in the casting process, since most of them lack the necessary fluidity for pouring. Ethyl cellulose is the principal material used; when casting, pressure is applied vertically on the molds.

Casting is recommended for preparing short rods, tubes, and various shapes that are to be used in subsequent machining operations or carving. Machined surfaces have a dull white appearance which may be removed by tumbling with wood blocks and abrasive particles or by buffing. Costume jewelry and novelties are cast because of the pleasing color combination that can be obtained and the fact that frequent style changes do not justify the preparation of expensive die equipment. Other examples of cast products are knobs, clock and instrument cases, handles, drilling jigs, and punches and dies for sheet-metal fabrication in the airplane industry.

Blowing. Many plastic materials can be formed into thin hollow articles by blowing when the material is in a soft pliable condition. The starting material may be either in sheet form or extruded as tubing. A single sheet can be expanded into a spherical shape by clamping and sealing the edges onto a mold and then introducing air pressure into the mold. Two flat sheets can be clamped in a split mold and pressure introduced between them. The sheets are forced to conform to the outlines of the mold, and heat seals the edges of the two sheets. Sections of extruded tubing are convenient forms to use in this process, as they require sealing only at the end. Steam, hot water, or air may all be used as means of creating an internal pressure. Articles made by blowing include toys, Christmas-tree ornaments, atomizer bulbs, bottle-shaped containers, and numerous synthetic-rubber articles such as hot-water bottles.

Laminated plastics. Laminated plastics consist of sheets of paper, fabric, asbestos, wood, or similar materials which are first impregnated or coated with resin and then combined under heat and pressure to form commercial materials. These materials are hard, strong, impact-resisting, unaffected by heat or water, and have desirable properties for numerous electrical applications. The final product may consist of either a few sheets or over a hundred, depending on the thickness and properties desired. Although most laminated stock is made in sheet form, rods and tubes as well as special shapes are available. The material has good machining characteristics which permit its fabrication into gears, handles, bushings, furniture, and many other articles.

In the manufacture of laminated products the resinoid material is dissolved by a solvent to convert it into a liquid varnish. Rolls of paper or fabric are then passed through a bath for impregnation, as shown in Figure 13. This is a continuous operation, and, as the sheet leaves the resinoid bath, it goes through a drier, which evaporates the solvent, leaving a fairly stiff sheet impregnated with the plastic material. To facilitate lamination, the sheets are then cut into convenient sizes and stacked together in numbers sufficient to make up the desired thickness of the final sheet. Each group is assembled between polished metal plates at top and bottom and is then stacked

Courtesy Bakelite Corporation.

FIG. 13.   Rolls of Paper or Fabric Passing through Resinoid Bath for Impregnation.

in a hydraulic press, as shown in Figure 14. Under the action of heat and pressure, a hard rigid plate having desirable properties for many industrial applications, is obtained.

Properties of laminates depend largely on the sheet material or filler used. Paper-base materials are used often in electrical applications because of their excellent electrical characteristics and their ability to be held to close tolerances.* Fabric-base materials are stronger and tougher and hence better for stressed applications. Gears made of a canvas base are quiet in operation and have proved

* According to NEMA (National Electrical Manufacturers Association) there are 12 grades of laminates which are standard for most applications. Six grades have paper base, four have fillers of cotton fabric, one has an asbestos paper filler, and one uses asbestos fibers.

very satisfactory in many applications. Asbestos and fiber-glass cloth are recommended for heat-resisting and low water-absorption uses. Thin sheets of wood are now being laminated to produce a light material equal in strength to some metals and resistant to moisture. These sheets, produced with a smooth surface, can be formed into panels of desired shape without expensive machining

Courtesy Bakelite Corporation.

FIG. 14. Press for Baking Laminated Fabric Sheet Bonded with Phenolic Resin.

operations. Safety glass is in effect a laminated plastic product, since thermoplastic layers are used between the glass sheets to give it the nonshattering characteristics. In addition to the latter application, many other materials, including rubber, metal, rayon, and spun glass, are used in the manufacture of laminated products.

A great amount of plastic laminated-sheet stock is fabricated by punching and blanking dies, as shown in Figure 15. This illustration shows a progressive die which is punching and blanking radio insula-

Courtesy Bakelite Corporation.

FIG. 15.   Punching Radio Insulation from Bakelite Laminated Punch State.

Courtesy Bakelite Corporation.

FIG. 16.   Stamped and Punched Pieces of Laminated Plastics Are Readily Produced in Great Variety.

tion plates. Linen, canvas, and paper-base stock are used for this type of fabrication because of their strength, close dimensional control, and excellent electrical properties. Figure 16 shows a variety of blanked and punched articles typical of the special shapes that are possible. Thin sheets may be blanked cold, but for sheets over 1/8 inch thick, it is advisable to preheat the stock to a temperature of about 280 F. Plastic materials in sheet, rod, and tube form may also be processed by the usual machining tools. Planing, shaping, lathe turning, threading, milling, and screw-machine operations can all be performed on this material. Cutting, rake, and clearance angles on tools differ from steel and should be checked with manufacturers' recommendations before any machining operation is attempted with plastics.

Transparent Plastics

An important development in the manufacture of plastics is a clear acrylic resin which has excellent light transmitting power as well as resistance to moisture. This material is chemically known as methyl methacrylate, but is better known by the commercial names Lucite (E. I. Du Pont de Nemours and Company) and Plexiglas (Rohm and Haas Company). Even though the source of raw materials for this product, namely coal or petroleum, air, and water, are quite abundant, this product is more expensive than most plastics because of the elaborate chemical processes involved in its manufacture. Its use is limited to applications where transparency is necessary; for example, airplane windows and covers for instruments and meters.

Transparent Plexiglas sheets are cast in special individual molds and are removed from the molds with a polished surface. These sheets can be heated in an oven and bent to any shape desired, or can be processed by the usual machining operations. In Figure 17 is shown the top turret of a bomber as it is being removed from a forming mold. The plastic sheet is first heated to a temperature of about 300 F and the edges clamped to the bed of the press. A vacuum is created below the sheet to stretch and draw it into the chamber at the press. The forming die is then lowered from above into position, and the vacuum released, causing the sheet to snap back against the forming die and conform to its shape. The lower part of the die is seldom used in such forming, as without it there is less tendency to mar the sheet, and cooling is accelerated. Because it is less than half the weight of glass, is strong, and is permanently transparent,

Plexiglas is used for gun turrets, observation hatches, and cockpit enclosures on all types of fighting planes. Through these transparent plastic parts, the crew can see their objective on raids or sight a machine gun if attacked. It is interesting to note that the transparent

Courtesy Rohm & Haas Company.

Fig. 17. Removing a Plexiglas Top Turret From Forming Chamber.

acrylic resins may also be processed by extrusion, compression, and injection molding. Many articles such as brush backs, cosmetic containers, ornaments, rods, tubes, gage covers, models, lenses, and containers are made by these processes.

Lucite, a methyl methacrylate resin, is also made into a molding powder suitable for compression, casting, or injection technique. Cast Lucite has a clearness exceeding that of glass, an ability to carry light around curves, freedom from color, lightness of weight, and weather resistance. Converted into a clear and practically unbreakable solid, it is used in the decorative arts, scientific equipment, including self-illuminating dental and surgical instruments, highway reflectors, and aircraft enclosures. Figure 18 shows one of the in-

Courtesy E. I. Du Pont de Nemours, Inc.

FIG. 18.   Inspection of Lucite Sheeting.

spections of Lucite sheeting after the casting operation.   These sheets can be heated and shaped to three-dimensional forms for airplane turrets, cowlings, and cockpit enclosures.

In addition to methyl methacrylate, transparent articles can also be made from the cellulose compounds, polyvinyl chloride acetate, polyvinyl chloride, polyvinylidene chloride, polyvinyl butyral, polystyrene, and polyvinyl alcohol.   Films, sheets, containers, and many other parts are made from these resins.   The selection of the respective resin depends on the properties desired and the manner in which it can be fabricated.

Synthetic Rubber

Attempts to synthesize natural rubber have been made for many years because of the commercial utility and numerous applications of this material.   The fact that many highly industrialized nations had no source of raw rubber under their control was also a contributing factor in developing research along this channel.   Recent research has developed many synthetic materials possessing some of the characteristics of rubber, but out of the group there are only five which have been accepted as synthetic rubbers.* They are Thiokol, Neoprene, Buna S, Buna N, and Butyl. Of this group Buna S, also known by Government terminology as GR-S, is produced in the largest quantity and is particularly adapted for tire

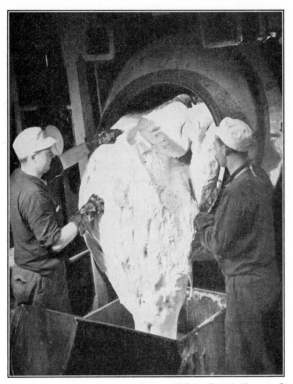

Courtesy E. I. Du Pont de Nemours, Inc.

FIG. 19. Neoprene Being Dumped from a Polymerization Kettle.

use. It is a copolymer of butadiene and styrene and can be cured to any degree of hardness desired. Buna N, a copolymer of butadiene and acrylonitrile, is exceedingly resistant to oils and has wide use in the manufacture of oil hose, bullet-sealing fuel cells, and similar products. The organic polysulfides, known as Thiokols, are also very resistant to gasoline, oils, paints, as well as sunlight, and are

* A. Black, " Recent Developments in Engineering Materials," *Mechanical Engineering,* April 1945.

used in the manufacture of many types of hose, coated fabrics, gaskets, and fuel tanks.

The chloroprene polymer, known as Neoprene, is produced from the basic raw materials: coal, limestone, water, and salt. Calcium carbide, a product of coal and limestone, when added to water, forms acetylene gas (C_2H_2). This gas, in combination with hydrogen chloride, forms chloroprene which is changed to Neoprene by polymerization. Figure 19 shows a batch of Neoprene being dumped from a polymerization kettle and illustrates the rubberlike consistency of this material. Neoprene has good resistance to oils, heat, and sunlight, and is used for such articles as conveyor belts, shoe soles, protective clothing, insulation, and hose. Butyl rubber, one of the most recent synthetics, has excellent ageing qualities and is resistant to attack by chemicals and ozone. For this reason it is useful for chemical tubing, life rafts, gas masks, and tires.

Many of the synthetic rubbers are stable under conditions which cause rapid deterioration of natural rubber. None of them have the same chemical formula as natural rubber, but they do have many of the same characteristics. Synthetics differ in physical and chemical properties, each serving best in its own field. Final properties may be controlled some by the kind of fillers added. In general, the synthetic rubbers are slightly inferior to natural rubber in physical properties, and they cost more; but they are superior in their resistance to oils, chemicals, sunlight, and heat.

Molds for Plastics

Molds for both the compression and injection processes are made of steel and are heat-treated. The production of these molds demands the same type of machine work and the usual precision required for dies used in pressure casting. There are, however, some differences in construction because of varying characteristics in the materials being processed. Ample draft and fillets should be provided to facilitate removing the article from the mold. Ejector pins are usually provided for this purpose and should be located at points where the pin marks are not noticeable. Like metals, plastic materials shrink upon cooling, and some allowance must be provided. Shrinkage varies according to the type of material and method of processing, but is usually 0.003 to 0.009 inch per inch.

Compression molds are made in *hand* and *semiautomatic* types. Each type might be further subdivided into *positive*, *semipositive*, and *flash* designs. The hand molds are charged and unloaded on a bench. Heating and cooling are accomplished by plates on the presses which are provided with the necessary circulating facilities. The semiautomatic molds are fastened rigidly to the presses and are heated or cooled by adjacent plates. Work is ejected automatically from the molds as they open. Both of these types are made in either single- or multiple-cavity molds.

Further classification of hand and positive molds may refer to the method of confining the plastic material in the mold. A positive mold is one that entirely confines the material in the mold. An example of this is a cylinder with a close-fitting plate at the bottom and a plunger which enters the cylinder to compress the powder. The thickness of the part being molded is controlled by the amount of powder charged. Molds of this type are commonly used in mounting metallurgical specimens. Semipositive molds are similar except that the plunger enters only a slight amount at the end of the stroke, and there is some possibility of overflow. This type of mold is used more than the other two in compression molding. The flash-type mold has no telescoping action whatsoever, and the material is not confined until the end of the stroke. As a result there is a slight fin, a few thousandths of an inch thick, around the lands of the mold where the closing is made.

A flash-type mold for a base to be made of thermoplastic material is illustrated in Figure 20. The small insert is first put in place, and then the granulations are charged slightly in excess of the weight of the part to be molded. Steam is circulated through the channels shown in both the upper and lower parts of the mold. As the mold is closed under pressure, the steam is turned off, and water is circulated through the same cavities to cool the mold. When the correct temperature is reached, the mold is opened and the part ejected by the pin shown at the bottom. This type of mold is especially adapted to thin flat articles.

Since most compression molding is done with thermosetting materials, alternate heating and cooling of the mold is not required. The molding temperature for these plastics is obtained from gas, steam, electricity, or heated liquids. Heat not only softens the plastic material so that it can be shaped to the mold, but also causes the chemical change which hardens it.

Injection molds are made in two pieces, one half being fastened to the fixed platen and the other half to the movable platen. Contact between the halves is made on accurately ground surfaces or lands surrounding the mold cavities. Neither half telescopes with the other, as is the case with many of the compression molds. The cavities should be centrally located with reference to the sprue hole in the fixed half so as to obtain an even distribution of material and

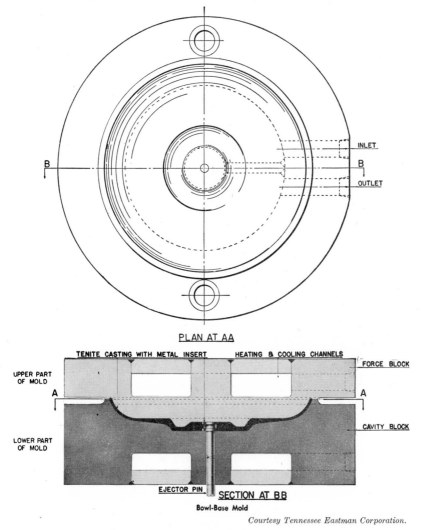

PLAN AT AA

TENITE CASTING WITH METAL INSERT   HEATING & COOLING CHANNELS
FORCE BLOCK
UPPER PART
OF MOLD
A   A
CAVITY BLOCK
LOWER PART
OF MOLD
EJECTOR PIN   SECTION AT BB
Bowl-Base Mold

INLET
B   B
OUTLET

Courtesy Tennessee Eastman Corporation.

Fig. 20.   Bowl Base — Flash-Type Compression Mold.

pressure in the mold. For locating purposes, guide pins are used which are similar to those used on metal-press dies. These pins are fastened in the fixed half of the mold and enter hardened bushings in the movable part of the mold.

Mold cavities can extend into both halves of the mold. It is best,
however, to have the outside of the molded part in the fixed half,
providing the shape is suitable for this arrangement. In the cooling
process the plastic material tends to shrink away from the cavity

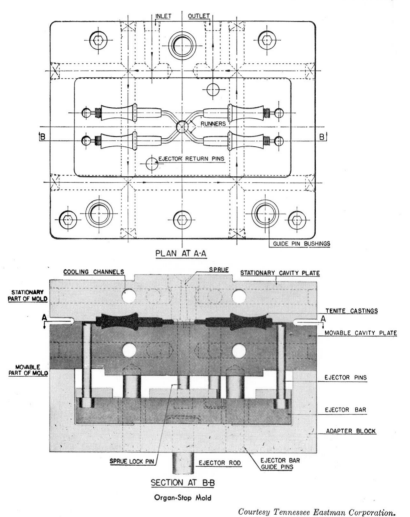

PLAN AT A-A

SECTION AT B-B

Organ-Stop Mold

Courtesy Tennessee Eastman Corporation.

Fɪɢ. 21.   Organ Stop — Four-Cavity Injection Molds.

walls and is withdrawn from this half as the mold opens. It is retained on the cores of the movable half until the ejector mechanism
operates.

Injection molds have cooling cavities in both halves to permit maintaining a uniform temperature for chilling the molded part, since most materials fabricated by this process are thermoplastic. The material is forced into the mold from the heater cylinder under pressure ranging from 2 to 20 tons per square inch and ejected from the mold at a temperature of approximately 125 F. Ejection of the parts occurs as the mold opens and is accomplished by either ejector pins or stripper plates.

An illustration of a typical injection-mold design is shown in Figure 21. This is a four-cavity mold with half the cavity extended into each plate. After the mold is filled, it opens with the sprue lock pin pulling the sprue. As the mold continues to open, the ejector rod is contacted, which pushes forward the plate holding the ejector pins. This, in turn, pushes forward the ejector pins until the molded parts are free from the cavity. These pins are forced back into position as the mold closes.

Any cores required in injection molding are placed on the movable half of the mold. The normal shrinkage of the molded part tends to cling to the cores, causing it to withdraw freely from the stationary half as the mold opens. Vents, to permit the escape of entrapped air, are extremely small and are usually located at the point farthest away from the sprue hole.

REVIEW QUESTIONS

1. List the principal materials used in the manufacture of plastic articles.
2. Distinguish between thermosetting and thermoplastic materials.
3. Name the various methods used in forming plastic materials.
4. Describe the process of injection molding.
5. What is meant by polymerization?
6. Illustrate by sketch how transfer molding is done.
7. What are the advantages or important characteristics of products made from plastic materials?
8. What are plasticizers, and why are they used in molding compounds?
9. Name four synthetic or natural resins, and state some outstanding characteristic or use of each.
10. What are preforms, and why are they used?
11. Describe the extruding process, and state what kind of plastic materials are used.
12. What kind of molds are used in the casting of plastics?
13. What various materials are laminated with plastic material? Name one final product for each.
14. Describe the process of laminating fabric or paper materials.
15. What raw materials are used for manufacturing transparent plastics?
16. What processes are used for making transparent plastics?
17. How does synthetic rubber differ from natural rubber?

18. How is Neoprene made? What are the raw materials used?
19. List the different types of molds used for making plastics.
20. What is a positive mold? A flash-type mold?
21. Describe the type of mold used on an injection-molding machine.
22. Is venting necessary in molds for plastic materials? If so, how is it done?

BIBLIOGRAPHY

AKIN, R. B., " Vinyl Resins in War and Peace," *Chemical Industries*, April and May, 1943.

ASTM Standards on Plastics, 1943, American Society for Testing Materials.

Bakelite Molding Technique, Bakelite Corporation, 1935.

Bakelite Molding Plastics, Bakelite Corporation, 1940.

BELL, L. M. T., *The Making and Moulding of Plastics*, Chemical Publishing Company.

DEARLE, D. A., *Plastic Molding*, Chemical Publishing Company, 1941.

DELMONTE, J., *Plastics in Engineering*, Penton Publishing Company, 1940.

DuBois, J. H., *Plastics*, American Technical Society, 1942.

LEHMANN, GEORGE P., " Proper Molds for Plastics," *American Machinist*, June 25, 1941.

LOUGEE, E. F., *Plastics from Farm and Forest*, Plastic Institute, Chicago, 1943.

Plastics Catalog, 1944, Plastics Catalogue Corporation.

" The Processing of Plastics," *American Machinist*, June 28, 1939.

RAHM, L. F., *Plastic Molding*, McGraw-Hill Book Company, 1933.

SASSO, J., *Plastics for Industrial Use*, McGraw-Hill Book Company, 1942.

SIMONDS, H. R., and ELLIS, C., *Handbook of Plastics*, D. Van Nostrand Company, Inc., 1943.

SIMONDS, H. T., *Industrial Plastics*, Pitman Publishing Company.

Tenite Molding, edition 5A-1940, Tennessee Eastman Corporation.

Tenite Specifications, edition 6B-1941, Tennessee Eastman Corporation.

THAYER, G. B., *Plastic Mold Designing*, American Industrial Publisher, 1941.

CHAPTER 6

HEAT TREATMENT OF STEEL

Heat treatment is the operation of heating and cooling a metal or alloy in its solid state. Steel responds to this treatment in a unique fashion: its physical properties can be greatly changed according to the procedure involved. Therefore, the purpose of heat treating is to enhance certain desired properties in steel. For example, a tool that has been machined can be made hard to resist cutting action and abrasion, whereas another part, already hard, can be softened so that further machine work can be done. With the proper treatment, internal stresses may be removed, grain size reduced, toughness increased, or a hard surface produced on a ductile interior. To use the correct treatment for desired properties, the analysis of the steel must be known; for small percentages of certain elements, notably carbon, greatly change the physical properties.

The following discussion applies principally to the ordinary commercial steels known as carbon steels, in which carbon is the controlling element. Alloy steels are those which owe their properties in a marked degree to the presence of one or more elements other than carbon. The elements used in the manufacture of alloy steels include nickel, chromium, manganese, molybdenum, tungsten, silicon, vanadium, copper, and cobalt. When one alloying element is used together with iron and carbon, the alloy is known as a *ternary* steel; steels containing two special elements are called *quaternary* steels. Their greatly improved physical properties enable alloy steels to fulfill many important commercial applications not possible with carbon steels.

The treatments discussed in this chapter apply only to steel. Bronze given similar treatments will not react the same, because of the nature of the elements it contains. For example, if the bronze were in a hardened state from cold working, it could be softened by heating followed by any rate of cooling. For steel, on the other hand, the rate of cooling is a controlling factor. Rapid cooling from above the critical point results in a hard structure, while very slow cooling has the opposite effect. To understand these changes, a

151

knowledge of the structure of steel and its various constituents is necessary.

Constituents of Steel

In most commercial steels of the low- and medium-carbon types, the total sum of all elements other than iron is not more than 1%. The usual elements found within this 1% are carbon, manganese, silicon, sulfur, and phosphorus. Of these, the carbon is the most important so far as its influence on physical properties is concerned. It does not exist in the steel as free carbon, but as an iron carbide known as *cementite* (Fe_3C). It is the carbon in this carbide state that gives the steel its strength, hardness, and ability to respond to

FIG. 1. Pearlite, a Laminated Structure of Ferrite and Cementite.
Magnification ×750

heat treatment. Cementite is the hardest constituent of steel, and its volume is about 15 times that of carbon. The upper limit for carbon is 2.0%, although few steels have greater than 1.5%. Above 2.0% the alloys are in the cast-iron range.

If a specimen of annealed medium-carbon steel, such as SAE 1040, is polished to a mirror surface and then etched lightly with a dilute solution of hydrochloric acid, grain boundaries and certain constituents can be observed under high-power magnification. In the micrograph shown in Figure 5c two such constituents are in evidence. The light area is pure iron and is called *ferrite* (from Latin word ferrum, " iron ").

The dark area is a constituent known as *pearlite*, so called because of its lustrous appearance under the microscope. Pearlite exists in most cases as a lamellar constituent composed of alternate layers of ferrite and cementite. High magnification of ×500 or over clearly shows this construction. This constituent under high magnification may be seen in Figure 1. There are some treatments which convert the pearlite into round globules or spheres of carbides. This process is known as *spheroidizing,* and it is usually done to facilitate machining.

Iron–Carbon Diagram

If a piece of 0.20% C steel is uniformly heated, a temperature–time curve similar to Figure 2 will be obtained. It is clearly evident

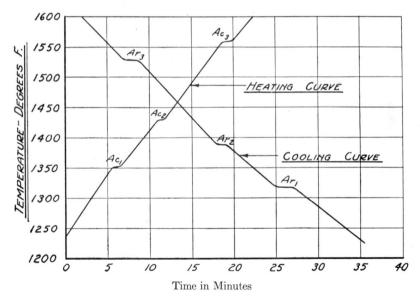

Fig. 2.   Time–Temperature Heating Curve for SAE 1020 Steel.

that at three temperatures there is a definite change in the heating rate. In a similar fashion these same three points again show upon cooling, but occur at slightly lower temperatures. These points are known as critical points and on heating are designated by the symbols Ac_1, Ac_2, and Ac_3. The letter " c " is the initial letter of the French word *chauffage*, meaning " to heat." The points on the cooling curve are designated by Ar_1, Ar_2, and Ar_3, the " r " being taken from the word *refroidissiment*, meaning " to cool." Certain changes which take place in the steel at these critical points are called *allotropic changes*. Although the chemical content of the steel remains the same, its properties are changed. Principal among these are changes in electrical resistance and atomic structure, and

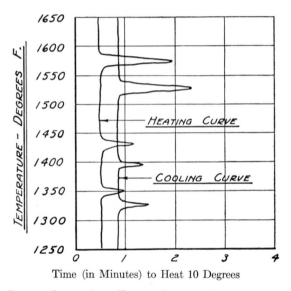

Time (in Minutes) to Heat 10 Degrees

FIG. 3.   Inverse-Rate Heating Curve for SAE 1020 Steel.

loss of magnetism. These critical points should be known, as most heat-treating processes require heating the steel to a temperature above this range. It must also be kept in mind that these points are not fixed points, since they vary with the rate of heating and cooling. Steel cannot be hardened unless it is heated to a temperature within or above the upper critical range.

As the critical points frequently do not show up clearly on a temperature–time curve, the data may be plotted another way which accentuates these points. Such a curve is shown in Figure 3 and is called an inverse-rate heating and cooling curve. The abscissa in this case is heating rate, or number of seconds required to heat or cool the steel 1 degree. The curve then becomes a vertical line except at the critical points where the heating or cooling rates show marked change.

If such curves were drawn for a piece of 0.40% C steel, it would be noted that there were only two arrest or critical points; and, for

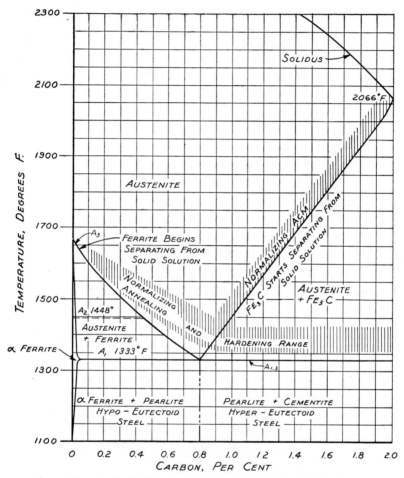

Fig. 4.   Partial Iron–Iron-Carbide Equilibrium Diagram.

0.80% C steel, there would only be one such point.   If all these points were plotted on a temperature–percentage-carbon curve, a diagram similar to Figure 4 would be obtained.   This diagram, which applies only under slow cooling conditions, is known as a partial iron–carbon or iron–iron-carbide diagram.   The latter term is frequently used. because the carbon in steel is nearly always in the carbide form.

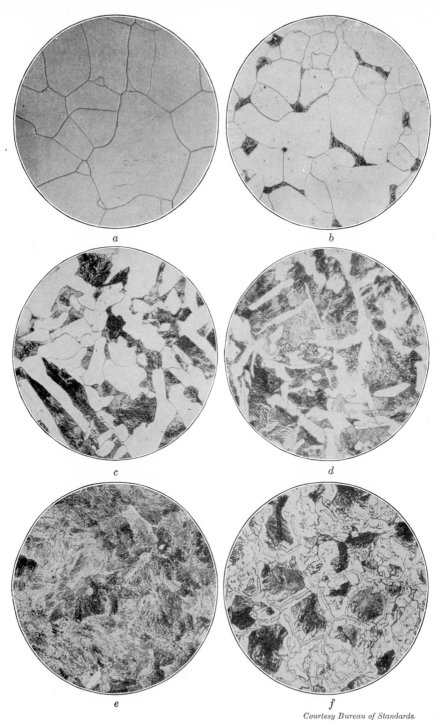

FIG. 5. Microphotographs of Iron–Carbon Alloys with Increasing Carbon.

(a) High-Purity Iron. (b) 0.12% Carbon. (c) 0.40% Carbon.
(d) 0.62% Carbon. (e) 0.79% Carbon. (f) 1.41% Carbon.

The importance of this diagram is evident, for by reference to it the proper quenching temperature for any carbon-content steel may be readily obtained.

It has been stated that there is a change in the atomic structure of the ferrite as it is heated. Below the Ac_1 line it is known as ferrite and has a cubic structure known as a " body-centered " cubic lattice. These cubes are exceedingly small, being less than one ten-millionth of an inch. One of the principal characteristics of ferrite is that it is strongly magnetic. As the temperature increases, this iron changes to another form, known as β iron, and finally above the upper critical point it has changed completely to another form known as γ iron. The crystal or cubes of gamma iron have a face-centered cubic lattice arrangement and are nonmagnetic. Iron in this form has the ability to absorb carbon. This solid solution of gamma iron and iron carbide, called *austenite*, is the constituent into which all steel must be converted before it can be hardened.

If a low-carbon steel is heated to the point where it is made up entirely of austenite and then allowed to cool slowly, ferrite will be rejected after passing the Ar_3 line. The remaining constituent will be pearlite. As the carbon content of the steel increases, the amount of ferrite rejected will be less and less until at 0.80% C the steel will be made up of 100% pearlite. This steel is known as a *eutectoid* steel. Such a steel is one that changes its composition in a solid state at the lowest temperature. Steels having less carbon than the eutectoid steel are called *hypoeutectoid* steels; those having more carbon are known as *hypereutectoid* steels. Since hypereutectoid steels cannot have any more pearlite, the additional carbon that is rejected in slow cooling is in the form of free cementite (Fe_3C). This rejection of cementite starts at the Acm line in the structural diagram and continues until the Ar_{123} line is reached. The photomicrograph in Figure 5f has as its constituents pearlite and cementite. This specimen has been cooled slowly and has a carbon content of 1.41%.

So far as steel treating is concerned, the partial diagram shown in Figure 4 is sufficient, as 2.0% is the limit of carbon content in steel. If the diagram is extended to include the cast irons having carbon contents up to 6.67%, it will appear as shown in Figure 6.

Effect of Carbon on Microstructure of Steel

In Figure 5 are shown six photomicrographs of iron with increasing amounts of carbon. The first photomicrograph is of high-purity iron, the structure being made up entirely of ferrite. As the percentage of carbon increases, the dark areas which are pearlite rapidly increase in size until at 0.79% C the entire area is almost all of this constituent. Above the eutectoid point, 0.80% C, two constituents

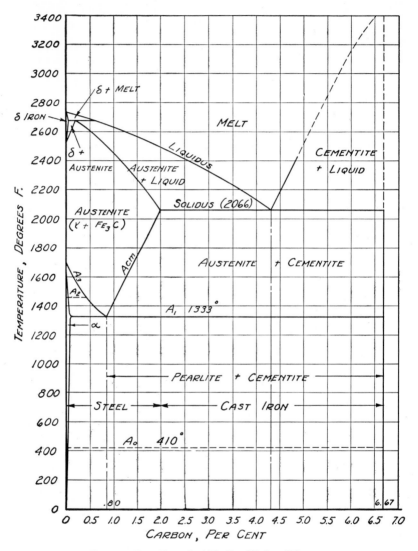

FIG. 6. Iron–Iron-Carbide Equilibrium Diagram.

again appear, as shown in the specimen containing 1.41% C. In this photomicrograph the light area is made up of the iron carbide called cementite. These iron–carbon alloys have all been cooled slowly to give the constituents just described. Those obtained by rapid quenching are described later.

Hardening

Hardening is the process of heating a piece of steel to a temperature within or above its critical range and then cooling it rapidly. If the carbon content of the steel is known, the proper temperature to heat the steel may be obtained by reference to Figure 4, the iron–carbon diagram. However, if the composition of the steel is unknown, a little preliminary experimentation may be necessary to determine the range. A good procedure to follow is to heat and quench a number of small specimens of the steel at various temperatures and observe the results, either by hardness testing or by microscopic examination. When the correct temperature is obtained there will be a marked change in hardness and other properties. Another method, which is quite simple, is to heat up the steel slowly and to check it periodically with a small magnet attached to a brass rod. As soon as the steel reaches the point where it will no longer attract the magnet, the temperature is within or above the critical range, and the steel may be quenched.

In any heat-treating operation the rate of heating is important. Heat flows from the exterior to the interior of steel at a definite maximum rate. If heated too fast, the outside becomes hotter than the interior, and uniform structure cannot be obtained. If a piece is irregular in shape, a slow rate is all the more essential to eliminate warping and cracking. The heavier the section, the longer must be the heating time to achieve uniform results. Even after the correct temperature has been reached, the piece should be held at that temperature for a sufficient period of time to permit its thickest section to attain a uniform temperature. The American Society for Steel Treating has worked out definite recommendations for heating rates based on both size and composition of the steel.

The hardness obtained from a given treatment depends upon the quenching rate, the carbon content, and the work size. In alloy steels the kind and amount of alloying element also have an influence on the hardness.

A very rapid quench is necessary to harden low- and medium-carbon steels. Quenching in a bath of water is considered to be rapid cooling and is common practice for low- and medium-carbon steels. For high-carbon and alloy steel oil is generally used as the quenching medium, because its action is not so severe as water.

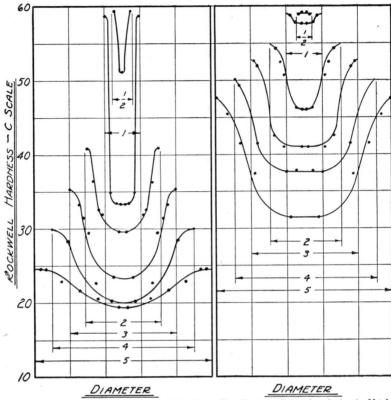

Courtesy M. A. Grossman, "Principles of Heat Treatment," American Society for Metals.

FIG. 7.   Hardness Distribution   FIG. 8.   Hardness Distribution
Curve SAE 1045.   Curve SAE 6140.

Various commercial oils, such as mineral oil, have different cooling speeds and consequently impart different hardnesses to steel on quenching. For extreme cooling, brine or water spray is most effective. Certain alloys can be hardened by air cooling, but for ordinary steels such a cooling rate is too slow to give any appreciable hardening effect. Large parts are usually quenched in an oil bath. This quenching medium has the advantage of cooling the part down to ordinary temperatures rapidly and yet is not too severe. It should be remembered that the temperature of the quenching medium must be kept uniform to achieve uniform results. Any quenching bath used in production work should be provided with means for cooling.

Steel with low carbon content will not respond appreciably to hardening treatments. The predominating constituent of such steel is ferrite, which is soft and not changed by the treatment. As the carbon content increases up to the eutectoid point, the ability of the steel to be hardened also increases. Above this point the hardness can be increased only slightly, because steels above the eutectoid point are made up entirely of pearlite and cementite in the annealed state. Pearlite responds best to heat-treating operations; any steel composed mostly of this constituent can be transformed into a hard steel. The small amount of excess cementite found in hypereutectoid steels will not greatly increase the hardness of these steels.

As the size of parts to be hardened increases, the surface hardness decreases, even though all other conditions have remained the same. This is evident if it is remembered that there is a limit to the rate of heat flow through steel. No matter how cool the quenching medium may be, if the heat in the inside of a large piece cannot escape faster than a certain critical rate, there is a definite limit to the hardness that may be obtained. For similar reasons it is evident that the inside of a piece of steel would be softer than the outside. Hardness-penetration curves for two types of steel are shown in Figures 7 and 8. In both cases six bars of steel were quenched, the diameters varying from 0.5 to 5 inches. After quenching, the bars were cut in two and hardness readings taken across the diameter as shown in Figure 9. These curves clearly show the effect of size on hardness as well as the possible variations that may be obtained in a given specimen. Although there is not much difference in the carbon content of the two steels, there is considerable difference in the two sets of curves. This difference may be attributed to the 0.94% chromium and the 0.17% vanadium in the SAE 6140 steel that are not present in the SAE 1045 steel. Alloys make it possible to harden small pieces uniformly from the outside to the interior. Also, much greater surface hardness is obtained on alloy steel than on similar sizes of carbon steel. Finally, alloy steel has greater hardenability; that is, it will harden at a slower cooling rate than will carbon steel. Therefore, it may be quenched in oil instead of water.

Fig. 9. Manner of Taking Rockwell Readings to Find Distribution of Hardness.

Constituents of Hardened Steel

It has been previously stated that austenite is a solid solution of iron carbide in gamma iron. All carbon steels are composed entirely of this substance above the upper (Ac$_3$) critical point. The appearance of austenite under the microscope is shown in Figure 10 at a magnification of 100. Extreme quenching of a steel from a high temperature will preserve some of the austenite at ordinary temperatures. This constituent is about one half as hard as martensite, but has excellent wear-resisting properties and is nonmagnetic. Austenitic structure for carbon steels is not recommended, as it is unstable and highly strained.

If a hypoeutectoid steel is cooled down slowly, the austenite is transformed into ferrite and pearlite. Steel having these constituents

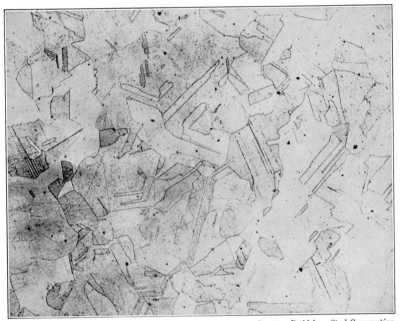

Courtesy Bethlehem Steel Corporation.

FIG. 10. Austenite, Magnification ×100.

is soft and ductile. Faster cooling will result in a different constituent, and the steel will be harder and less ductile. A rapid cooling, as a water quench, will result in a martensitic structure, which is the hardest structure that can be obtained. Cementite, although somewhat harder, is not present in its free state except in hypereutectoid steels and then only in such small quantities that its influence on the hardness of the steel can be ignored.

The essential constituent of any hardened steel is martensite. A. Martens, a German scientist, first recognized this constituent about 1878. Martensite is obtained by rapid quenching of carbon steels and is the transitional substance formed by the rapid decomposition of austenite. It is a mixture of very small crystals of alpha iron with carbon, which is probably in the form of cementite. Under the microscope it appears as a needlelike constituent, as may be seen in Figure 11. It is not clearly known why martensite is so hard, although a number of theories have been proposed. One theory is that the carbon is in a finely divided state and that the crystal structure is distorted and highly strained. The hardness of marten-

Courtesy Bethlehem Steel Corporation.

Fig. 11.   Martensite.   Magnification ×100.

site depends upon the amount of carbide present and varies from Rockwell C45 to C67. It cannot be machined, is quite brittle, and is strongly magnetic.

If steel is quenched at a rate slightly less than the critical rate, a dark constituent with somewhat rounded outlines will be obtained. This constituent is frequently associated with martensite, as may be seen in the photomicrograph shown in Figure 12. The name recently given to this constituent is *fine pearlite*, although for a long time it has been known as *troostite*. Under the microscope at usual magnifications, it appears as a dark unresolved mass, but at very high magnification a fine lamellar structure can be seen. The structure consists of fine cementite particles imbedded in a matrix of ferrite, which is the composition of the pearlite found in slowly cooled steels. Fine pearlite is less hard than martensite, having a Rockwell C hardness varying from 34 to 45, but it is quite tough and capable of resisting considerable impact. Steels with this constituent can be machined only with difficulty.

Fine pearlite may also be obtained by reheating a martensitic

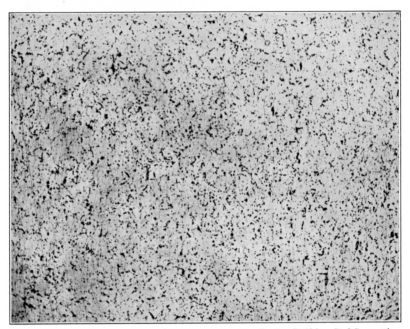

Courtesy Bethlehem Steel Corporation.

Fig. 12.   Fine Pearlite (Troostite).   Magnification ×100.

steel to a temperature of 400 to 600 F. The structure obtained consists of a fine dispersion of carbide particles in ferrite and has in the past been known as *secondary troostite.*

Figure 13 shows a photomicrograph of a structure obtained by quenching at a less rapid rate than for fine pearlite (troostite) or by reheating hardened steels to temperatures of 600 to 1000 F. This structure is now known as pearlite (formerly called *sorbite*), since its structure is definitely laminated when resolved at high magnification. At moderate magnification it has a granular appearance with no definite form or arrangement. It consists of a fine aggregate of cementite in alpha iron and is very strong and tough. Steel with this structure can be readily machined. The term *tempered martensite* is frequently given to this constituent when it is obtained by reheating a martensitic steel.

Maximum Hardness of Steel

One of the most valuable properties of steel is its ability to be hardened. The maximum hardness obtainable in a given piece of

Courtesy Bethlehem Steel Corporation.

FIG. 13. Pearlite (Sorbitic). Magnification ×100.

steel depends on the carbon content. This applies equally well to alloy steels. Although various alloys such as chromium and vanadium increase the rate and depth-hardening ability of alloy steels, their maximum hardness will not exceed that of a carbon steel having the same carbon content. This fact is illustrated in the curve shown in Figure 14, where Rockwell C hardness is plotted against percentage of carbon. This curve shows the maximum hardness that is possible for a given carbon percentage. To obtain maximum hardness, the carbon must be completely in solution in the austenite when quenched. The *critical quenching rate*, which is the slowest rate of cooling that will result in 100% martensite, should be used. Finally austenite must not be retained in any appreciable percentages, as it is considerably softer than martensite.

The curve in Figure 14 is made up of test points from both alloy and carbon steels, and it may be seen that there is little variation in the results. However, the same quenching rate cannot be used for both alloy and carbon steels of the same carbon content. A quenching rate on SAE 1050 steel that results in a martensitic struc-

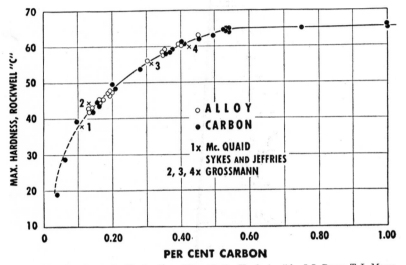

Courtesy American Society for Metals.   From "Quantitative Hardening," by J. L. Burns, T. L. Moore, R. S. Archer, *Trans. ASM*, Vol. XXVI, 1938.

Fig. 14.   Maximum Hardness versus Carbon Content.

ture would be too fast for SAE 2350.   In the latter case some austenite would be retained, and a lower hardness would result.   The maximum hardness obtained in any steel represents the hardness of martensite and is approximately Rockwell C65.

Annealing

The primary purpose of *annealing* is to soften hard steel so that it may be machined or cold-worked. This is usually accomplished by heating the steel to a temperature slightly above the critical temperature, holding it there until the temperature of the piece is uniform throughout, and cooling at a slow rate. This process is known as *full annealing* because it wipes out all trace of previous structure and refines the crystalline structure in addition to softening the metal. Annealing also relieves internal stresses previously set up in the metal and removes gases trapped in the metal during the initial casting.

By reheating hardened steel to above the critical range, the constituents are changed back into austenite, and slow cooling then provides ample time for complete transformation of the austenite into the softer constituents. For the hypoeutectoid steels these constituents are pearlite and ferrite. It may be noted by referring to the equilibrium diagram that the annealing temperature for hypereutectoid steels is lower, being slightly above the A_1 line. There is no reason to heat above the Acm line, as it is at this point that the precipitation of the hard constituent cementite is started. All martensite is changed into pearlite by heating above the lower critical range and slowly cooling. Any free cementite in the steel is unaffected by the treatment.

The temperature to which a given steel should be heated in this process depends on its composition, and for carbon steels can be obtained readily from the partial iron–iron-carbide equilibrium diagram shown in Figure 4. The heating rate should be consistent with the size and uniformity of sections so that the entire part is brought up to temperature as uniformly as possible. If the heating time is to be very long, it is advisable to protect the surface from oxidation either by controlling the furnace atmosphere or by packing the part in some material such as sand or cast-iron turnings. When the annealing temperature has been reached, the steel should be held there until conditions are uniform throughout. This usually takes about 45 minutes for each inch of thickness of the largest section. For maximum softness and ductility the cooling rate should be very slow, such as allowing the parts to cool down with the furnace. The higher the carbon content, the slower must be this rate. Low-carbon steels may be cooled more rapidly. Various rates can be obtained by burying the parts in lime or sand or cooling them in air.

The process of *normalizing*, considered a form of annealing, consists of heating the steel about 100 F above the critical range and cooling in still air to room temperature. This process is principally used with low- and medium-carbon and alloy steels to make the grain structure more uniform, to relieve internal stresses, or because the treatment results in desired physical properties. Most commercial steels are normalized after being rolled or cast.

Spheroidizing is the process of changing the pearlite from a lamellar to a spheroid structure. If a hard steel is slowly reheated to a temperature just below the critical range and then slowly cooled, this structure will be obtained. The globular structure of the carbide obtained by this treatment gives improved machinability to the steel.

Process annealing practiced in the sheet and wire industry between cold-working operations consists of heating the steel to a temperature a little below the critical range and then cooling slowly. This process is more rapid than the spheroidizing process and results in the usual pearlitic structure. It is similar to the tempering process but will not give so much softness and ductility as a full anneal. Also at the lower heating temperature there is less tendency for the steel to scale or decarburize.

Tempering

Steel which has been hardened by rapid quenching is brittle and not suitable for most uses. By *tempering* or " drawing," the hardness and brittleness may be reduced to the desired point for service conditions. As these properties are reduced, there is also a decrease in tensile strength and an increase in the ductility and toughness of the steel. The operation consists of the reheating of hardened steel to some temperature below the critical range, followed by any rate of cooling. Although this process softens steel, it differs considerably from annealing in that the process lends itself to close control of the physical properties and in most cases does not soften the steel to the extent that annealing would.

Tempering is possible because of the instability of the martensite, the principal constituent of hardened steel. At about 400 F this constituent will start to break down to the softer constituents. Low draws from 300 to 400 F do not cause much decrease in hardness and are used principally to relieve internal strains. As the tempering temperatures are increased, the breakdown of the martensite takes place at a faster rate, and at 600 F the change to fine pearlite (secondary troostite) is very rapid. Fine pearlite is less hard than martensite, but is very tough and can be machined only with difficulty. If heating is continued, this structure will start changing to pearlite (sorbite) at about 600 F. This constituent, although much softer than fine pearlite, is tough and has considerable resistance to impact. Increasing tempering temperatures to 1200 F causes further decrease in hardness and increase in toughness.

In the process of tempering some consideration should be given to time as well as temperature. Although most of the softening action occurs the first few minutes after the temperature is reached, there

PHYSICAL PROPERTIES CHART

S. A. E. 10-40

(Average Values)

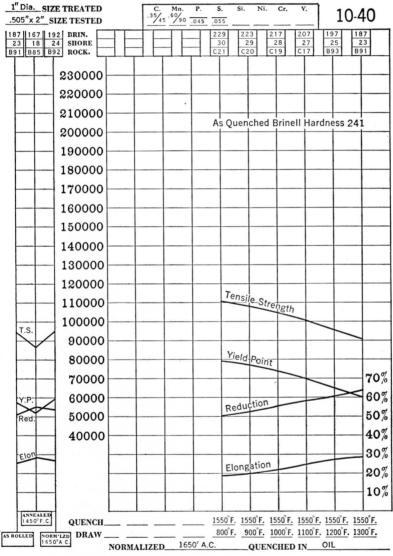

Courtesy Bethlehem Steel Corporation.

FIG. 15. Physical Properties of SAE 1040.

is some additional reduction in hardness if the temperature is maintained for a prolonged time. Usual practice is to heat the steel to the desired temperature and hold it there only long enough to have it uniformly heated.

The properties of a given steel vary considerably according to the heat treatment used. Figure 15 shows a typical set of property curves for SAE 1040 steel, giving the tensile strength, hardness, percentage elongation, and percentage reduction in area for various tempering temperatures. The influence of tempering (drawing) upon the physical properties of the steel is clearly shown by the curve.

Case Hardening

The oldest known method of producing a hard surface on steel is case hardening or *carburizing*. This process, in brief, is merely heating iron or steel to a red heat in contact with some carbonaceous material. Iron, at temperatures close to and above its critical temperature, has an affinity for carbon. The carbon enters the metal to form a solid solution with iron and converts the outer surface into a high-carbon steel. The process is the same as the old cementation process for making steel, except that it is stopped before the carbon penetrates very far into the interior of the metal.

The steel used for this process is usually a low-carbon steel of about 0.15% C which does not respond appreciably to heat treatment. In the course of the process, the outer layer is converted into a high-carbon steel with a carbon content ranging from 0.9% C to 1.2% C. By proper heat treatment such steel will have an extremely hard surface on the outside and a soft ductile center.

Figures 16 and 17 show the appearance of a carburized case obtained by heating the specimen in contact with potassium cyanide salt at a temperature of 1650 F. The specimens were water-quenched after being heated in the bath for 5 and 30 minutes, respectively. The light outside constituent is martensite, and the darker one beneath is fine pearlite.

This process is merely one of changing the carbon content of the surface steel, which makes it possible to obtain different physical properties in a given piece of steel. A steel with varying carbon content, and consequently different critical temperatures, requires consideration in the heat-treatment procedure. The case has a much lower critical temperature than the interior. Since there is some grain growth in the steel during the prolonged carburizing treatment the work should be first heated up to the critical temperature of the core and cooled, which refines the core structure. The steel should then be reheated to the critical range of the case and quenched to produce a hard fine structure. A third tempering treatment may be used to reduce strains.

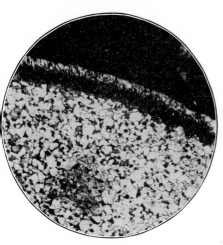 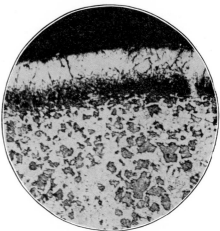

Both from University of Texas Shop Laboratories.

FIG. 16. SAE 1015 Carburized in Cyanide Bath 5 Minutes, Followed by Water Quench.

FIG. 17. SAE 1015 Carburized in Cyanide Bath 30 Minutes, Followed by Water Quench. Magnification ×200.

Nitriding

Nitriding is somewhat similar to ordinary case hardening, but it uses a different material and treatment to create the hard-surface constituents. In this process the metal is heated to a temperature of around 950 F and held there for a period of time in contact with ammonia gas. Nitrogen from the gas is introduced into the steel, forming very hard nitrides which are finely dispersed through the surface metal.

It has been found that nitrogen has greater hardening ability with certain elements than with others; hence, special nitriding alloy steels have been developed. Aluminum in percentages of 1 to 1½ has proved to be especially suitable in steel, as it combines with the gas to form a very stable and hard constituent. The temperature of heating ranges from 850 to 1200 F, although 960 to 975 F is the temperature range generally used.

The nitriding process develops extreme hardness in the surface of steel. This hardness ranges from 900 to 1100 Brinell, which is considerably higher than that obtained by ordinary case hardening. Nitriding steels, by virtue of their alloying content, are stronger than ordinary steels and respond readily to heat treatment. It is recommended that these steels be machined and heat-treated before nitriding, as there is no scale or further work necessary after this process. Fortunately, the structure and properties are not affected appreciably by the nitriding treatment; and, since no quenching is necessary, there is little tendency to warp, develop cracks, or change condition in any way. The surface effectively resists corrosive action of water, salt-water spray, alkalies, crude oil, and natural gas.

This process is used on many automotive, airplane, and Diesel-engine wearing parts, as well as on numerous miscellaneous parts, such as pump shafts, gages, drawing dies, gears, clutches, and mandrels. Its use is limited by the time and expense necessary for the treatment and by the higher cost of nitriding steels.

Induction Hardening

Metals have been heated and melted by induced electric currents for some time. It is only recently, however, that this means of heating has been employed in surface hardening. The process here described is commonly known as the *Tocco process** and was developed for the purpose of surface hardening crankshaft bearings and other similar wearing surfaces. It differs from ordinary case-hardening practice in that the analysis of the surface steel is not changed, the hardening being accomplished by an extremely rapid heating and quenching of the wearing surface which has no effect on the interior core metal. A surface hardness of approximately 58 to 62 Rockwell C is obtained.

An inductor block acting as a primary coil of a transformer is placed around, but not touching, the journal to be hardened. A high-frequency current, usually 2000 cycles, is passed through this block, inducing a current in the surface of the bearing. The heating effect is due to induced eddy currents and hysteresis losses in the surface material. As the steel is heated to the upper critical range, the heating effect of these losses is gradually decreased, thereby eliminating any possibility of overheating the steel. The inductor block surrounding the heated surface has water connections and numerous small holes in its inside surface, and, as soon as the steel has been brought up to the proper temperature, it is automatically spray-quenched under pressure.

* Developed by The Ohio Crankshaft Company.

An important feature of this method of hardening is its rapidity of action, since it requires only 5 seconds to heat the steel to a depth of 1/8 inch. Another advantage is that only a small percentage of the weight of the object to be treated is heated to the necessary high temperature. Obviously, this procedure eliminates warping to a great extent and consequently necessitates only a small allowance

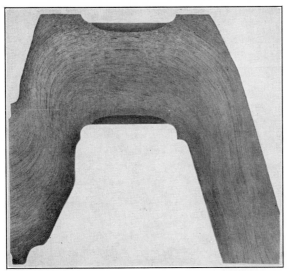

Courtesy The Ohio Crankshaft Company.

FIG. 18. Section of an Induction-Hardened Crankpin Bearing.

for grinding to the finished size. The local heating does not affect any previous treatment given to the core or cause trouble at fillets. Medium-carbon steel has proved very satisfactory for parts treated in this manner, and the nature of the process has practically eliminated the necessity of using costly alloy steels. Figure 18 illustrates the local heating obtained in a hardened crankpin bearing which has been induction-hardened.

Austempering

The conventional method of hardening and tempering steel just described may be illustrated by the temperature–time curve in Figure 19. Rapid quenching of the steel from above the critical temperature gives a hard martensitic structure which by tempering is changed to a structure having the desired properties. Control of the properties is obtained by varying the tempering time and temperature. The process of *austempering* is depicted in Figure 20. Steel is heated to a temperature above the critical range until it is all converted to austen-

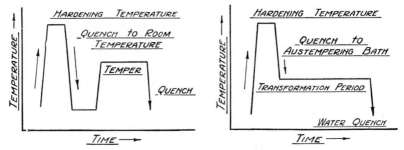

FIG. 19. Customary Quench and Temper Process.

FIG. 20. Austempering Process.

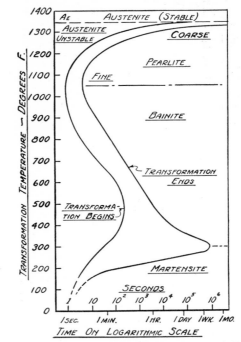

Three Figures Courtesy Carnegie-Illinois Steel Corporation.

FIG. 21. The Process of Austenite Transformation at Constant Temperature — Eutectoid Steel.

ite. It is then quenched into a molten salt or metal bath held at some predetermined temperature ranging from 350 to 800 F and held there for a definite period of time until the transformation is complete.

This structure, called *bainite,* is named after E. C. Bain, of the United Steel Corporation, for his outstanding work on the transformation of austenite at various constant temperatures. Under the microscope it is darker than martensite and has a fine needlelike structure. For the same hardness it is tougher and more ductile than tempered steel and is less likely to crack or warp during the heat-treating process. Figure 22 shows five microphotographs at high magnification in the progress of transformation from austenite to bainite.

The foundation for the austempering process is based upon investigation of austenitic transformation at constant temperatures. It is depicted diagrammatically by the S curve shown in Figure 21. This figure is drawn for steel of eutectoid composition, but is similar to curves of other carbon and alloy steels. Because of the long intervals required for some

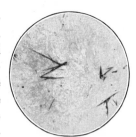

Transformation Begins — 400 Seconds.

25% Transformed -- 500 Seconds.

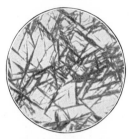

50% Transformed — 850 Seconds.

75% Transformed — 900 Seconds.

Transformation Completed — 2500 Seconds.

Courtesy J. A. Vilella, United States Steel Corporation.

FIG. 22. Microphotographs at High Magnification, Showing the Progress of Transformation from Austenite to Bainite.

transformations, the transformation temperatures are plotted against time on a logarithmic scale. The distance between two curves indicates the time required for complete transformation from austenite to the final constituent. This constituent will depend upon the rate of quench and the temperature at which it is held. Above the upper hump in the curves it will be pearlite; from 350 to 800 F, bainite; and below the lower hump, martensite. These constituents vary in structure within each range, and the change is gradual from one to the other. Because of the curves extending far to the left at the upper hump, steel must be rapidly quenched to the desired transformation temperature to clear the curve and eliminate the formation of some pearlite. Most alloying elements in steel have the effect of shifting the curves to the right, thus allowing more time to harden the steel fully without hitting the upper bend of the curve. This increase in the hardenability of the steel permits the hardening of thicker sections than would otherwise be possible. Complete austempering is limited to steels of a size that can be quenched at a rate that will not cut through this portion of the curve. However, large sections will be benefited by having a hard tough surface of bainite with fine pearlite at the interior. This process is widely used for treating such articles as small tools, springs, lock washers, firearm parts, link chains, fishing rods, shovels, and miscellaneous machinery parts.

Grain Size

All steel is crystalline in structure, and the size of these crystals or grains has an important effect on the quality of the steel. Molten steel upon cooling starts solidifying at many small centers or nuclei, the atoms in each group orienting themselves in the same direction. The irregular grain boundaries seen under the microscope after polishing and etching are the outlines of each group of atomic cells that have the same orientation. The size of these grains depends upon a number of factors, the principal ones being the composition of the steel and the heat treatment it has received.

It has long been known that coarse-grain steels are weaker and more brittle than those having a fine grain; however, they have better machinability and greater depth-hardening power. The fine-grained steels, in addition to being tougher, are more ductile and have less tendency to distort or crack during heat treatment. Control of grain size is possible through regulation of composition in the initial manufacturing procedure, but after the steel is made the control is through proper heat treatment.

When a piece of low-carbon steel is heated, there is no change in the grain size up to the Ac_1 point. As the temperature increases through the critical range, the ferrite and pearlite are gradually transformed to austenite, and at the upper critical point Ac_3 the average grain size is a minimum. Further heating of the steel causes an increase in the size of the austenitic grains, which in turn governs the final size of the grains when cooled. Quenching from the Ac_3 point would result in a fine-grained structure, whereas slow cooling or quenching from a higher temperature would give a coarser structure. The final grain size depends entirely on the prior austenitic grain size in the steel at the time of quenching.

All steels do not start growing large crystals immediately upon being heated above the upper critical range, and such steels can be heated to some higher temperature with little change in their structure. A temperature known as a *coarsening temperature* is eventually reached, and grain-size increase becomes rapid. This is characteristic of medium-carbon steels, many alloy steels, and steels which have been deoxidized with aluminum. The coarsening temperature is not a fixed temperature and may be changed by prior hot or cold working and heat treatment.* Normalizing lowers this temperature. Low-carbon cold-rolled steel will also show increased grain structure at the surface upon being reheated at lower temperatures. Hot work on steel is started at temperatures well above the critical range with the steel in a plastic state and has the effect of refining the grain structure and eliminating any coarsening effect due to the high temperature. Hot forging or rolling should not continue below the critical temperature.

The principal method of determining grain size is by microscopic examination, although it may be roughly estimated by examination of a fracture. For microscopic determination it is necessary that the grain boundaries be clearly outlined by some constituent. Low-carbon steels have ferrite precipitated from the austenite upon slow cooling, and the outline of these grains can be clearly brought out by polishing and etching. In estimating the former austenitic grain size for low-carbon steels, the area of the pearlite formed from the remainder of the austenite must also be considered. For medium-carbon steels the former austenitic grain size would be represented roughly by the pearlitic area plus one-half the surrounding ferrite. Hypereutectoid steels will have the grain boundaries outlined by the

---

* M. A. Grossman, "Grain Size in Metals with Special Reference to Grain Growth in Austenite," *Trans. ASM*, Vol. 22, no. 10, 1934.

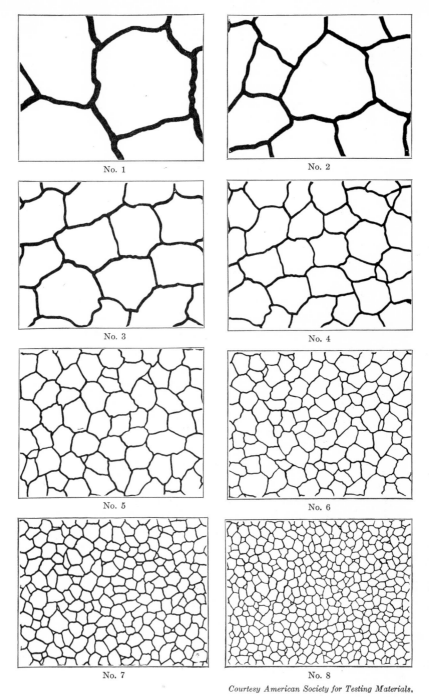

No. 1

No. 2

No. 3

No. 4

No. 5

No. 6

No. 7

No. 8

Courtesy American Society for Testing Materials.

Fɪɢ. 23.   Tentative Grain-Size Chart for Classification of Steels.
Magnification × 100.

178

cementite that is precipitated. A carburizing test developed by H. W. McQuaid and E. W. Ehn is widely used for studying grain size. The steel is heated to a temperature of 1700 F in contact with a carbonaceous material, is held at that temperature for 8 hours, and then slowly cooled. A high-carbon case results, and the grains are well outlined by the precipitated carbide. The size of the grains is obtained by comparing the photomicrograph at 100 magnification with a series of grain-size charts. These standard grain sizes shown in Figure 23 are numbered 1 to 8. Any grain structure similar to the first four numbers is considered to be coarse-grained. These

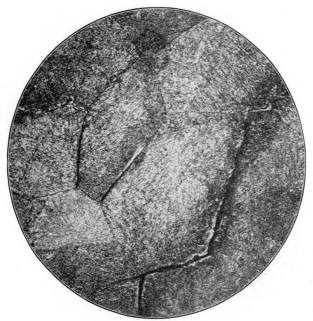

Fig. 24.   Crystalline Separation and Excessive Grain Size.
Magnification ×300.

charts for designating grain size have been standardized by the American Society for Testing Materials.

The McQuaid–Ehn carburizing test determines the austenitic grain size at 1700 F and indicates whether or not that temperature is above or below the coarsening temperature. It is especially adapted to carburizing steels. There is a definite correlation between the grain size and the hardening ability of the steel, a steel with large grain size having greater depth-hardening ability. This test has the advantage in that the grains are clearly outlined by the carbide. All steels do not require this test for grain-size determination, and the desired results may frequently be obtained by other heat-treatment procedures.

An example of a large-grained steel is shown in the photomicrograph in Figure 24. This specimen has been heated to an excessively high temperature, resulting in large grain growth and some crystalline separation. Steel which has been " burnt " shows this separation owing to oxidation at the grain boundaries, and such structure cannot be remedied by heat treatment. It can be rendered fit for commercial use only by remelting

Courtesy Westinghouse Electric Corporation.

FIG. 25.   Electrical-Resistance Heat-Treating Furnace.

Heat-Treating Furnaces

A furnace for heat treating consists of a well-insulated heating chamber provided with combustion space or other heating facilities and means for measuring temperatures. The fuel used may be oil, gas, or electricity, depending on the type of furnace in question, temperature desired, and cost. Because of its close control and ease of operation, electricity is widely used for small furnaces in the heat treatment of tools and similar parts.

The simplest type of heat-treating furnace is the box-shaped *stationary-hearth furnace.* Parts to be treated are placed on the hearth, and the heating flames or elements are spaced around the heating chamber in such a way that the hot gases do not contact the work. This, known as a muffle furnace, is a desirable arrangement for uniform heating of parts to be treated.

An electrical-resistant box-type furnace, with its accessory control equipment used for batch heat treatment of gears and other engine

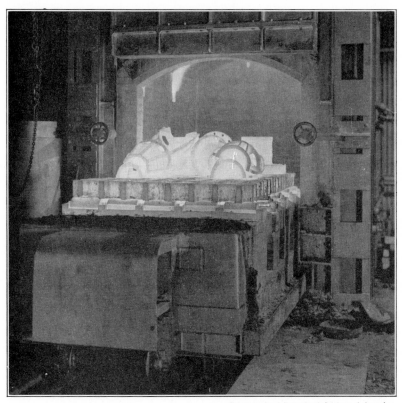

Courtesy Steel Founders' Society of America.

Fig. 26.   Normalizing Steel Castings in Car-Type Furnace.

parts, is shown in Figure 25. This furnace has a controlled atmosphere to permit clean hardening of steels without decarburization.

As furnaces increase in size for the heating of large-run production jobs and large parts, the stationary-hearth type of furnace is not suitable, and movable-hearth furnaces are employed. In this group the *car-type* furnace, as shown in Figure 26, is used for heavy work, frequently requiring the use of handling equipment for loading and unloading. The furnace shown is a gas-fired furnace used in normalizing miscellaneous steel castings. Another furnace similar to this one utilizes trucks instead of cars.

The *tunnel type* of furnace is used where many similar parts are to be treated. It is continuous in its operation: pieces are charged onto a metal chain or apron conveyor at one end, move slowly through the heating oven, and are discharged into a quenching bath at the other end. Some continuous furnaces employ a pusher mechanism on a stationary hearth and slowly push or roll the work through the furnace. Still another design is the rotating horizontal hearth in which the parts are loaded and unloaded at the same point after completing the heating cycle.

Certain heat-treating operations are best carried on in *bath-type* furnaces, since uniform temperatures can be maintained around the parts and no oxidizing atmosphere is present. Oil, lead, or various salts are melted in a pot surrounded by heating elements. All this is enclosed in a well-insulated furnace wall and, in addition, suitable temperature-control equipment is provided for furnace regulation.

REVIEW QUESTIONS

1. What elements other than iron are found in most commercial carbon steels?
2. Distinguish among steel, wrought iron, and cast iron.
3. What are the principal microconstituents found in all annealed steels?
4. What changes take place in steel at the critical points? How are these points determined?
5. Sketch a partial iron–iron-carbide diagram, and indicate correct temperatures for all lines.
6. What microconstituents would be found in annealed SAE 1030 steel? 1.1C steel?
7. What is meant by the following terms: Eutectoid, pearlite, cementite, austenite, and alpha ferrite?
8. Describe the process known as hardening.
9. What is martensite? How does it appear under the microscope?
10. What determines the maximum hardness that can be obtained in a piece of steel?
11. What is the purpose of annealing and how is it done?
12. Distinguish between normalizing and spheroidizing.
13. Describe the process of case hardening.
14. What is nitriding, and what advantages does the process have over carburizing?
15. How is induction hardening accomplished?
16. What is the difference between austempering and ordinary tempering?
17. How is grain size determined by the McQuaid–Ehn test?
18. List the different types of heat-treating furnaces used.

BIBLIOGRAPHY

BOYLSTON, H. M., *Iron and Steel*, 2d edition, John Wiley & Sons, 1936.

BULLENS, D. K., *Steel and Its Heat Treatment*, 4th edition, Vols. I and II, John Wiley & Sons, 1938, 1939.

CAMP, J. M., and FRANCIS, C. B., *The Making, Shaping, and Treating of Steel*, 5th edition, Carnegie-Illinois Steel Corporation, 1940.

COONAN, F. L., *Principles of Physical Metallurgy*, Harper & Brothers, 1943.

DAVENPORT, E. S., and BAIN, E. C., *Transformation of Austenite at Constant Subcritical Temperatures*, AIME Technical Publication 348, 1930.

DOAN, G. E., *The Principles of Physical Metallurgy*, 1st edition, McGraw-Hill Book Company, 1935.

DOWDELL, R. L., JERABEK, H. S., FORSYTH, A. C., and GREEN, C. H., *General Metallography*, John Wiley & Sons, 1943.

HEYER, R. H., *Engineering Physical Metallurgy*, D. Van Nostrand Company, 1939.

KELLER, J. F., *Lectures on Steel and Its Treatment*, 2d edition, American Society for Steel Treating, 1930.

LIDDELL, D. M., and DOAN, G. E., *The Principles of Metallurgy*, 1st edition, McGraw-Hill Book Company, 1933.

Metals Handbook, American Society for Metals, 1939.

NEWTON, J., *An Introduction to Metallurgy*, John Wiley & Sons, 1938.

PALMER, F. R., *Tool Steel Simplified*, 1st edition, Carpenter Steel Company, Maple Press Company Printers, 1937.

Physical Properties of Bethlehem Steels, Catalog — 147, Bethlehem Steel Company, 1938.

ROSENHOLTZ, J. L., and OESTERLE, J. F., *The Elements of Ferrous Metallurgy*, 2d edition, John Wiley & Sons, 1938.

SACHS, G., and VAN HORN, K. R., *Practical Metallurgy*, American Society for Metals, 1940.

SISCO, F. T., *Modern Metallurgy for Engineers*, Pitman Publishing Company, 1941.

STOUGHTON, B., *The Metallurgy of Iron and Steel*, 4th edition, McGraw-Hill Book Company, 1934.

STOUGHTON, B., and BUTTS, A., *Engineering Metallurgy*, 3d edition, McGraw-Hill Book Company, 1938.

TEICHERT, E. J., *Ferrous Metallurgy*, 3 Vols., McGraw-Hill Book Company, 1944.

WILLIAMS, R. S., and HOMERBERG, V., *The Principles of Metallography*, 4th edition, McGraw-Hill Book Company, 1939.

CHAPTER 7

POWDER METALLURGY

Powder metallurgy is the art of producing commercial products from metallic powders by pressure. Heat may or may not be used in the process; if it is, the temperature is kept below the melting point of the powder. The application of heat during the process or subsequently is known as *sintering* and results in bonding the fine particles together, thus improving the strength and other properties of the finished product. Sintering is also defined as " the process by which solid bodies are bonded by atomic forces."* Products made by powder metallurgy are frequently alloyed or contain nonmetallic constituents. Such combinations may be made to improve the bonding qualities of the particles, but more frequently the purpose is to improve certain properties or characteristics of the final product. For example, cobalt or other metals are necessary in the bonding of tungsten carbide particles, whereas graphite is added with bearing-metal powders to improve the lubricating qualities of the finished bearing. Electric contacts are alloyed, because, in addition to having good heat and electrical conductivity, they must also be wear-resistant and somewhat refractory.

The manufacture of products from powders is not a new art. In some applications it dates back many centuries. Much of the early experimentation had to do with metals such as platinum and tungsten which could not be readily melted and processed. Early in the 19th century Wollaston experimented with platinum powders and was able to produce articles of platinum by a pressing and sintering treatment. The tungsten filament developed by Coolidge in 1909 was the result of research for a metallic filament to replace the then common carbon filament. Shortly after the first World War, the development of porous bearings and cemented carbides added much impetus to the further expansion of powder metallurgy.

Important Characteristics of Metal Powders

The manufacture of metal powders is closely associated with the final product, since the particle size, shape, and size distribution

---

* P. E. Wretblad and J. Wulff, " Sintering," *Powder Metallurgy*, ASM, 1942, p. 36.

have definite effects on the characteristics and physical properties of the compacted product. For this reason, powders are produced according to certain specifications; and a powder that may be desirable for one product may not prove entirely satisfactory for another. No standard form of specifying powders has been developed, but in general the points covered in specifications are structure or shape, fineness, particle-size distribution, flowability, chemical properties, and apparent specific gravity.

The *shape* of a powder particle, depending largely on how it was produced, may be spherical, ragged, dendritic, flat, angular, or otherwise. While the shape is important, the manufacturer is limited in his selection by the method through which a given metal can be produced. *Fineness* may be determined by passing the powder through a standard sieve or by microscopic measurement. Standard sieves ranging from 100 mesh to 325 mesh are used for checking sizes and also for determining particle-size distribution within that range. *Particle size distribution* is also important, since it has considerable influence in determining the flowability and apparent density, as well as the final porosity, of the product. Once it is established for a product it cannot be varied appreciably without affecting the size of the compact. *Flowability* is that characteristic of a powder which permits it to flow readily and conform to the mold cavity. It can be expressed as the rate of flow through a definite orifice. *Chemical* properties have to do with the purity of the powder, amount of oxides permitted, and the percentage of other elements allowed. The *apparent density* or specific gravity of a powder may be expressed in grams per cubic centimenter. This should remain constant to insure that the same amount of powder is fed into the mold each time.

Methods of Producing Powders

Metal powders, because of their individual physical and chemical characteristics, cannot all be manufactured the same way. The procedures vary widely, as do the sizes and structures of the particles obtained from the various processes. Machining results in coarse particles and is used principally for producing magnesium powders. Milling processes utilizing various types of crushers, rotary mills, and stamping mills, break down the metals by crushing and impact. Brittle materials may be reduced to irregular shapes of almost any fineness by this method; however, the process is also used in pigment manufacture for ductile materials. With ductile materials, flake particles are obtained, an oil being used in the process to keep them from sticking together. *Shotting* is the operation of pouring molten metal through a sieve or orifice and cooling by dropping into water. Spherical or pear-shaped particles are obtained by this process. Most metals can be shotted, but the size of the particles is too large in many instances. Atomization, or the operation of metal spraying, is an excellent means of producing powders from many of the low-temperature metals such as lead, aluminum, zinc, and tin. The particles are irregular in shape and are produced over quite a range of size distribution. A few metals can be converted into small particles by rapidly stirring the metal while it is cooling. This process, known as *granulation*, depends on the formation of oxides on the individual particles during the stirring operation. Electrolytic deposition is a common means for processing copper, iron, tantalum, silver, and several other metals. The characteristic structure obtained by this method is dendritic, and the apparent density is low. Reducing metal oxides in powder form by contact with a reducing gas at temperatures below the melting point is an economical method for some metals. Tungsten, iron, molybdenum, nickel, and cobalt are all produced commercially by this process.

Various other methods involving precipitation, condensation, and other chemical processes have been developed for producing powdered metals. These methods as well as some of those previously mentioned are not widely used, but prove satisfactory for some metals. Production costs for metal powders should gradually be reduced as further research develops the processes and the demand for powder-compacted products increases.

Pressing to Shape

Powder for a given product must be carefully selected to insure economical production and to obtain the desired properties in the final compact. If only one powder is to be used in the product, no additional processing or blending will be necessary before pressing, unless the particle-size distribution is not correct. In some cases various sizes of powder particles are mixed together to change such characteristics as flowability or density; but most powder is produced with sufficient particle-size variation to make mixing unnecessary. Mixing or blending becomes necessary in production when the powders are alloyed or when nonmetallic particles are added. Any mixing or processing of the powder must be done under conditions which will not permit oxidation or defects to develop.

Powders are pressed to shape in steel dies under pressures ranging from a few thousand to 200,000 pounds per square inch. Because the soft particles can be pressed or keyed together quite readily, powders which are plastic do not require so high a pressure as the harder powders to obtain adequate density. Quite obviously, the density and hardness increase with the pressure; but in every case there is an optimum pressure above which little advantage in improved properties can be obtained. Furthermore, owing to the necessity for strong dies and large capacity presses, production costs increase with high pressures.

Many of the commercial presses developed for other materials are adaptable for use in powder metallurgy. Both mechanical and hydraulically operated presses are used. Most presses are so designed that their operation from the filling of the cavity with powder to the ejection of the finished compact is entirely automatic. To accomplish this, rotary table presses which have a series of die cavities each provided with top and bottom punches are frequently used. In the course of production the table indexes around and the oper-

Courtesy Kux Machine Company.

Fig. 1. Rotary Press for Compressing Powdered Metals into Solid Form.

ations of filling, pressing, and ejecting the product are accomplished at the various stations. A rotary press of this type is shown in Figure 1. This press has a $2\frac{1}{2}$-inch maximum tablet diameter and a $2\frac{1}{4}$-inch depth of fill. Die fills and pressures can be quickly adjusted and when once established remain constant. Experience has proved that more uniform compacting is obtained if pressure is applied at both top and bottom simultaneously.

In Figures 2 and 3 are shown press setups for compacting small bushings and pinions from metal powders. Many products similar to these are entirely completed by the pressing operation and require no further processing. Others require a heat treatment or sintering operation to increase the strength and improve the crystalline structure. Still another method used is to combine the pressing and heating operations. Although this procedure has certain advantages as to improved properties, it is more difficult to accomplish.

Courtesy Chrysler Corporation — Amplex Division.

Fig. 2.   From Metal Powder to Briquetted Bearing in 2 to 6 Seconds.

Sintering

The operation of heating a " green compact " to an elevated temperature is known as *sintering*. As previously stated, it is the process by which solid bodies are bonded by atomic forces. Baeza[*] states that by the application of heat the particles are pressed into more intimate contact and the effectiveness of surface-tension reactions is increased. Plasticity is increased, and there is a possibility that better mechanical interlocking is produced by building a fluid network. Also, any interfering gas phase present is removed by the heat. The temperatures used in sintering are usually well below the

* W. J. Baeza, *Powder Metallurgy*, Reinhold Publishing Company, 1943.

melting point of the principal powder constituent, but may vary over a wide range up to a temperature just below the melting point. Tests have proved that there is usually an optimum sintering temperature for a given set of conditions, with nothing to be gained by going above this temperature. Aside from the temperature, other

Courtesy Moraine Products Division of General Motors Corporation.

Fig. 3.   Pressing Small Pinions from Powdered Metal.

factors in sintering are time and atmosphere. The time element varies with different metals, but in most cases the effect of the heating is complete in a very short time, and there is no economy in prolonging the operation. Atmosphere is nearly always important, as the product, being made up of small particles, has a large surface area exposed. The problem is to provide a suitable atmosphere of some reducing gas or nitrogen to prevent the formation of undesirable oxide films during the process. Textile-machine bearings entering a sintering furnace are shown in Figure 4. In the fabricating of powder metal parts, dimensional accuracy should be maintained so that no subsequent machining is necessary. There is always some dimensional change in the operation of sintering; it may be either a growth or a shrinkage.* What happens depends on the shape and

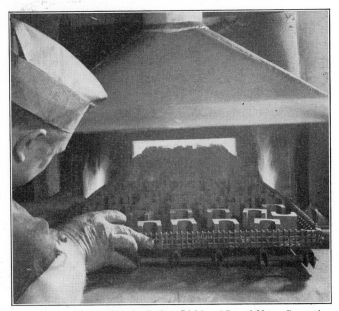

Courtesy Moraine Products Division of General Motors Corporation.

FIG. 4. * Textile-Machine Bearings Entering a Sintering Furnace.

particle-size variation of the powder, the powder composition, sintering procedure, and briquetting pressure. Accurate size is maintained by compensating for the change in making the green compact and then maintaining uniform conditions.

Hot Pressing and Sizing

Considerable effort has been made to combine the operation of pressing and sintering. Such combination, of course, has obvious advantages. Experiments with several metals have demonstrated that this method can produce compacted products with improved strength and hardness over the usual methods of production. Factors to overcome for successful operation are atmospheric control, method

* R. P. Koehring, " Sintering Atmospheres for Production Purposes," *Powder Metallurgy*, ASM, 1942.

of applying heat, and suitable dies to resist high-temperature wear and creep.

Products requiring close tolerances may necessitate a final coining or sizing operation. Coining is usually a cold-working operation. In addition to providing close dimensional tolerance, it gives the product increased strength and density.

Advantages and Limitations

Powdered metal is an expensive raw material as compared with other materials used for making similar products. In addition, the tools and dies are expensive; and the total production cost may be high as compared with other processes. However, these facts are frequently outweighed by the advantages obtained in the products. The use of powder metallurgy is rapidly increasing today, and many products are being made better and more cheaply than by other manufacturing methods. Some of the advantages obtained by this process are as follows:* †

1. Many products such as sintered carbides and porous bearings cannot be produced by any other method or process. This applies also to a number of products from alloys containing both metallic and nonmetallic powders. It is also possible to mold layers of different metal powders to form bimetallic products.

2. It is possible to produce parts with controlled porosity, such, for example, as is found in self-lubricating bearings made from nonferrous powders and graphite.

3. Large-scale production of many small parts can compete favorably with machined parts because of the close tolerances and surface finish that are obtained. On parts up to 2 inches in diameter, tolerances of ±0.001 inch or less can be maintained.

4. Product of extreme purity can be made, as it is possible to obtain powders in a very pure state. In the operation of pressing there is little chance for impurities to enter.

5. The process is economical in material, since there are no losses of material in the fabrication, and in most cases the dimensional accuracy is so close that no material allowance is needed for machinery.

6. Labor cost is low: skilled mechanics are not required to operate presses or other necessary equipment.

7. A wide range of physical properties is possible with any given material. These can be controlled by varying the die pressure, particle size, sintering temperature, or by introducing alloying elements.

* E. Schumacher and A. G. Souden, " Powder Metallurgy," *Metals & Alloys,* November 1944.

† P. Schwarzkopf and C. G. Goetzel, " Processing Trends in Powder Metallurgy," *Iron Age,* September 19, 1940.

Powder metallurgy has certain limitations which will restrict its use, particularly with those products that can be made economically by other manufacturing processes. In addition there are certain other limitations such as pertain to the mechanical equipment, the thermal characteristics of the powder, safety, and design. Some of these limitations are as follows:

1. Metal powders are expensive and in some cases difficult to store without some deterioration. Prices on powders should gradually decrease as the demand increases and the methods of production improve.

2. Equipment costs are high. Presses with capacities up to 60 tons per square inch are required for certain products. Dies operating in these presses must be accurately machined and capable of withstanding high pressures and temperatures. Sintering furnaces present problems of temperature and atmospheric control. These facts preclude the use of this process for short-run jobs.

3. The size of powder-fabricated parts is controlled by the capacity of the presses available and also by the compression ratio of the various powders. Compression ratios of different powders vary considerably. A compression ratio of 4:1 means the mold depths must be four times the finished compact. Since pressure is not distributed uniformly in a powder as it is in a liquid, the final product is not so likely to be uniform in density. This will affect the shape and dimensional accuracy when a sintering operation is necessary.

4. Intricate designs in products are difficult to attain, since there is no flow of the metal particles during compacting. Abrupt changes in thickness must be avoided; and it is not possible to mold undercuts, internal threads, and grooves. Uniform density is difficult to attain in long pieces.

5. Some thermal difficulties appear in sintering operations, particularly with the low-melting powders such as tin, lead, zinc, and cadmium. Most oxides of these metals cannot be reduced at temperatures below the melting point of the metal; hence, if such oxides exist, they will have detrimental effects on the sintering process and result in an inferior product.

6. Some powders in a finely divided state present explosion and fire hazards, and precaution must be taken to keep dust out of the air. Such metal powders include aluminum, magnesium, zirconium, and titanium.

7. A completely dense product is not possible if the sintering operation is carried out. However, porosity can be reduced materially if the heating accompanies the pressing operation.

Metal Powder Products*

Many metals are now available for use in powder-metal parts, and the number of products made by this process is steadily increasing.

* H. E. Hall, "Development in Metal Powders and Products," *Powder Metallurgy*, ASM, 1942.

Some parts are produced by powder metallurgy, because they cannot be produced by any other method, whereas others are made in this manner because of special properties or characteristics inherent to the process. Finally there are those parts which can be made by other methods but are economically produced by this process. A representative selection of machine parts made from a wide variety of metal powders is shown in Figure 5. It is interesting to note the

Courtesy Chrysler Corporation — Amplex Divisions.

FIG. 5. Machine Parts Made from a Wide Variety of Metal Powders.

intricate shape and design of the parts, most of which are made complete without the necessity of machining. Some of the prominent powder-metal products are as follows:

Cemented Carbides. Tungsten carbide particles are mixed with a cobalt binder, pressed to shape, and then sintered at a temperature above the melting point of the matrix metal. The metal cobalt binds the carbide particles together and gives strength and toughness to the final product. Cemented carbides are used for cutting tools, dies, and various wear-resistant applications.

Motor brushes. Brushes for motors are made by mixing copper with graphite in sufficient quantities to give the compact adequate mechanical strength. Tin or lead may also be added in small quantities to improve wear resistance. In the sintering operation, the compact is heated in a reducing atmosphere.

Porous bearings. Most bearings are made from copper, tin, and graphite powders, although other metal combinations also are used. After sintering the bearings are sized and then impregnated with oil by a vacuum treatment. Porosity in the bearings can be controlled readily and may run as high as 40% of the volume.

Metallic filters. Porous metal filters having greater strength and shock resistance than ceramic filters are made with porosities up to 80%. Bronze and nickel are common metals used for this purpose.

Gears and pumps rotors. Gears and pump rotors are made from powdered iron mixed with sufficient graphite to give the product the desired carbon content. Parts are produced with close dimensional accuracy requiring a minimum of machining. A porosity of around 20% is obtained in the process; and after the sintering operation the pores are impregnated with oil to promote quiet operation. The physical properties of iron-powder parts are close to those of ordinary gray cast iron.

Magnets. Excellent small magnets can be produced from iron, aluminum, nickel, and cobalt when combined in powder form. Alnico magnets made principally from iron and aluminum powders are superior to those cast. A finer-grain structure is obtained, there are no internal defects, and the magnets are produced with close dimensional tolerances.

Contact parts. Electric-contact parts lend themselves well to powder-metallurgy fabrication, since it is possible to combine several metal powders and still maintain some of the principal characteristics of each. Contact parts must be wear-resistant and somewhat refractory and at the same time must have good electrical conductivity. Many combinations such as tungsten–copper, tungsten–cobalt, tungsten–silver, silver–molybdenum, and copper–nickel–tungsten have been developed for electrical applications.

Numerous other parts including clutch faces, tungsten filaments, diamond cutting wheels, brake bands, laminated metals, and welding rod are produced by powder metallurgy. Likewise many small machine parts such as micrometer frames, bearing retainers, V blocks, gages, commutator segments, and packing glands are economically

made. Numerous parts for war matériel have been accepted by the
Ordnance Department, one interesting application being the use of
copper and brass rotating bands on projectiles. There are many
other uses for powdered metals which are not pressed to shape, as
for example the use of paint pigments and other protective coatings.
Aluminum powder is used in thermit welding; and in the field of
pyrotechnics and explosives powdered aluminum and magnesium are
both prominent. The addition of powdered metals to plastics increases their strength and contributes other metallic properties.

REVIEW QUESTIONS

1. Define powder metallurgy.

2. What is meant by sintering, and how is it accomplished?

3. What characteristic should be included in specifying a metal powder?

4. How is the size of metal powders determined and how are the sizes
specified?

5. What is meant by the term "particle-size distribution"?

6. Name and describe five methods of producing powders.

7. Describe the usual steps in producing a metal powder part.

8. In the operation of sintering, what factors must be taken into consideration?

9. What factors determine the sintering temperatures to use?

10. What are the advantages and limitations of hot pressing?

11. What are the advantages claimed for powdered metal parts?

12. List the limitations of this manufacturing process.

13. What precautions must be taken in designing a powder metallurgy part?

14. Name three products made by powder metallurgy that cannot be made
by other processes.

15. What advantages do bearings made by this process have over cast bearings?

16. How does powder metallurgy serve the electrical industry?

17. What types of presses are used in pressing compacts?

BIBLIOGRAPHY

BAEZA, W. J., *A Course in Powder Metallurgy*, Reinhold Publishing Company,
1943.

BALKE, C. C., "Powder Metallurgy — Some Theoretical Aspects," *Iron Age*,
April 17, 1941.

GOETZEL, C. G., "Sintered, Forged, and Rolled Iron Powders," *Iron Age*, October
1, 1942.

LENEL, F. V., "Powder Metallurgy," *Mechanical Engineering*, July 1943.

SCHUMACHER, E. E., and SOUDEN, A. G., "Some Aspects of Powder Metallurgy,"
Bell System Technical Journal, Vol. XXIII, October 1944.

SCHWARZKOPF, P., and GOETZEL, C. G., "Processing Trends in Powder Metallurgy," *Iron Age*, September 19, 1940.

SKAUPY, F., *Principles of Powder Metallurgy*, Philosophical Library, 1944.

VICTOR, M. T,. and SORG, C. A., "Design of Powder Metallurgy Parts," *Metals
& Alloys*, March 1944.

WULFF, J., *Powder Metallurgy*, American Society for Metals, 1942.

CHAPTER 8

WELDING AND ALLIED PROCESSES

Welding is the fusion or uniting of two pieces of metal by means of heat. Many welding processes have been developed which differ widely in the manner in which the heat is applied and in the technique employed. Some processes require hammering, rolling, or pressing to effect the weld; others bring the metal to a fluid state and require no pressure. The term *plastic welding* is given to the first group, whereas the latter methods are known as *fusion* welding. No additional metal is required in the plastic processes, but in most fusion processes additional weld metal is required. Welds are also made by casting, in which case the metal is heated to a high temperature and poured into the cavity between the two pieces to be joined. In this method the heat in the weld metal must be sufficient to cause it to fuse properly with the parent metal.

Soldering is the uniting of two pieces of metal with a different metal, which is applied between the two in a molten state. The usual metal for this purpose is a low-melting alloy of lead and tin. *Brazing* is a similar process in which the metal parts are joined by either a copper, silver, or aluminum alloy. Strictly speaking, neither soldering nor brazing is a welding process, since the parts being united are not brought to a fluid or plastic state. The strength of such joints is limited to the strength of the third metal used and is naturally weaker than properly welded joints. The brazing process known as " bronze welding " is generally more reliable for cast iron than is fusion welding.

Welding as a commercial process is comparatively recent. The only commercial process that dates back into the past century is forge welding. Most of the developments of modern welding have come about since the first World War. This development was brought about by the demands of industry for more rapid means of fabrication and assembly of metal parts. Welding processes are employed extensively in the manufacture of automobile bodies, aircraft, high-speed railroad cars, machine frames, structural work, tanks, and general machine-repair work. In the oil industry welding is extensively used at refineries and in pipe-line fabrication. During the war the largest single use for welding has been in shipbuilding; in peacetimes it is the fabrication of metal structures. Welding has also had its influence in the casting industry, as many machine parts are now made up of steel members welded together. Such construction has the advantage of being lighter and stronger than cast iron. Gas cutting has likewise had its influence on forged products. Many parts are now accurately cut from thick steel plates, thus saving the cost of expensive dies. There is hardly an industry today which is not affected in some way by welding and cutting processes.

The first welding processes were all limited to low-carbon steel and wrought iron. All such materials are easily welded and have a wide welding range. As the carbon content increases or as alloying elements are added, the welding range decreases, and good welds become increasingly difficult. However, the development of new electrodes and new welding techniques has greatly altered our concept of what is weldable material. Practically all alloy steels can now be welded if proper equipment and materials are used. Cast iron at first presented serious difficulties, because of its low ductility, poor fusion, and tendency to crack on cooling. Most of these difficulties are now overcome by proper methods and the selection of suitable welding materials. Such nonferrous metals as brass, bronze, Monel metal, aluminum, copper, and nickel can all be successfully welded, although special precaution must usually be taken to prevent oxidation.

Principal Welding Processes

The principal commercial welding processes are listed in Table 4. All welding processes may be classified under two general heads, plastic welding and fusion welding. Plastic welding is, as the name implies, the uniting of two pieces of metal by pressure without the addition of weld metal. The old form of blacksmith-forge welding comes under this classification as do the modern forms of electrical-resistance welding. In this process the metal is heated until it is in a plastic state, when the weld is effected by applying pressure.

Fusion welding includes those processes which have the metal in a fluid state. In uniting two pieces of metal by this method, weld metal is added but no pressure is applied. The surface of the metals is in a fluid condition. Examples of this form of welding are the gas and electric-arc processes.

TABLE 4

WELDING PROCESSES

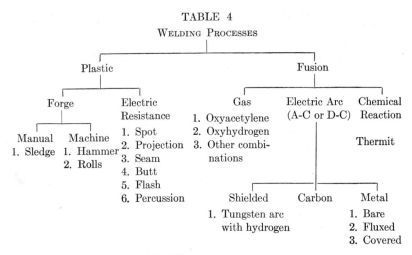

General Conditions for Welding

All surfaces to be welded should be thoroughly cleaned and freed from foreign matter by wire brushing, machining, or sand blasting. All impurities tend to weaken a weld, causing the metal to be either brittle or filled with gas and slag inclusions. They also cause poor cohesion of the metals.

Tendencies towards oxidation increase with temperature. At the high temperatures used in many welding processes, the oxidation of the weld metal is likely to have serious weakening effects in the weld. In some processes this influence is counteracted by the use of a flux which removes the oxides and permits perfect cohesion of the metals. In the electric-arc process the flux is coated on the electrodes and when dissolved forms a protective coating of slag over the weld metal as well as a nonoxidizing atmosphere. In gas welding and forge welding it is usually added in powder form. Other processes eliminate any oxidation tendencies by creating a nonoxidizing atmosphere at the point where the welding is done. Hydrogen is used for this purpose with the atomic hydrogen process.

Inasmuch as oxidation takes place rapidly at high temperature, speed in welding is important. Some processes are naturally quicker than others, but in any event the work should be done as rapidly as possible.

Forge Welding

Forge welding was the first form of welding used and for many centuries the only one in general use. Briefly, the process consists of heating the metal in a forge to a plastic condition and then uniting it by pressure. The heating is usually done in a coal- or coke-fired forge, although modern installations frequently employ oil or gas furnaces. The manual process is naturally limited to light work, as all forming and welding is accomplished with a hand sledge. Before the weld is made, the pieces are first formed to a correct shape, so that when they are welded they will unite at the center first. As they are hammered together from the center to the outside edges, any oxide or foreign particles will be forced out. The process of preparing the metal is known as *scarfing*.

Forge welding is naturally rather slow, and there is considerable danger of an oxide scale forming on the surface. The tendency to oxidize can be counteracted somewhat by using a thick fuel bed and by covering the surfaces with a fluxing material which dissolves the oxides. Many special fluxes have been prepared; however, borax in combination with sal ammoniac is commonly used. Heating must be slow on account of unequal section thickness. As soon as the desired uniform temperature is reached, the pieces are removed to the anvil and hammered together

For this type of welding, low-carbon steel and wrought iron are recommended, as they have a large welding-temperature range. This range decreases rapidly as the carbon content increases. High-carbon steels and alloy steels require considerably more care in controlling temperatures and producing the welds.

Large work may be welded in hammer forges driven by air or steam. Such equipment is especially valuable for forming and shaping work in the plastic state and has the additional advantage of refining the grain size when worked above the critical temperature of the metal. Welded steel pipe is made mechanically by running the preheated steel strips through rolls which form it to size and apply the necessary pressure for the weld.

Electrical-Resistance Welding

In this process a heavy electric current is passed through the metals to be joined, causing a local heating, and the weld is completed by applying pressure. This process dates back to the latter part of the 19th century, being first used by Elihu Thompson. When the current passes through the metal, the greatest resistance is at the point of contact; hence, the greatest heating effect is at the point where the weld is to be made. Alternating current is generally used coming to the machine with the usual commercial voltages. A transformer in the machine reduces the voltage to 4 to 12 volts and raises the amperage sufficiently to produce a good heating current. The amount of current necessary is 30 to 40 kva per square inch of area to be united, based on a time of about 10 seconds. For other time intervals the power varies inversely with the time. The necessary pressure to effect the weld will vary from 4000 to 8000 pounds per square inch.

Resistance welding does not compete in any way with the electric-arc process. It is essentially a production process adapted to the joining of light-gage metals which can be lapped. Usually the equipment is suitable for only one type of job, and the work must be moved to the machine. The process is especially adapted to quantity production, and its use includes a large amount of the welding done at the present time. It is the only process which permits a pressure action at the weld, while permitting an accurately regulated heat application. Also, the operation is extremely rapid.

The weldability of a given metal depends to some extent on its melting point. Practically all metal can be welded by resistance welding, although some few, such as tin, zinc, and lead, can be welded only with great difficulty.

In all resistance welding the three factors which must be given consideration are expressed in the formula: $Heat = I^2RT$, where I is the welding current in amperes, R the resistance of the metal being welded, and T the time. The amperage of the secondary or welding current is determined by the transformer. To provide possible variation of the secondary current, the transformer is equipped with a regulator on the primary side to vary the number of turns on the primary coil. This may be seen in Figure 1A.

The timing of the welding current is very important. There should be an adjustable delay after the pressure has been applied until the weld is started. The current is then turned on by the timer and held a sufficient time for the weld. It is then stopped, but the pressure remains until the weld cools, thus eliminating any tendency for the electrodes to arc and also protecting the weld from discoloration. The pressure on the weld may be obtained manually, by mechanical means, by air pressure, by springs, or by hydraulic means. Its application must be controlled and co-ordinated with the application of the welding current.

Resistance welding may be subdivided into six separate heads: (1) Spot welding, (2) projection welding, (3) seam welding, (4) butt welding, (5) flash welding, and (6) percussion welding.

Spot welding. In this form of resistance welding two or more sheets of metal are held together between metal electrodes (or

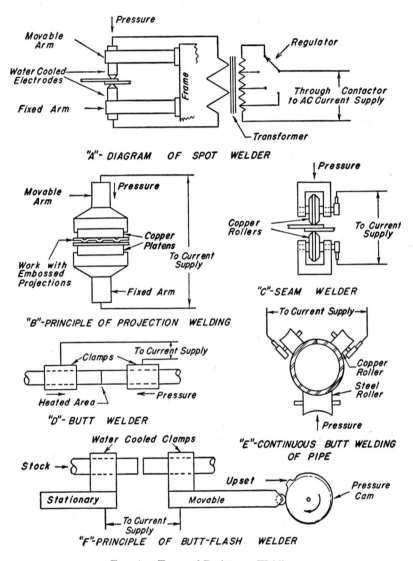

"A"- DIAGRAM OF SPOT WELDER

"B"-PRINCIPLE OF PROJECTION WELDING

"C"-SEAM WELDER

"D"- BUTT WELDER

"E"-CONTINUOUS BUTT WELDING OF PIPE

"F"-PRINCIPLE OF BUTT-FLASH WELDER

FIG. 1.   Types of Resistance Welding.

" dies ") as shown in Figure 1A.   A low-voltage current of sufficient strength, and definitely timed, is passed between the electrodes, causing the metal in contact to be rapidly raised to welding temperature. As soon as the temperature is reached, the pressure between the electrodes squeezes the metal together and completes the weld. The pressure is then released, and the work is either removed from the machine or moved so that another portion may be welded. Spot welding is probably the simplest form of resistance welding and for ordinary sheet steel does not present much of a problem. However, good welds require sheet steel which is free from scale or foreign substances. Such films cause variations in surface resistance and tend to increase the heating effect of metal in contact with the electrodes. Surface imperfections, variations in weld strength, and electrode pickup are defects to be expected if sheet surfaces are not

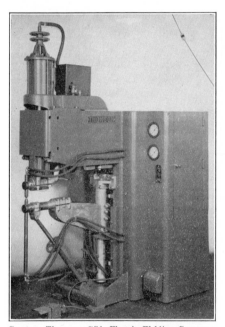

Courtesy Thompson-Gibb Electric Welding Company.

FIG. 2.   Direct-Pressure Spot-Welding Machine.

properly prepared. It may be noted that in spot welding there are three zones of heat generation: one at the surface between the two sheets, and the other two at the contact surfaces of the sheets with the electrodes. The center surface reaches a fusing temperature first, since the heat is rapidly conducted from the outside surfaces by the water-cooled electrodes.

Machines for single-spot welding are built in both stationary and portable types. Stationary machines may be further classified as rocker-arm and the direct-pressure types. The rocker-arm type is the simplest and cheapest but is generally limited to machines of small capacity. This machine is so designated because the motion for applying pressure and raising the upper electrode is done by rocking the upper arm. The larger machines usually employ direct straight-line motion of the upper electrode. This arrangement permits them to be used also for projection welding. A 250-kva air-operated machine of this type is shown in Figure 2.

As assemblies to be welded increase in size, it is not always possible to bring them to a machine. Portable spot welders, connected to the transformer by long cables, and capable of being moved to any desired position are then used. Welding jig assemblies where all welds cannot be made by a single machine setup are also served best by portable welders. Figure 3 illustrates the use of portable spot-welding guns for welding different operations on demolition bomb

Courtesy Progressive Welder Company.

Fig. 3. Portable Spot-Welding Guns Welding Different Operations on
Demolition-Bomb Fins.

fins. A wide variety of welding guns is made, the principal difference among them being in the manner of applying the pressure.

For production work, multiple-spot-welding machines have been developed which are capable of producing two or more spots simultaneously. In machines of this type several direct welds can be made either from one or more transformers. In some cases a system known as *indirect welding* is used, where two electrodes are in series, and the current passes through a heavy plate underneath the sheets and between the electrodes. Figure 4 shows a duplex machine spot welding reinforcing rings to 40-mm ammunition boxes. This is a special-purpose machine designed to accommodate the particular assembly to be spot-welded.

Spot welding lends itself readily to production work because of its simplicity and the rapidity with which welds can be made. Welds have been made at the rate of 300 per minute. Most ferrous and nonferrous alloys may be spot-welded, but each metal requires special manipulation and current regulation. Spot welding has a wide application, from thin foils, in which low currents are used, to heavy plates requiring high current and long time. Theoretically, any material of

Fig. 4. Duplex Hydromatic Progressive Welder Spot Welding Reinforcing Rings to 40-Mm Ammunition Box.

reasonable thickness can be spot-welded; however, as the thickness increases, the required tremendous force necessary becomes a limiting factor. One-half inch is the usual upper limit. Spot welding is widely used in the manufacture of automobiles, refrigerators, and metal toys, and in numerous other metal-stamping assemblies.

Projection welding. One of the latest developments of spot welding, termed projection welding, is shown diagrammatically in Figure 1*B*. Material to be welded in this manner is first put through a punch press, which presses small projections or buttons into the metal. These projections are made with a diameter on the face equal to the thickness of the stock and project above the stock about 60% of its thickness. Such projection spots or ridges are made at all points where a weld is desired. One advantage of this form of welding is that 1 to 30 welds may be made at one time. The size and location of these spots are readily controlled according to the desired strength of the joint. It is also very economical so far as dies are concerned, since only flat surfaces are necessary. This means reduced wear, burning, and electrode maintenance.

Projection welding requires two to three times the pressure needed in spot welding, and for most applications requires about twice the current. It is possible to weld all metals by this method as readily as by other resistance-welding processes. The production of fabricated-sheet-metal parts is a good example of projection welding.

Seam welding. Seam welding, shown in Figure 1*C*, consists in making a continuous weld or joint on two overlapping pieces of metal. Such welds are made by passing the metal between two electrode rollers or dies which transmit current, and to which mechanical pressure is applied. In some cases only a single roller is used in conjunction with a flat track. This method is in effect a continuous spot-welding process, as the current is not on continuously but regulated by the timer on the machine. The spots can be spaced close together to form a continuous seam or can be regulated so that they are several inches apart. The welding time is measured in cycles (1 cycle = 1/60 second), and electronic-tube controls are used to provide current interruptions. To produce a pressure-tight seam requires 8 to 18 spots per inch, depending on the type and gage of the material.

An air-operated rocker-arm-type seam and roll-spot-welding machine is shown in Figure 5. The maximum pressure obtainable between the electrodes runs as high as 3500 pounds at an air pressure of 90 pounds per square inch. Continuous rotation of the upper wheel can be adjusted up to 6 rpm, or intermittent rotation can be obtained if desired. This machine is capable of producing 858 spot welds per minute on two 0.040-inch thicknesses, which is 66 inches per minute with 13 spots per inch.

This form of welding is used a great deal in the manufacture of metal containers, automobile mufflers, stove pipes, refrigerator cabinets, and gasoline tanks. Advantages of this type of fabrication include neater design, saving of material, tight joints, and low cost of construction.

Butt welding. This form of welding, illustrated in Figure 1*D*, consists in the gripping together of two pieces of metal that have the same cross section and pressing them together while heat is being generated in the contact surface by electrical resistance. Although pressure is maintained while the heating takes place, at no time is the temperature sufficient actually to melt the metal. The joint is upset somewhat by the process, but this defect can be eliminated by subsequent rolling or grinding. Both parts to be welded should be

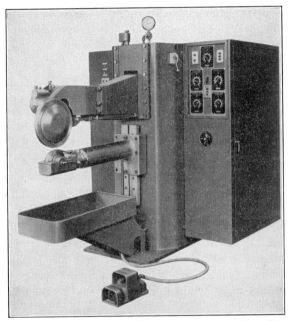

Courtesy Sciaky Bros.

FIG. 5.   Electrical-Resistance Rocker-Arm Seam-Welding Machine.

of the same resistance in order to have uniform heating at the joint. If two dissimilar metals are to be welded, the metal projecting from the die holders must be in proportion to the specific resistance of the materials to be welded. The same treatment must be used where materials of different cross section are butt-welded.

Butt welding is not limited to pipe or small sections. Round bars with a cross-section area as high as **70** square inches have been successfully welded. There have also been machines developed that will butt-weld 16-gage plates **120** inches wide. For such light work the current must be higher than for solid work and must be applied more quickly and with greater pressure. To illustrate, a solid round bar with a cross-sectional area of 20 square inches requires a transformer capacity of 40 kva for each square inch of area and a pressure of 5000 pounds per square inch. A seam butt weld 60 inches long of 18-gage metal has a cross-sectional area of about 3 square inches. Such a weld requires a transformer capacity of 125 kva per square inch area, and the upset pressure should be about 7000 pounds per square inch.

Practically all metals that can be spot- or projection-welded may be butt-welded, with the strength of the weld being about equal to the strength of the metal being welded. This type of welding is especially adapted to rods, pipes, small structural shapes, and many other parts of uniform section. Figure 1*E* illustrates a special type of butt-seam welding used in pipe manufacture. The action is similar in this case to seam welding.

Flash welding. Butt and flash-butt welding are similar in their application, but differ somewhat in the manner of heating the metal. Figure 1*F* illustrates diagrammatically the operation of this type of machine. The stock is clamped by dies as in ordinary butt welding. As soon as the metal is clamped, the current is turned on, and the two joints are brought together by means of the cam control. Just before the two parts come together, there is vigorous arcing which melts down any unevenness of the surface and rapidly brings it to a plastic state. The operating cam is adjusted so that the upset portion engages the moving platen at the proper time, and the two surfaces are forced together under high pressure, thus completing the weld. This system requires a very heavy current but takes a relatively short time. A small fin or projection left at the joint can be easily removed. An important advantage of the flash-welding process is that the plate edges do not have to be specially prepared.

Welding of small areas is usually done by the butt-welding method and those of large area by the flash-butt method; however, there is no clear demarcation between the two. The shape of the piece and nature of the alloy are frequently the determining factors. Areas ranging from 0.002 square inch to 50 square inches have been successfully welded by flash welding. With this process, less current is required than in ordinary butt welding; there is less metal to remove around the joints; the metal that forms the weld is protected from atmospheric contamination; and the operation consumes little time. Brass and copper cannot be successfully welded by this process. Flash welding is widely used for tubular furniture, rear-axle housings, steel rims, sheets in body manufacture, steel forgings, and rolled sections.

Percussion welding. This is a recent development in welding, and, like the flash-weld process, relies on arc effect for heating rather than on the resistance in the metal. Pieces to be welded are held apart, one in a stationary holder and the other in a clamp mounted in a slide and backed up against heavy spring pressure. When the movable clamp is released it moves rapidly, carrying with it the piece to be welded. When the pieces are about 1/16 inch apart there is a sudden discharge of electric energy, causing intense arcing over the surfaces and bringing them to a high temperature. The arc is extinguished by the percussion blow of the two parts coming together with sufficient force to effect the weld.

The electric energy for the discharge is built up in one of two ways. In the electrostatic method, energy is stored in a capacitor, and the parts to be welded are heated by the sudden discharge of a heavy current from the capacitor. The electromagnetic welder uses the energy discharge caused by the collapsing of the magnetic field linking the primary and secondary windings of a transformer or other inductive device. In either case intense arcing is created, which is followed by a quick blow to make the weld.

The action of this process is so rapid (about 0.1 second) that there is little heating effect in the material adjacent to the weld. Heat-treated parts may be welded without being annealed. Parts differing in thermal conductivity and mass may be successfully joined, as the heat is concentrated only at the two surfaces. Some applications are welding Stellite tips to tools, copper to aluminum or stainless steel, silver contact tips to copper, cast iron to steel, lead-in wires on electric lamps, and zinc to steel. Butt welds are made without any upset or flash at the joint. The principal limitation of the process is that only small areas (up to 1/4 square inch) of nearly regular sections can be welded. Thin sheets of equivalent area cannot be joined by this process. The equipment is expensive, since it must be extremely rugged, provided with accurate holding fixtures, and equipped with elaborate electric timing devices and large transformer capacity.

Gas Welding

Gas welding includes all the processes in which gases are used in combination to obtain a hot flame. Those commonly used are acetylene, natural gas, and hydrogen in combination with oxygen. Oxyhydrogen welding was the first gas process to be commercially developed. The maximum temperature developed by this process is 3600 F. The most-used combination is the oxyacetylene process, which has a flame temperature of 6300 F.

Oxygen is produced by two main processes: electrolysis and liquefying air. Electrolysis separates water into hydrogen and oxygen by passing an electric current through it. Most of the commercial oxygen is made by liquefying air and separating the oxygen from

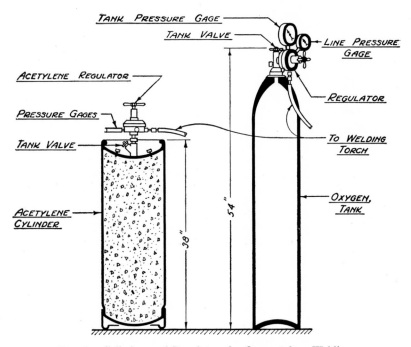

FIG. 6.   Cylinders and Regulators for Oxyacetylene Welding.

the nitrogen. It is stored in steel cylinders, as shown in Figure 6, at a pressure of 2000 pounds per square inch. Hydrogen is produced either by the electrolysis of water or by passing steam over coke.

Acetylene gas (C_2H_2) is obtained by dropping lumps of calcium carbide in water. The gas bubbles up through the water, and the remainder of the calcium carbide is converted into slaked lime. The reaction that takes place in an acetylene generator is

$$CaC_2 + 2H_2O = Ca(OH)_2 + C_2H_2$$

| Calcium carbide | Water | Slaked lime | Acetylene gas |

The calcium carbide used for making this gas is a hard gray stone-like material formed by smelting calcium with coal in an electric furnace.   This material is crushed, sized, and stored in air-tight steel drums prior to its use.   Acetylene gas can be obtained either from acetylene generators, which generate the gas by mixing the carbide with the water, or it can be purchased in cylinders ready for use. Because this gas may not be safely stored at pressure much over 15 pounds per square inch, it is stored in combination with acetone. Acetylene cylinders are filled with a porous filler saturated with acetone in which the acetylene gas can be compressed.   These cylinders hold 300 cubic feet of gas at pressures up to 250 pounds per square inch.

A cross section of a typical welding torch is shown in Figure 7.   It consists of a series of brass tubes through which and into which the gases are conducted and finally mixed, valves for controlling the volumes of acetylene and oxygen, and a copper tip from which the gas mixture is burned.   Regulation of the proportion of the two gases is of extreme importance, as the characteristics of the flame may be varied.

Three types of flame that can be obtained are *reducing, neutral,* and *oxidizing.*   Of the three, the neutral flame has the widest application in welding and cutting operations.   This flame occurs with approximately a one-to-one mixture of oxygen and acetylene.   There are two sharply defined zones, an inner luminous cone surrounded by an outer envelope flame which is only faintly luminous and slightly bluish in color.   The maximum temperature of 6300 F is obtained at the tip of the inner luminous cone.

When there is an excess of acetylene used, there is a decided change in the appearance of the flame.   In this flame there will be found three zones instead of the two just described.   Between the luminous cone and the outer envelope there is an intermediate cone of whitish color, the length of which is determined by the amount of the excess acetylene.   This flame, known as a reducing or carbonizing flame, is used in the welding of Monel metal, nickel, certain alloy steels, and many of the nonferrous, hard-surfacing materials, such as Stellite and Colmanoy.

If the torch is adjusted to give excess oxygen, a flame similar in appearance to the neutral flame is obtained, except that the inner luminous cone is much shorter, and the outer envelope appears to have more color.   This, the oxidizing flame, may be used in fusion welding of brass and bronze, but it is undesirable in other applications.

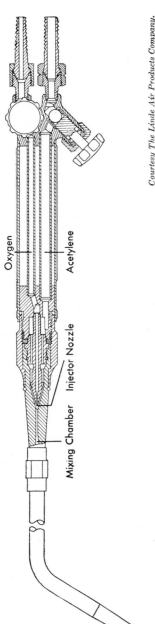

Oxygen

Acetylene

Mixing Chamber   Injector Nozzle

FIG. 7.   Cross Section of an Oxyacetylene Welding Blowpipe.

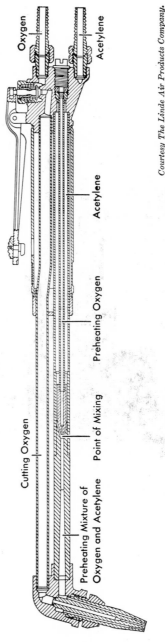

Oxygen

Acetylene

Acetylene

Preheating Oxygen

Point of Mixing

Cutting Oxygen

Preheating Mixture of
Oxygen and Acetylene

FIG. 8.   Cross Section of an Oxyacetylene Cutting Blowpipe.

The advantages and uses of gas welding are numerous. The equipment necessary for welding is comparatively inexpensive and requires little maintenance. It is portable and can be used with equal facility out in the field and in the factory. With proper technique practically all metals may be welded. There is also the added advantage that the equipment can be used for cutting as well as welding. The process is especially adapted to the welding of sheet metal, to flame hardening, and to the application of many hard-facing materials. It finds wide use in airplane welding maintenance and job-shop welding, brazing, pipe-line welding, and many other industrial applications. However, this process is not all-inclusive and will not solve all welding problems, as discussions of other processes will indicate.

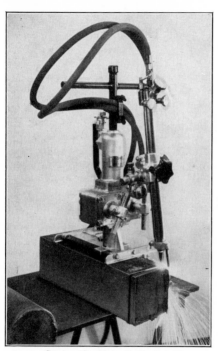

Courtesy *The Linde Air Products Company.*

FIG. 9. Cutting a Square Billet with a Portable Bar-Cutting Machine.

Gas Cutting

The cutting of steel with a flame-cutting torch has developed into a very important production process. A simple hand torch for flame cutting is shown in Figure 8. It differs from the welding torch in that it has several small holes for preheating flames surrounding a central hole through which pure oxygen passes. The preheating flames are exactly like the welding flames and are intended only to preheat the steel prior to the cutting operation. The principle upon which flame cutting operates is based on the fact that oxygen has an affinity for iron and steel. At ordinary temperatures this action is slow, but eventually an oxide in the form of rust materializes. As the temperature of the steel is increased, this action becomes much more rapid. If the steel is heated to a red color and a jet of pure oxygen is blown on the surface, the action is almost instantaneous, and the steel is actually burned into an iron oxide slaglike appearance. About 1.3 cubic feet of oxygen are required to burn up 1 cubic inch of iron. This action is illustrated in Figure 9, where a square steel billet is being cut with a portable machine. This machine will cut cold rounds or squares of any size from 2 to 10 inches. In the operation shown the blowpipe starts and finishes at an angle, while a constant clearance is maintained at all times between the cutting nozzle and the billet. A similar cutting operation is shown in Figure 10 in preparing plate edges for welding fabrication. Cuts in several planes can be made simultaneously, with accuracy comparable to preparation by machine tools, but at less cost. Metal plates up to 30 inches thickness can be cut by this process.

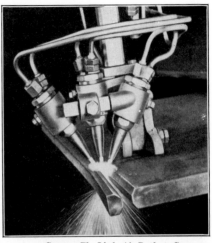

Courtesy The Linde Air Products Company.

FIG. 10.   Preparing Plate Edges Preparatory to Welded Fabrication.

Many cutting machines have been developed which automatically control the movement of the torch to cut any desired shape. Such a machine is shown in Figure 11 cutting a bearing cap out of 8-inch stock. Motion is transferred from the tracing unit to the blowpipe by means of two carriages. The lower carriage runs on rails along the edges of the tracing table, while the upper carriage moves on rails above the lower carriage and provides the transverse movement. Electrically driven, the drive unit automatically controls the movement of the machine at a proper cutting speed in accordance with the templet shape. Multiple cutting can be accomplished by mounting several blowpipes on a cross bar. An example of multiple cutting of tank sprockets is shown in Figure 12. This machine operates on the pantograph principle so that one or several cutting torches are made to conform exactly to the movement of the tracing device. The latter may be a hand-guide, or a spindle roller which is held against the templet by the operator.

Many parts which previously required shaping by forging or casting are now cut to shape by this process. Flame-cutting machines, replacing many machining operations where accuracy is not paramount, are widely used in the shipbuilding industry, structural fabrication, maintenance work, and in the production of numerous items made from steel sheets and plates.

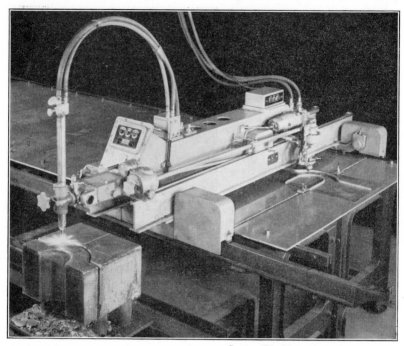

Courtesy The Linde Air Products Company.

FIG. 11.   Cutting a Bearing Cap out of 8-Inch Stock.

Flame machining.*   Flame machining is the term used to describe the operation of removing metal with a cutting blowpipe. It differs from ordinary flame cutting in that it does not sever the main body of metal, but merely removes metal as is done in machining operations. The torch is held at a small angle to the work surface and, as it progresses, cuts out a groove instead of penetrating. The process is rapid and requires no power, and the work setup need not be rigid. On the other hand, the surface finish is not good, and close dimensional accuracy cannot be attained. However, in many rough machining operations involving the removal of a large amount of metal, this method of cutting should be given consideration.

* E. L. Cady, Flame Cutting and Machining Methods, *Metals & Alloys*, May 1945.

Courtesy Air Reduction Sales Company.

FIG. 12.   Multiple Cutting of Tank Sprockets.

Brazing and Soldering

The processes of brazing and soldering, which unite metals by means of a different metal, are shown in Table 5.   In these processes, joints are made without pressure, the joining metal merely being introduced into the joint in a liquid state and allowed to solidify. Both of these processes have wide commercial use in the uniting of small assemblies and electrical parts.

TABLE 5

BRAZING AND SOLDERING

Dipping	Furnace	Gas Torch	Electric	Soft Soldering
1. Metal	1. Gas	1. Oxyacetylene	1. Resistance	1. Soldering iron
2. Chemical	2. Electric	2. Oxyhydrogen	2. Induction	2. Wiping
			3. Arc	

Brazing. In the process of brazing, a nonferrous alloy is introduced in a liquid state between the pieces of metal to be joined and allowed to solidify. The filler metal has a melting temperature somewhat over 1000 F, lower than the melting temperature of the parent metal. Special fluxes are required to remove surface oxide and to give to the filler metal the fluidity necessary to wet completely the joint surfaces. The brazing metals and alloys commonly used are as follows:

1. Copper — melting point 1982 F.

2. Copper alloys — brass and bronze alloys having melting points ranging from 1600 F to 2250 F.

3. Silver alloys — melting temperatures ranging from 1250 F to 1550 F.*

4. Aluminum alloys — melting temperatures ranging from 1025 F to 1785 F.†

In the brazing of two pieces of metal, the joint must first be cleaned of all oil, dirt, or oxides, and the pieces properly fitted together with appropriate clearance for the filler metal. Mechanical or chemical cleaning may be necessary in the joint preparation in addition to the flux used during the process. Borax, either alone or in combination with other salts, is commonly used as a flux. The final selection of a flux should be determined by handbook or manufacturers' recommendations, according to the filler material used.

According to Table 5 there are four methods used in heating the metal to complete a joint. They are:

1. Dipping the assembled parts in a bath of filler metal or flux. When dipped in a flux bath, which is held at a temperature sufficient to melt the filler metal, the assembly must be held together securely in a jig and the joint preloaded with the brazing alloy.

2. In furnace brazing the assemblies are held in position by jigs and are introduced into a controlled-atmosphere furnace maintained at the proper temperature to melt the brazing metal. These furnaces may be either gas- or electric-heated, and may be of the batch or continuous type. In Figure 13 is shown a group of assemblies ready for furnace brazing. This method is rapid and lends itself to quantity-production jobs.

3. Torch brazing is similar to oxyacetylene welding. Heat is applied locally by an oxyacetylene or oxyhydrogen torch, and the filler metal, in wire form, is melted into the joint. Flux is applied by immersing the wire.

4. In electric brazing the heat may be applied by resistance, by induction, or by an arc. Of these methods, the first two are most often used because of their speed and accurate temperature control.

* *Metals Handbook*, American Society for Metals, 1939, p. 1211.

† *Welding and Brazing of Alcoa Aluminum*, Aluminum Company of America, p. 99.

To facilitate speed in brazing, the filler metal is frequently prepared in the form of rings, washers, rods, or other special shapes, to fit the joint being brazed. This insures having the proper amount of filler metal available for the joint as well as having it placed in the correct position.

Joints in brazing may be of the lap, butt, sleeve, or scarf types, or of various shapes obtained by curling, upsetting, or seaming processes. The strength of the joint* is determined principally by the strength of the brazing material used, although other influencing factors are the strength of the parent metal, amount of clearance, cleanness of the joint, and method of heating and cooling.

Advantages of the brazing process include its ability to effect joints in materials difficult to weld, in dissimilar metals, and in exceedingly thin sections of metal. In addition, the process is rapid and results in a neat-appearing joint requiring a minimum of finishing. Brazing is used for the fastening of pipes and fittings, tanks, carbide tips on tools, radiators, heat exchangers, electrical parts, and the repair of castings.

Courtesy Aluminum Company of America.

FIG. 13.   Batch Brazing of Aluminum Gasoline Tanks in Furnace.

Soldering. Soldering differs from brazing in that lower-temperature filler metals are used in the joint, and there is no tendency for the soldering metal to alloy with the parent metal. Lead and tin alloys having a melting range from 300 to 700 F are principally used in soldering, and the strength of the joint is determined by the adhesive qualities and the strength of these alloys. Although any heating method used in brazing can be employed in this process, much soldering is done with the common soldering iron which is

* *Materials and Processes,* edited by J. F. Young, John Wiley & Sons, 1943, p. 438.

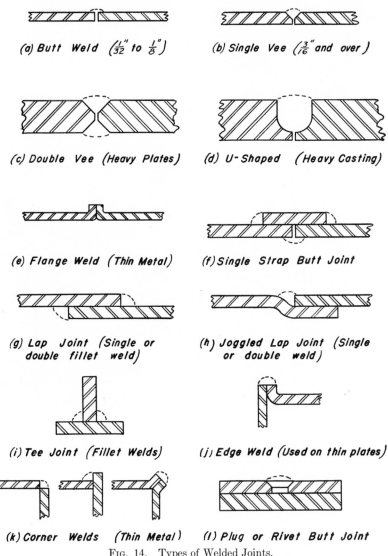

FIG. 14. Types of Welded Joints.

especially suitable for small parts and light-gage metal. Heat is supplied by the iron, and solder is fed to the joint in the form of wire. Cleaning of the joint surface is equally as important in soldering as in brazing, and a flux is necessary. Electric connections, wire terminals, and similar small parts are typical of the joints made by soft soldering. A form of soldering known as *wiping* is used in making connections of lead pipe.

Welded Joints

Both gas and arc welding (Figure 14) use the same type of joints. The six principal types are *butt, lap, edge, corner, plug,* and *tee.* Some of these types, such as butt welds, may be further subdivided, as they vary in form according to the thickness of the material. Joints for forge welding differ in their manner of preparation and do not resemble those shown in the figure. Lap and butt joints are the principal types used in resistance welding. In general, resistance-welded joints must be prepared more accurately and must be considerably cleaner than in other processes.

Electric-Arc Welding

Arc welding is a fusion process in which the metal is heated to a state of liquefaction permitting it to flow into a solid joint. Contact is first made between the electrode and the work to create an electric circuit, and then, by separating the conductors, an arc is formed. The electric energy is converted into intense heat in the arc, which attains a temperature around 7000 F. A filler metal obtained by melting a welding electrode in the arc is fused with the molten base metal to make the weld.

Both direct current and alternating current may be used for arc welding, direct current being preferred for most purposes. A d-c welder (see Figure 15) is simply a motor–generator set of constant-energy type (constant potential may also be used), having the necessary characteristics to produce a stable arc. There should not be too great a current surge when the short circuit is made, and the machine should compensate to some extent for varying lengths of the arc. D-c machines are built in capacities up to 600 amperes having an open-circuit voltage of 40 to 95 volts. A 200-ampere machine has a rated current range of from 40 to 250 amperes, according to the standard of the National Electrical Manufacturers Association. While welding is going on, the closed-circuit voltage is 18 to 25 volts. Having the electrode as the negative terminal is known as *straight polarity,* while *reverse polarity* has the electrode as the positive terminal.

The first methods of arc welding employed only carbon electrodes. One of the first processes, invented by Zerener, used two carbon electrodes. Contact was made between the two electrodes, and weld metal was supplied by a separate rod as in the oxyacetylene process. This equipment was bulky and not suitable for high currents. A second process, utilizing a single carbon electrode of negative polarity,

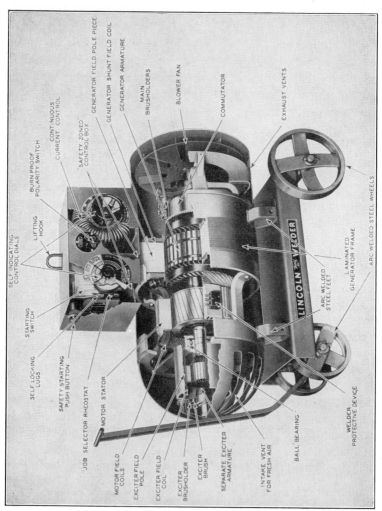

Fig. 15.   Lincoln D-C Arc-Welding Machine.

is considerably simpler. In this case the arc is created between the carbon electrode and the work, and any weld metal needed is supplied by a separate rod. Such an arc is easy to start, as there is no tendency for the electrode to stick to the metal. Straight polarity

Courtesy Westinghouse Electric Corporation.

FIG. 16. A Team of Welders Seals the Joints of a Frame for a Turbogenerator. The Big Circular Ladder in the Foreground is a Frame for a 4500-Horsepower Electric Motor.

must always be used, as a carbon arc is unstable when held on the positive terminal. There is also a tendency for vaporized carbon or carbon monoxide to contact the weld metal and form hard spots.

Carbon-arc welding is used principally for welding copper, bronze, galvanized steel, and, to some extent, cast iron. It may also be used effectively for rough cutting of metals.

Shortly after the development of the process just described, Slavianuff discovered that, by the use of a metal electrode with the proper current characteristics, the electrode itself could be melted down to supply the necessary weld metal. A basic patent for this process was issued in 1889, and it is this process that is in general use today. An example of this type of welding is shown in Figure 16 in the welding of a frame for a turbogenerator. In the fabrication of large weldments it is desirable to position the work so that flat or down-hand welding is possible. This is frequently done by mounting the work on a large table which can be rotated or moved to a position favorable to the operator. However, when this is not possible, vertical or overhead welding can be used.

One of the principal advantages of arc welding is that there is little tendency for grain growth in the metal adjacent to the weld. This is due to the heat of the arc's being concentrated in a very small area and the rapidity of the process. Arc welding is faster than the oxyacetylene process, as a result of the higher temperature developed by the arc. Furthermore, it is specially adapted to overhead and position welding. It successfully welds most ferrous and nonferrous alloys and has many applications, for example, the welding of pressure vessels, structural fabrication, production work, maintenance, manufacture of machine frames, and ship construction.

Electrodes. The three types of metal electrodes (or " rods ") are bare, fluxed, and heavy coated. Bare electrodes have a limited use for the welding of wrought iron and mild steel. Straight polarity is generally recommended. Improved welds may be made by applying a light coating of flux on the rods by a dusting or washing process. The flux assists both in eliminating undesirable oxides and in preventing their formation. However, the heavily coated arc electrodes are by far the most important ones used in all types of commercial welding. Over 95% of the total manual welding that is being done today is with coated electrodes.

Figure 17 is a diagrammatic sketch showing the action of an arc using a heavy-coated electrode. In the ordinary arc with bare wire the metal is affected to some extent by the oxygen and nitrogen in the air. This causes oxides and nitrides to be formed in the weld metal, both of which are undesirable. The effect of heavy coatings on electrodes is to provide a shield around the arc to eliminate such conditions and also to cover the weld metal with a protective slag coating which prevents oxidations of the surface metal during cooling. Welds made from rods of this type have superior physical characteristics. Manufacturers' recommendations should always be followed in the selection of an electrode for a given job.

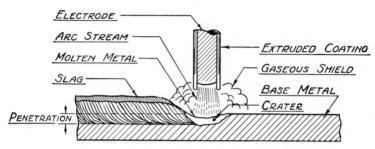

FIG. 17. Diagrammatic Sketch of Arc Flame.

Electrode coatings. Electrodes coated with slagging or fluxing materials are particularly necessary in the welding of alloys and nonferrous metals. Some of the elements in these alloys are not very stable and are lost if there is no protection against oxidation. Heavy coatings also permit the use of larger welding rods, higher current, and greater welding speeds. In summary, the coatings do the following things:

1. Provide a protecting atmosphere.
2. Provide slag of suitable characteristics to protect the molten metal.
3. Facilitate overhead and position welding.
4. Stabilize the arc.
5. Add alloying elements to the weld metal.
6. Perform metallurgical refining operations.
7. Reduce spatter of weld metal.
8. Increase deposition efficiency.
9. Remove oxides and impurities.
10. Influence the depth of arc penetration.
11. Influence the shape of the bead.
12. Slow down the cooling rate of the weld.

These functions are not common to all coated electrodes, since the coating put on a given electrode is largely determined by the kind of welding it has to perform. It is interesting to note that the coating composition is also a determining factor as to electrode polarity. By varying the coating, rods may be used with either the positive or negative terminal or may work equally well either positive or negative. Properly coated electrodes make possible a weld metal having equal physical properties with the base metal.

Many coating compositions have been developed to accomplish these results. In general, they may be classified as organic and inorganic coatings, although in some cases both types might be used. Inorganic coatings can be further subdivided into flux compounds and slag-forming compounds. These are some of the principal constituents used:

1. Slag-forming constituents — SiO_2, MnO_2, and FeO. Al_2O_3 is sometimes used, but it makes the arc less stable.
2. Constituents to improve arc characteristics — Na_2O, CaO, MgO, and TiO_2.
3. Deoxidizing constituents — graphite and wood flour.
4. Binding material — sodium silicate and asbestos.
5. Alloying constituents to improve strength of weld — V, U, Ce, Co, Mo, Al, and Zr.

A-c arc welding. A-c welders are growing in popularity because of certain advantages inherent in this type of equipment and because of the development of electrodes suitable for this process. For ordinary welding there is little difference in the quality of the welds made by a-c and d-c equipment. The a-c machines consist principally of static transformers which are simple pieces of equipment having no moving parts. Their efficiency is high, their loss at no load is negligible, and their maintenance and initial costs are low. Welders of this type are built in five sizes specified by NEMA, and are rated at 150, 300, 500, 750, and 1000 amperes. For welding requiring 750 amperes or higher, a-c equipment is preferred. The fact that there is less magnetic flare of the arc or " arc blow " than with the d-c equipment is important in the welding of heavy plates or fillet welding. For jobs requiring medium- or small-amperage current loads the a-c equipment is limited by the type of electrodes required. Most of the nonferrous metals and many of the alloys cannot be welded with a-c equipment, because electrodes have not been developed for this purpose.

The equipment required for a-c welding is shown in Figure 18. In this case the operator is welding light tubular sections with a high-frequency stabilized a-c welder. The superimposed high-frequency arc stabilizer permits striking and maintaining a steady arc on current settings as low as 10 amperes. This type of equipment is particularly useful in the welding of thin-wall tubular sections and light sheet-metal work.

If we compare a-c and d-c welders, the welding speed, the quality of the welds, and the ease in welding is the same. However, d-c machines are still to be preferred, as this type of machine permits the selection of the proper polarity for the welding electrode. Because the alternating current is continually reversing with every cycle, the correct polarity is being used only half the time. This eliminates

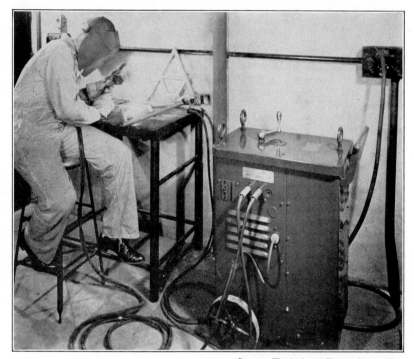

Courtesy Westinghouse Electric Corporation.

FIG. 18. Welding Light Tubular Sections with High-Frequency Stabilized A-C Welder.

the successful use of carbon electrodes and many of the metal electrodes. Also, a-c welders operate at slightly higher voltages, and hence the danger of shock to the operator is increased. In spite of these limitations, there is a growing demand for the latter type of welder in fabrication and maintenance shops, sheet-metal work, and jobbing machine shops.

Automatic-arc welding. In production shops, where a large number of similar joints have to be made, automatic arc welding is satisfactory and rapid. The electrode is manipulated by a " welding head " which feeds it to the work at the proper rate as it is consumed. The rate of feed is controlled by the voltage of the arc, which varies as its length changes.　Such machines may be stationary, with the work moving past the electrode, or the welding head may be mounted so as to move over the work.　Automatic-welding heads are available for bare and lightly coated electrodes, shielded arc welding of either

Courtesy General Electric Company.

Fig. 19.　Gantry-Type Machine Arc Welder.　Capacity 40 Feet Long, 18 Feet Wide, and 11 Feet High.　Welding Speed 2¼ × 8 Inches per Minute.　Arms Being Welded on Waterwheel Generator for Boulder Dam.

coiled or short-length electrodes, carbon electrodes, inert gas-shielded arc welding, and flux-shielded welding using bare electrodes.

Prerequisites* for economic machine welding are (1) a sufficient volume of production to justify expensive equipment, and (2) a uniform product.　Assemblies must fit together readily, and each piece must be of the same size and similar in contour.　Often jigs and handling devices are necessary to position the work.　If these conditions are fulfilled, machine welding will be economical.　The use

* R. F. Wyer, " Progress in Automatic Arc Welding," *Machinery*, November–December 1944.

Courtesy The Linde Air Products Company.

Fig. 20. "Unionmelt" Welding Head Shown Welding the Transverse Seam on a 52-Inch Pipe.

of automatic-welding machines results in increased welding speed and uniform quality of the weld. In addition, the operator is relieved of tedious work, since he does not have to maintain the proper arc length and travel speed.

Figure 19 shows a thyratron-controlled gantry-type machine welder operating on a large part for a vertical waterwheel generator. This equipment uses a lightly coated electrode which is started and fed to the work automatically. The arc voltage is the regulating characteristic employed by the thyratron tube to maintain a constant arc length.

Another automatic-arc-welding process, called Unionmelt, is illustrated in Figure 20. A bare electrode is fed through the welding

Fig. 21.   Automatic-Welding Machine Mounted on a Tractor.   Used for Welding Deck Plates on an LST.

head into a highly resistant conductive medium supplied as a granulated material and known as Unionmelt. This material is laid down along the seam to be welded by the welding head, and the entire welding action takes place beneath it. As this medium does not conduct electricity when cold, a small wad of steel wool is placed beneath the electrode to start the action. A portion of the Unionmelt is fused into a slag coating which covers the weld, and the unused material is picked up by suction, as shown in the figure. As in most of the other systems, the rod feed is governed by voltage control of the arc. This process has the advantage of fairly high-speed welding; furthermore, any commercial thickness of metal can be welded by one pass with no loss of weld metal. Vertical and overhead welding cannot be done with this equipment.

The Lincolnweld automatic-arc-welding machine is illustrated in Figure 21. This process is similar to the Unionmelt method in that both weld under granulated flux and feed a metallic wire automatically into the arc; however, each uses a different flux, and there are several differences in the equipment, particularly the control. The tractor model shown is used for such large work as ship plates and large tanks. Equipment on the tractor includes welding head, operator's control station, flux hopper, and electrode reel. The travel speed of this machine can be varied from 8 to 50 inches per minute. A bare electrode is fed through the flux while direct current, supplied by the welder, produces the arc between the electrodes and the joint. As the weld is made, it is covered with melted flux which forms a slag coating protecting the metal from contact with air. Automatic-arc-welding machines are indispensable in production welding and are particularly useful in the fabrication of large assemblies such as ships, tanks, and railroad cars.

Atomic hydrogen arc welding. In this process a single-phase a-c arc is maintained between two tungsten electrodes, and hydrogen is introduced into the arc. As the hydrogen enters the arc, the molecules are broken up into atoms which recombine into molecules of hydrogen outside the arc. This reaction is accompanied by the liberation of an intense heat, attaining a temperature of about 7200 F. Weld metal may be added to the joint in the form of welding rod, the operation being very similar to the oxyacetylene process. The atomic hydrogen process differs from other arc-welding processes in that the arc is formed between two electrodes, rather than between one electrode and the work. This makes the electrode a rather mobile tool, as it can be moved from place to place without being extinguished. The complete atomic hydrogen arc-welding equipment is shown in Figure 22.

The outstanding advantage of this process is the protection which the hydrogen offers in shielding the electrode and the molten metal from oxidation. The hydrogen also acts to intensify and concentrate the heat. Metal of the same analysis as the metal being welded can be used, and many alloys, difficult to weld by other processes, can be successfully treated. The welds are clean, smooth, and free from scale, and they respond to heat treatment in the same way as the parent metal when weld metal of the same composition is used. The process has wide use in die repair; it sucessfully welds heat-resisting alloys; it has proved to be an excellent means of applying

Courtesy General Electric Company.

Fig. 22. Complete Atomic Hydrogen Arc-Welding Equipment.

carbides and many other hard-surfacing alloys; and it is widely used in production work where special ferrous and nonferrous alloys are used.

Gas-shielded arc welding. The only commercial method of welding magnesium is the recently developed " Heliarc " process, in which a standard d-c machine is used with a tungsten electrode surrounded by an envelope of helium gas. The arc is maintained between the single tungsten electrode, and the work with the required filler metal is supplied by a separate welding rod. The torch is constructed with a gas nozzle surrounding the tungsten electrode so that the helium gas, as it leaves the nozzle, completely envelops the tip of the electrode and the work beneath it. Magnesium is always welded with reverse polarity.

Argon gas can also be used with the same torch equipment, but at prevailing prices helium is the more economical. In the welding of aluminum alloys argon seems to give much better results. No flux is required and an a-c source of power is used. Both gases are successfully used in automatic-arc-welding machines.

Thermit Welding

Thermit welding is the only process employing an exothermal chemical reaction for the purpose of developing a high temperature. It was discovered by Dr. H. Goldschmidt and patented March 16, 1897, but was not used extensively for welding until several years later. It is based on the fact that aluminum has a great affinity for oxygen and can be used as a reducing agent for many oxides. The usual Thermit metal consists of finely divided aluminum and iron oxides, mixed at a ratio of about 1 to 3 by weight. The iron oxide is usually in the form of roll scale. This mixture is not explosive and can be ignited only at a temperature of about 2800 F. A special ignition powder (usually magnesia ribbon or barium peroxide) is used to start the reaction. The chemical reaction requires only a few seconds and attains a temperature of around 4500 F. The mixture reacts according to the chemical equation:

$$8Al + 3Fe_3O_4 = 9Fe + 4Al_2O_3$$

The resultant products are iron and an aluminum oxide slag, which floats on top and is not used. Other reactions also take place, as most Thermit metal is alloyed with manganese, nickel, or other metals.

Figure 23 illustrates the method of preparing the material for such a weld. Around the break where the weld is to be made, a wax pattern of the weld is built up. A refractory sand is packed around the joint, and necessary provision is made for riser and gates. A preheating flame is used to melt and burn out the wax, to dry the mold, and to bring the joints to a red heat. The reaction is then started in the crucible, and, when it is complete, the metal is tapped and allowed to flow into the mold. As the weld-metal temperature is approximately twice the melting temperature of steel, it readily fuses in the joint. Such welds are sound, because the metal solidifies

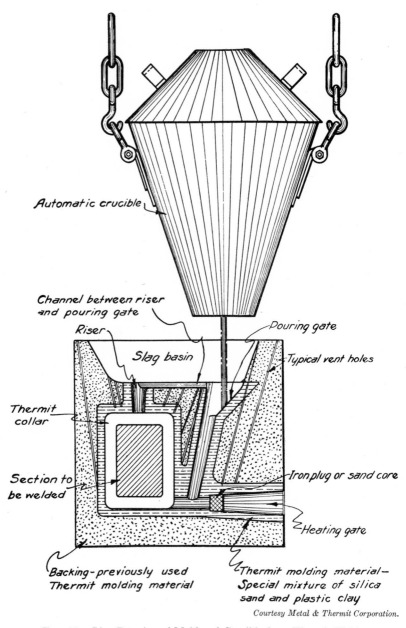

Automatic crucible

Channel between riser
and pouring gate

Riser

Slag basin

Pouring gate

Typical vent holes

Thermit
collar

Section to
be welded

Iron plug or sand core

Heating gate

Backing—previously used
Thermit molding material

Thermit molding material—
Special mixture of silica
sand and plastic clay

Courtesy Metal & Thermit Corporation.

FIG. 23.   Line Drawing of Mold and Crucible for a Thermit Weld.

from the inside toward the outside, and all air is excluded from around the mold.

There is no limit to the sizes of welds that can be made by this process which is used for the repair of large parts which would be difficult to weld by other processes. Other uses are welding of large crankshafts, machine frames, steel-mill rolls, railroad rails (see Figure 24), ship-rudder frames, and large shafts. In addition to repairing

Courtesy Metal & Thermit Corporation.

FIG. 24. Thermit Equipment for Welding Railroad Rails.

old parts, it may be used for building up worn surfaces and, in production work, for welding together forgings that would otherwise require expensive dies or patterns. It can be also used for cast iron and many ferrous alloys by controlling the composition of the Thermit metal.

Hard Surfacing

Hard surfacing is the process of applying to a wearing surface some metal or treatment which renders the surface highly resistant to abrasion. Such processes vary a great deal in their technique. Some apply a hard surface coating by fusion welding; in others no material is added, and the surface metal is changed by heat treatment or by contact with other materials. With the development of the processes many new hard-surfacing materials were discovered. The research for these hard materials has been especially keen during recent years, owing to the great demand of industry for longer-life products.

TABLE 6

. METHODS OF PRODUCING HARD SURFACES

1. Heat Treatment
    A. Carburizing — heating in contact
        1. With solids as charcoal
        2. With liquids as KCN
        3. With gases as CO
    B. Special case-hardening processes
        1. Nitriding — contact with NH_3 gas
        2. Chapmanizing — contact with liquid containing N and C
        3. Dry cyaniding — contact with gases
        4. Ni-carbing — contact with gases
    C. Induction hardening — electric heating and rapid quenching
    D. Flame hardening — heating with torch and rapid quenching

2. Metal Spraying — Applying with Air Pressure
    A. High-carbon steel
    B. Stainless steel
    C. Other alloys

3. Metal Plating — Electrolytic Deposit of Chromium and Other Hard Elements

4. Fusion Welding Processes
    A. Overlay process — welding with
        1. Ferrous alloys
            (a) High-carbon steel
            (b) Steel alloys
        2. Nonferrous alloys
            (a) Chiefly of chromium, cobalt, molybdenum, and tungsten
    B. Diamond substitutes
        1. Cemented, cast, or sintered carbides of tungsten and other elements
            (a) Inserts
            (b) Screen sizes in tubes
            (c) Screen sizes and binder cast into rods
            (d) Screen sizes loose or with gelatin binder
            (e) " Sweat-on " — paste containing very fine hard particles
    C. Casting or spinning process — chiefly nickel borides

The several properties required of materials subjected to severe wearing conditions are hardness, abrasion resistance, and impact resistance. Hardness is easily determined by several known methods, and an accurate comparison of metals for this property can be obtained readily. Tests for wear or abrasion resistance have not been standardized, and it is difficult to obtain comparative results. In general, experience has shown that, to obtain proper results, wear testing must simulate the service conditions for each type of hard-facing material. The statement that " the wear resistance of a material is a function of the method by which it is measured " has been confirmed by both practical experience and research. Considering all the factors involved, hardness is probably the best criterion of wear resistance. Ability to withstand wear and abrasion usually increases as the hardness of the metal increases.

Table 6 is a classification of the various processes used for obtaining a hard surface. Obviously, there is a great difference in the hardness that can be obtained from these methods. The classification does not include heat-treating methods which produce a hard interior surface.

Flame Hardening

Flame hardening, like the induction-hardening process, is based on rapid heating and quenching of the wearing surface. The heating is accomplished by means of an oxyacetylene flame, which is applied for a sufficient length of time to heat the surface above the critical temperature of the steel. Integral with the flame head are water connections which cool the surface by spraying as soon as the desired temperature is reached. By proper control, the interior surface is not affected by the treatment, the depth of the case being a function of the heating time and flame temperature. Figure 25 shows an etched cross section of a gear tooth and the hardened areas.

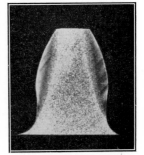

Courtesy The Linde Air Products Company.

FIG. 25. Section through a Gear Tooth Showing Structure Obtained by Flame Hardening.

There are several methods employed in this process. In the stationary method of spot hardening, both torch and work are stationary, and the effect is local. Progressive hardening refers to cases where the flame and work move with respect to one another, as in the case of rail hardening. As the flame progresses, the work is immediately quenched behind the flame. Spinning or rapidly rotating circular work may be used, employing one or more flames. As soon as the work is brought up to the proper temperature, it is quenched while rotating. This method is usually applied to fairly small work when the heating time is short. Spinning may also be used in connection with a progressive movement of the torch along the side of the work. Hand operation is not recommended, as it does not produce uniform results.

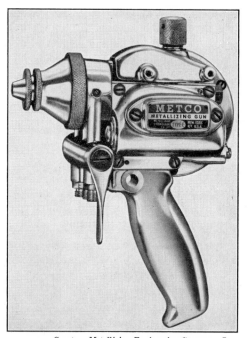

Courtesy Metallizing Engineering Company, Inc.

Fig. 26.   Metal Spray Gun.

The following advantages are claimed for this process: Hard surfaces with a ductile backing may be obtained, large pieces may be treated without heating the entire part, the case depth is easily controlled, the surface is free of scale, and the equipment is portable. This process may be used on most of the commonly used steels. The surface hardness will depend on the percentage of carbon and the alloying elements that may be present. Average values will vary from 350 to 650 Brinell.

Metal Spraying

The spraying of molten metal, a comparatively recent development, is rapidly becoming an important process in industry. Any
metal obtainable in wire form can be applied in this manner. The
wire is fed into the spray gun at a definite rate, where it is melted
by an oxyacetylene flame and then blown by compressed air to the
surface being coated. Such a spray gun is shown in Figure 26.
Another type of gun, developed in England, has a heated container

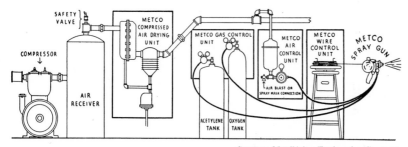

Courtesy Metallizing Engineering Company.

FIG. 27.   Arrangement of Equipment for Metal Spraying.

into which molten metal is poured at intervals; but otherwise it acts
in the same manner. A typical layout of metal spraying equipment
is shown in Figure 27.

Because the bond between the sprayed metal and the parent metal
is entirely mechanical, it is important that the surface of the metal
be properly prepared before spraying. The usual method of cleaning
and preparing the surface is by blasting with sharp silica sand or
angular steel grit. Cylindrical objects may be prepared by rough
turning on a lathe.

Either of these methods roughens the surface and provides the
necessary interlocking surfaces or keys to make the plastic metal
adhere to the surface. The molten metal is blown with considerable
force against the surface, causing it to flatten out and interlock with
surface irregularities and the adjacent metal particles. The sprayed
metal itself provides a suitable surface for successive coatings and
permits building up a layer of considerable thickness.

Obviously, there is some change in the physical properties of metal
applied in this manner. There is an increase in porosity and a
corresponding decrease in the tensile strength of the material. The
reason for this is that the bond is mechanical and not fusile, as for
welding. The compressive strength is high, and there is some in-

crease in hardness. Stainless-steel and high-carbon deposits will develop a Scleroscope hardness of 70 to 75, which corresponds to a Brinell hardness of 500 to 550. The wearing quality of sprayed metal is good, as evidenced by its wide use in building up worn shafts and other moving parts. All deposits can be finished satisfactorily by usual machining or grinding methods. In general, the metals retain most of their original properties.

Metal spraying is used principally to prevent corrosion. Zinc, cadmium, lead, and aluminum are the principal metals used. Because the coatings may be thicker and there is no trace of acids at their base, it is frequently claimed that metals applied by this method have greater resisting properties than if applied by plating.

Aside from building up worn surfaces, sprayed metal produces hard-wearing surfaces. A low-carbon shaft can be sprayed with 0.80% C steel, increasing the surface hardness materially. For greater hardness, 1.20% C steel may be applied successfully. There are two types of stainless steel in general use, ordinary 18–8 and a special air-hardening steel. The latter steel gives the maximum surface hardness obtained by this process. Sprayed metal, being air-quenched, is naturally harder than the original metal and has proved very satisfactory for wearing surfaces.

The success of this process is due largely to its economy and the rapidity with which the metal can be applied. There is no distortion in the parts being surfaced, nor are any internal stresses developed. Practically any metal can be applied to any other commercial metal and even to other base surfaces such as wood and glass.

Metal Plating

Electroplating has long served as a means of applying decorative and protective coatings on metals. For wear or abrasive resistance, the outstanding metal for plating metallic surfaces is chromium. For this use, coatings are seldom less than 0.002 inch thick, and may be considerably more. Coatings of this nature are not plated on a soft base metal, but directly onto the hard parent metal. If plated on a soft base metal, as copper or nickel, it will have greater corrosion-resisting power, but its resistance to abrasion and deformation will be much less. Hence, any measure of hardness or abrasive resistance is to some extent a function of the metal upon which it is plated as well as of the chromium deposit itself.

The process, electrolytic, consists in passing an electric current from an anode to a cathode (the cathode being the object upon which the metal is deposited) through a suitable chromium-carrying electrolytic solution in the presence of a catalyst. The catalyst does not enter into the electrochemical decomposition. A solution of chromic acid with a high degree of saturation is used as the electrolyte. The surfaces must be thoroughly polished and cleaned before operations start; and, since the rate of deposition is fairly slow, the work must remain in the tanks several hours for heavy plating.

Chromium has proved very satisfactory for wear-resisting parts because of its extreme hardness, which exceeds most other commercial metals. According to the Brinell scale, the hardness of plated chromium ranges from 500 to 900. This wide variation is due not to the metal but to methods and equipment used.

Fusion-Welding Processes

Where thick coatings of materials are required, it is necessary to use some form of welding. The electric-arc process is both simple and economical in its operation, and for most ferrous alloy materials it is very suitable. The acetylene process operates at a lower temperature than an electric arc and can be controlled more readily where excessive puddling of the material is not desirable. A certain amount of penetration or bonding is necessary when hard-facing materials are being applied, but in general it should be kept at a minimum so as not to dilute the weld metal with the parent metal.

The hard-facing materials are classified roughly as " overlay " and " diamond-substitute " types. The overlay materials include such metals as high-carbon steel, ferrous alloys of chromium and manganese, and numerous nonferrous alloys containing principally cobalt, manganese, and tungsten. The hardness of these materials varies considerably, ranging from around Rockwell C 40 to 70. According to the Mohs scale, the hardness seldom exceeds 8. The " diamond substitutes " are such materials as tungsten, boron, tantalum carbides, and chromium boride. These materials are among the hardest available and on the Mohs scale fall between 8.5 and 9.5. They cannot be applied by self-fusion but must be bonded to the parent metal with some lower-melting alloy.

High-carbon welding rod with a carbon content ranging from 0.9 to 1.1% is the most economical hard-facing material to apply from the standpoint of initial cost. Such rods may be applied equally well by either the electric-arc or the oxyacetylene method of welding. They form a tough surface of moderate hardness, ranging from Rockwell C 30 to 45. Hardest surfaces are obtained by rapid quenching;

and, as is the case with all martensitic deposits, not much additional hardness can be obtained by cold working. Their corrosion resistance is poor, but such coatings have a wide application where wear resistance is desired.

Increased hardness and wear resistance can be obtained by alloying steel with such elements as nickel, manganese, molybdenum, and chromium. The limit of hardness for such coatings is around Rockwell C 55, and, since many of the alloys result in austenitic deposits, their hardness can be increased by cold working. Corrosion resistance of most of these materials is good, as is their resistance to impact, and no heat treatment is required after application.

In the nonferrous group are included all rods which are made up of elements other than iron, but in some cases small percentages of iron may be present. The principal elements in this group are tungsten, chromium, molybdenum, and cobalt. The average room-temperature hardness of this group is about the same as for the ferrous-alloy group. A high percentage of this hardness is retained while at red heat, which adds greatly to their wear-resisting power. In severe abrasive work considerable heat is developed by friction, which acts on the minute areas of particles in contact. The effect of this heat is to soften the metal on these areas and cause them to wear away. However, if the metal in contact can retain a hardness at a relatively high temperature, it has a much greater resistance to wear than metals that do not have this property. In such cases the initial hardness is not a true criterion of the wear-resisting ability of the metal.

Both the electric-arc and oxyacetylene processes can be used in applying this material, the latter process being preferred. Better control of the deposit is obtained, and there is less dilution of the rod with the parent metal. There is also no loss of the expensive rod material by volatilization and spattering. Practically any carbon or alloy steel can be hard-surfaced with this material, and it is especially adapted for coating surfaces subject to severe abrasion and impact such as valve seats and oil drilling tools.

Diamond Substitutes

The so-called diamond substitutes constitute the hardest materials that are available for hard surfacing. These materials, generally spoken of as cemented carbides, include tungsten carbide, tantalum carbide, titanium carbide, boron carbide, and chromium boride, or a combination of these and other carbides with a suitable cementing agent. In tungsten carbide, which is one of the most common of the group, the usual analysis by percentage is tungsten 81.4, cobalt 12.7, carbon 5.3, and iron 0.6. The cobalt serves as a binder and adds to the ductility of the carbide. It may vary in percentage from 5 to 13. Tantalum carbide is 87% TaC, with 13% of some binder. Usually the binder is either a combination of molybdenum and iron or tungsten carbide and cobalt. Boron carbide contains about 78.2% boron and 21% carbon, with a trace of silicon and iron. It is usually known by the symbol B_4C. Many similar carbide materials are manufactured under special trade names, the compositions of which are not generally known and vary with manufacture. Some of the commonly known commercial carbide products are Carboloy, Firthite, Carmet, Widia, Borium, and Hastellite.

Carbide material cannot be applied as other hard-surfacing materials because of its high melting temperature and is therefore furnished either in the form of small inserts or in screen sizes. Inserts can be applied by a brazing or sweating-on process or placed in melted or puddled metal and then surrounded by metal from a steel or hard-surfacing welding rod. Screen sizes of crushed carbide particles can be applied conveniently by putting the particles in steel tubes. The steel sheath melts like an ordinary welding rod and fuses to the metal. The carbide particles do not melt but are distributed through the molten metal and are held fast when the metal cools. Screen sizes can also be applied by mixing the particles with a suitable binder and casting them into rods. These rods can be used conveniently like other hard-surface welding rods.

These materials all have hardnesses approaching that of a diamond, and on the Mohs scale they range from 9 to 9.5. This hardness is maintained to a large extent at a red heat. Because of such extreme hardness and brittleness, diamond substitutes do not have a high strength rating and are not suitable where severe shock and impact conditions exist. This difficulty is partially eliminated by properly supporting the elements with a tough binding material. Another characteristic of these surfacing materials is that they do not respond to heat treatment or cold working and retain their initial hardness under all conditions. They are not suitable for casting, although a few hard materials, principally boron alloys with an iron base, can be processed in this manner.

Cemented carbides are used principally for cutting-tool material and for parts subjected to extreme wearing conditions. Tools that are tipped with such inserts greatly outlast any other form of cutting tool.

REVIEW QUESTIONS

1. Classify all welding processes.

2. What kind of steel is most easily welded?

3. How is acetylene gas made? How is it stored?

4. How is oxygen made? Will pure oxygen ignite?

5. Distinguish between butt-resistance welding and flash welding.

6. How is projection welding done?

7. Describe the Thermit process. For what type of work is it used?

8. What are the advantages of the atomic hydrogen process?

9. Show by sketches the type joints to use in oxyacetylene welding for thin, medium, and heavy joints.

10. Define brazing and list the different ways it is done.

11. What are the common metals used in brazing?

12. Compare a-c and d-c welders as to quality of welds, welding speed, and range of application.

13. What kinds of arc-welding machines are in the shop? Is the current alternating or direct?

14. Describe the electrical-resistance process of welding.

15. What is flame machining, and how does it differ from flame cutting?

16. Show by sketch the position of acetylene torch and welding rod when welding. Indicate direction of travel.

17. What type of electrodes are recommended for hard surfacing?

18. How is it that steel is cut so readily with a torch?

19. Why are coated rods used in electric-arc welding?

20. How is brazing accomplished?

21. Describe the Unionmelt process of arc welding.

22. What are the prerequisites for economic machine welding?

23. What are the advantages of flame hardening?

24. What materials are known as " diamond substitutes "?

25. How should a surface be prepared for metal spraying?

26. How is magnesium welded?

BIBLIOGRAPHY

Arc Welding in Design, Manufacture, and Construction, James F. Lincoln Arc Welding Foundation, 1939.

Arc Welding Manual, General Electric Company, 1939.

Arc Welding Manual and Operators Training Course, Hobart Brothers Company, 1941.

BEGEMAN, M. L., " Hard-Surfacing Processes and Materials," *Mechanical Engineering,* Vol. 60, December 1938.

CADY, E. L., " Flame Cutting and Machining Methods," *Metal & Alloys,* May 1945.

CHAFFEE, W. J., *Arc Welding and How to Use It,* 3d edition, Hobart Brothers Company, 1938.

HOMERBERG, V. O., " Nitriding," *Iron Age,* Vol. 138, October 15, 1936.

HUBERT, E. H., *Manual of Electric Arc Welding,* 1st edition, McGraw-Hill Book Company, 1932.

Instructions in Oxy-Acetylene Welding and Cutting Processes, Air Reduction Company, 1940.

JOHNSON, J. B., *Airplane Welding and Materials,* 2d edition, Goodheart-Willcox Company, 1941.

MAMPLE, A. Z., " Soldering with Tin-Lead Alloys," *Metals & Alloys,* May 1945.

KERWIN, H., *Arc and Acetylene Welding,* McGraw-Hill Book Company, 1944.

Procedure Handbook of Arc Welding Design and Practice, 7th edition, Lincoln Electric Company, 1942.

Properties of Haynes Stellite, Haynes Stellite Company, 1936.

SHAPIRO, C. H., " Oxy-Acetylene Application of Hard Facing Materials to Oil Well Drilling Bits," *Metal Progress,* Vol. 35, April 1939.

The Oxwelder's Handbook, 15th edition, Linde Air Products Company, 1939.

The Welding Encyclopedia, 11th edition, Welding Engineer Publishing Company, 1943.

Welding Handbook, American Welding Society, 1942.

WYER, R. F., " Some Developments in Gas-Shielded Arc Welding," paper presented at Philadelphia Chapter of American Welding Society, May 1944.

CHAPTER 9

HOT FORMING OF METALS

The mechanical working of metal is the shaping of metal in either the cold or hot state by some mechanical means. This does not include the shaping of metals by machining or grinding, in which processes metal is actually machined off, nor does it include the casting of molten metal into some form by use of molds. In mechanical working processes, the metal is shaped by pressure — actually forging, bending, squeezing, drawing, or shearing to its final shape. In these processes the metal may be either cold- or hot-worked. For cold working of steel, normal room temperature is used; however, temperatures up to critical are included in this range. Hot working of steel takes place at temperatures above the critical range. At such high temperatures the metal is in a plastic state and is readily formed by pressures. The properties of the metal are actually improved by mechanical working, since grain refinement is caused by the process. Little grain refinement takes place below the critical range, the action there being principally one of grain distortion. Both hot and cold working produce a structure elongated in the same direction as that of rolling, drawing, or other form of mechanical work.

Metals which are cast from the furnace into ingot molds must be reduced to commercial shapes by mechanical force. This force is applied to the metal while it is hot, for the following reasons: To shape the product more easily into some commercial article; to improve the physical properties of the metal; and to impart a fine crystalline srtucture to the metal. Principal methods of hot-working metals are:

HOT WORKING — METAL IN A PLASTIC STATE

1. Rolling.

2. Forging
   (a) Hammer or smith forging.
   (b) Drop forging.
   (c) Machine or upset forging.
   (d) Press forging.
   (e) Forging rolls.

FIG. 1. Steel Being Poured in Molds to Make Ingots for Subsequent Rolling.

    3. Pipe welding
       (*a*) Butt welding of heated strips.
       (*b*) Butt welding by electrical resistance.
       (*c*) Lap welding.
       (*d*) Hammer welding.

    4. Piercing.

    5. Drawing or cupping.

    6. Extrusion.

Hot-Working Processes

Before metals are mechanically worked they are first cast into ingot molds of suitable form for subsequent operations. The mold

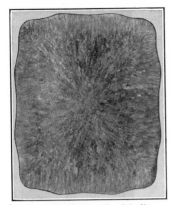

Cross Section of Medium-Carbon Killed-Steel Ingot Showing Crystalline Structure.

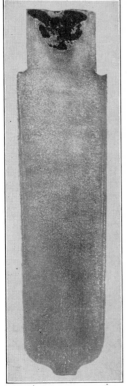

Courtesy Drop Forging Association.

Longitudinal Section of Steel Ingot.

FIG. 2.

may be either square or round in cross section, and the final casting may vary in size from a few hundred pounds to several tons. The kind of metal cast and the product desired are the principal governing factors in determining the ingot size. The operation of pouring ingot molds with steel from an open-hearth furnace is shown in Figure 1. The steel remains in these molds until solidification is about complete, and then the molds are removed. While still hot, the ingots are taken to the rolling mill and placed in gas-fired furnaces called *soaking pits*, where they remain until they have attained a uniform working temperature throughout.

The structure of the steel ingot is very coarse, because the grains grow while the metal is slowly cooling from a liquid to a solid state. This is illustrated in Figure 2, which shows a longitudinal and cross section of a medium-carbon killed-steel* ingot. The coarse dendritic crystalline structure may be clearly seen in the cross section. This coarse structure will be subsequently eliminated by the effects of hot working. The impurities in ingots tend to segregate in the shrink head during the process of solidification. Cutting off the end of the ingot, either before the rolling starts or shortly after it has started, largely eliminates this defect. Frequently as much as 25% of the ingot is scrapped in this fashion.

* Steel which has been deoxidized to the extent that gases are evolved during the process of solidification is known as "killed" steel.

Rolling. After the ingot has remained in the soaking pit at a temperature around 2200 F for a sufficient time to become uniformly heated, it is then removed by an overhead crane and placed on the rolling-mill table. Because of the large variety of finished shapes to be made, ingots are first rolled into such intermediate shapes as *blooms, billets,* or *slabs.* These shapes all have rounded corners and vary principally in cross-sectional area. A *bloom* has a square cross section with a minimum size of 6 by 6 inches. Figure 3 shows an ingot being rolled in a blooming mill. This mill is a two-high

Courtesy Bethlehem Steel Company.

FIG. 3.   Hot-Ingots Are First Rolled in the Blooming Mill until Reduced to Desired Size.

reversing mill, permitting the metal to pass back and forth through the rolls. At frequent intervals it is turned 90 degrees on its side to keep the section uniform and to refine the metal throughout. About 20 passes are required to reduce a large ingot into a bloom. Grooves are provided in both the upper and lower rolls to accommodate the various reductions in cross-sectional area.

A *billet* is smaller than a bloom, and may have any square section from 1½ by 1½ inches up to the size of a bloom. Although billets could be rolled to size in a large blooming mill, this is not usually done for economic reasons. Frequently they are rolled from blooms in a continuous billet mill consisting of about eight rolling stands in a straight line. The steel makes but one pass through the mill and emerges with a final billet size of approximately 2 by 2 inches. The billet is the raw material for many final shapes such as bars, tubes, and forgings. The diagram in Figure 4 illustrates the number of

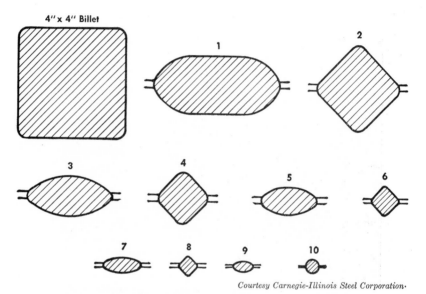

FIG. 4. Diagram Illustrates Number of Passes and Sequence in Reducing the Cross Section of a 4-by-4-Inch Billet to Round Bar Stock.

passes and sequence in reducing the cross section of a 4-by-4-inch billet to round bar stock.

Slabs may be rolled from either an ingot or a bloom. They have a rectangular cross-sectional area with a minimum width of 10 inches and a minimum thickness of $1\frac{1}{2}$ inch. Slabs constitute the material from which plates, skelp, and thin strips are rolled.

Many special rolling mills take the intermediary products just described and fabricate them into such finished articles as rails, structural shapes, plates, and bars. Such mills usually bear the name of the product being rolled and in appearance are similar to mills used for rolling blooms and billets. Materials which are commonly hot-rolled include steel, aluminum, copper, and magnesium.

Hammer or smith forging. This type of forging consists of hammering the heated metal either with hand tools or between flat dies in a steam hammer. Hand forging as done by the blacksmith is the oldest form of forging. It is largely used in repair or maintenance work and in the production of numerous small parts such as hooks, crowbars, chisels, cutting tools, U bolts, and similar articles, where only small quantities are involved. Much of this type of work is now done in power forges called steam or smith hammers. Machines of this type have an open-frame construction to allow plenty of room around the flat dies. Power is usually supplied by steam, although air is sometimes used. The force of the blow is closely controlled by the operator, and the work done is generally for the purpose of reducing the cross section of the metal. It is frequently a preliminary operation to press or drop forging in closed-impression dies.

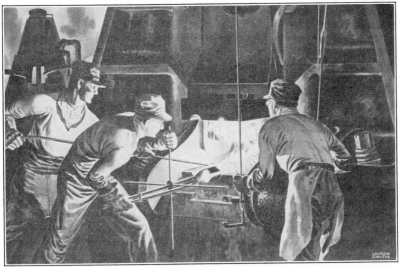

Courtesy Chambersburg Engineering Company.

Fig. 5. Drop Forging with Closed-Impression Dies.

Drop forging. Drop forging differs from smith forging in that closed-impression rather than open-face dies are used. The forging is produced by impact or pressure which compels the plastic metal to conform to the shape of the dies. In this operation there is drastic flow of the metal in the dies caused by the repeated blows on the metal. The ram, operating in a vertical plane, carries one half of the dies and the other half is held stationary on the anvil block.

Both are held in perfect alignment. The operations of drop forging a crankshaft with closed-impression dies is shown in Figure 5. A crankshaft is forged from a steel billet of proper length involving a number of successive forging operations. Each set of dies contains a different impression for the forming of the crankshaft, the final shape being obtained in the finishing dies.

Steps in the forging of a connecting rod are illustrated in Figure 6. The first step in the manufacture of this part is to shear off bar stock to proper length and bring it up to forging temperature in a furnace adjacent to the forge. The dies used in this operation, shown at (a), contain impressions for several operations. Preliminary hot working first proportions the metal for forming the connecting rod; then the fillering operation, illustrated at (b), reduces the cross-sectional area of the center while the edging gathers metal for the two ends of the rod. The blocking operation, which forms the rod into a definite shape, is shown at (c). This operation requires several blows of the hammer, causing the plastic metal to conform to the impressions in the die. Surplus metal called *flash* is forced around the edges of the forging. The appearance of the connecting rod after several blows in the finishing die is shown at (d). Flash around the edges of the finished forging is removed in a separate press by trimmer dies immediately after the finishing operation is completed. The completed connecting rod, ready for heat treatment, is shown at (e). At (f) is a macroetched cross section of a connecting-rod forging, showing the flow lines of the metal and the fiber structure obtained by the hammering action.

The two principal types of drop-forging hammers are the *steam hammer* and the *board hammer*. In the former the ram and hammer are lifted by steam, and the force of the blow is controlled by throttling the steam. These hammers work rapidly: over 300 blows a minute can be obtained. The capacities of steam hammers range from 500 to 50,000 lbs. In the board drop hammer several boards are attached to the ram, the lifting of which is accomplished by several friction rollers contacting the boards on both sides. When the top of the stroke is reached, the rollers are spread apart, allowing the boards and ram to fall by gravity. The force of the blow is dependent on the weight of the ram or head assembly, which seldom exceeds 8000 lbs.

Advantages of the drop-forging process are: The physical properties of the metal are improved by the severe mechanical working; the operation is rapid; many complicated parts can be forged to shape; a

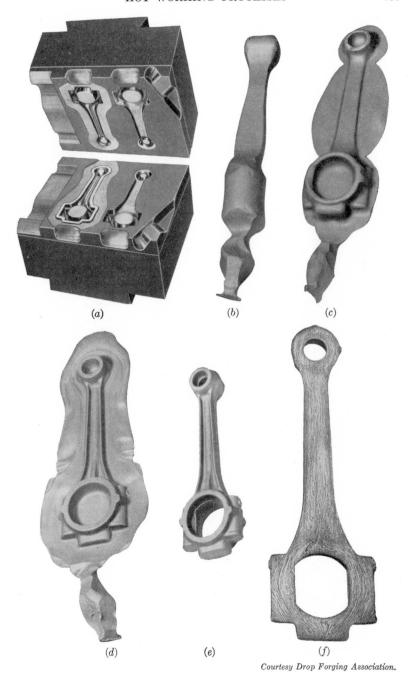

(a)

(b)

(c)

(d)

(e)

(f)

Courtesy Drop Forging Association.

Fig. 6. Steps in Drop-Forging Connecting Rod.

minimum amount of machining is necessary; and internal defects are eliminated. This method of fabrication is adapted to both carbon and alloy steels, wrought iron, copper-base alloys, aluminum alloys, and magnesium alloys. A limitation is the cost of the equipment and dies, which necessitates the making of a large number of parts. Casting or flame cutting may be employed if the advantages by hot working are not a paramount consideration.

Press forging. Press forging employs a slow squeezing action in deforming the plastic metal, as contrasted with the rapid-impact blows of a hammer. The squeezing action is carried completely to

Courtesy The Hydraulic Press Manufacturing Company.

Fig. 7.   A 1000-Ton Flat-Die Forging Press with Auto Floor-Type Manipulator.

the center of the part being pressed, thoroughly working the entire section. The size of the forging is no longer a limitation in press-forging work, since large presses up to capacities approaching 18,000 tons are now available.

In the press forging of such large parts as a heavy-artillery gun barrel, either flat dies or dies of simple construction are used. In Figure 7 is shown a 1000-ton flat-die forging press with the forging being handled by a 3-ton auto floor-type manipulator. By using mechanical means for locating and rotating the forging between the dies, the operation of the press is greatly simplified. This particular press is controlled by an oil-hydraulic operating system.

For small press forgings, closed-impression dies are used, and only one stroke of the ram is normally required to perform the forging operation. A close-up view of the bolster plate, dies, and forgings produced in such a press is shown in Figure 8. The work of pressing the conical-shaped blank (left) in the 1300-ton press is accomplished in two operations. The preliminary operation, performed in the dies to the left, compresses the slug (right) and roughly forms it to shape (center), so that there is a minimum of flow and abrasion to the finishing dies. Presses of this type are now widely used in the forming of nonferrous metals. Figure 9 shows a

Fig. 8. Pressing Cone Forgings for Oil-Well Drilling Head.

group of rather complicated press forgings made of wrought copper, brass, and bronze. Both magnesium and aluminum can also be processed in this fashion.

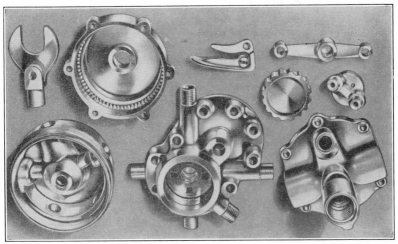

Fig. 9. Hot-Pressed Forgings Are Made in Wrought Copper, Brass, Bronze or Special Copper Alloys.

Another type of press forging which is a combination of piercing and extrusion is shown in Figure 10. The illustration shows an operator removing a hot steel slug from a 200-ton piercing press, the first operation in forging a 90-mm high-explosive shell. After one more extruding or drawing operation the forging is complete. The average production for forgings of this type is around 3000 forgings per 24-hour day.

Courtesy The Hydraulic Press Manufacturing Company.

FIG. 10. Removing a Hot Steel Slug from a 200-Ton Piercing Press, the First Operation in Forging a 90-Mm High-Explosive Shell.

In the forging press a greater proportion of the total work put into the machine is transmitted to the metal than in a drop hammer press. Much of the impact of the drop hammer is absorbed by the machine and foundation. Furthermore, press reduction of metal is faster and the cost of operation is consequently lower. Press forgings are more accurate than drop forgings, since less draft is required; however, many parts of irregular and complicated shapes can be more economically forged by drop forging.

Machine or upset forging. Upset forging entails gripping a bar of uniform section in dies and applying pressure on the heated end, causing it to be upset or formed to shape. A large machine of this type is shown in Figure 11. Here rear-axle drive shafts with flanged ends are being forged from round stock. Machines of this type are an outgrowth of smaller machines designed for heading bolts and making nuts.

Dies for these machines consist of two gripper dies, grooved to hold the stock during the various forging operations, and a header or punch assembly. A die for forging a socket wrench in four operations is shown in Figure 12. Such dies are not limited to upsetting

Courtesy The Ajax Manufacturing Company.

FIG. 11.   Five-Inch Upset Forging Machine in Operation — Forging Flanged
Rear-Axle Drive Shafts.

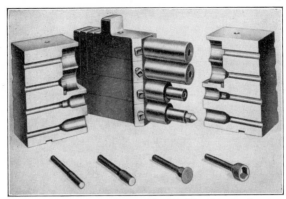

FIG. 12.   Dies for Upset Forging of Socket Wrench Consisting of Two Gripper
Dies and One Punch Assembly.

FIG. 13.   Forging Rolls Producing Automobile Rear-Axle Drive Shafts.

operations but may also be used for piercing, punching, trimming, extrusion, or bending. Parts forged by this process include axle shafts, pinions, axle housings, valve stems, engine cylinders, worm gears, and numerous other parts requiring upset ends.

Forging rolls. Forging rolls are being used in a wide variety of reduced, straight, and taper operations. A typical machine of this type producing automobile rear-axle drive shafts is shown in Figure 13. The rolls in this machine are semicylindrical in cross section and are grooved according to the type of forging or shaping to be done. When the rolls are in open position, the operator places the heated bar between them, retaining it with tongs. As the rolls re-

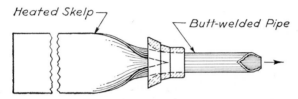

FIG. 14. Producing Butt-Welded Pipe by Drawing Skelp Through a Welding Bell.

volve, the bar is gripped in the roll grooves and pushed toward the operator. This operation is repeated several times, using other grooves, until the part is completely forged to shape. Gages and guiding grooves assist the operator in accurately placing the stock between the rolls. In this manner axles, gear-shift levers, blanks for eyebolts, and connecting rods, brake levers, aluminum propeller blanks, and leaf springs are forged.

Pipe welding. Pipe and tubular products may be produced by either welding or seamless processes. The butt-welding process uses heated strips of steel, known as *skelp*, the edges of which are first beveled slightly so they will meet when formed to a circular shape. In the intermittent process, one end of the skelp is trimmed to a V shape to permit it to enter the welding bell shown in Figure 14. When the skelp is brought up to welding heat, the end is gripped by tongs which engage a draw chain. As the tube is pulled through the welding bell, skelp is formed to a circular shape, and the edges are welded together. A final operation passes the pipe between sizing and finishing rolls to give it correct size and to remove all scale.

Continuous butt welding of pipe is accomplished by supplying the skelp in coils and providing means for flash-welding the coil ends

FIG. 15.   Skelp Emerging from Furnace at Right Is Formed into Continuous Butt-Weld Pipe.

together to form a continuous strip.   As the skelp enters the furnace, flames impinge on the edges of the strip to bring them to welding temperatures.   Leaving the furnace, the skelp enters a series of horizontal and vertical rollers, shown in Figure 15, which form it into pipe.   An enlarged view of the rollers, showing how the pipe is formed and sized, is shown in Figure 16.   As the pipe leaves the rollers, it is sawed into lengths which are finally processed by descaling and finishing operations.   Pipe is made in this fashion in sizes up to 3 inches in diameter.

FIG. 16.   Skelp Being Formed into Continuous Butt-Weld Pipe.

Electric butt welding of pipe necessitates cold forming of the steel plate to shape, prior to the welding operation.   The circular form is developed by passing the plate through a continuous set of rolls which progressively change its shape.   In Figure 17 is shown the plate for electric butt welding before forming and at various stages during the forming operation, indicating the degree of curvature imparted by each pass.   The welding unit through which the formed pipe passes is shown diagrammatically in Figure 18. It consists of three centering and pressure rolls to hold the pipe in position and two electrode rolls which supply the current

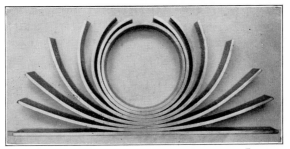

Courtesy Republic Steel Company.

Fig. 17.   Cross Section of Plate for Electric Butt Welding before Forming and at Various Stages during the Forming Operation, Indicating the Degree of Curvature Imparted by Each Pass.

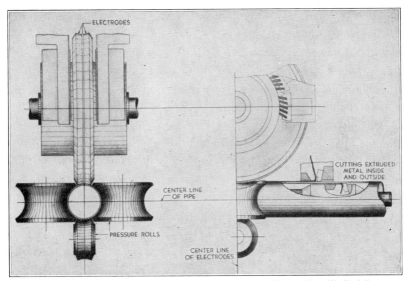

Courtesy Republic Steel Company.

Fig. 18.   The Operations of Electric Butt Welding and Flash Removal in the Manufacture of Steel Pipe.

to generate the heat.   Immediately after the pipe passes the welding unit, the extruded flash metal is removed from both inside and outside of the pipe.   Sizing and finishing rolls then complete the operation by giving the pipe accurate size and concentricity.   This process

is adapted to the manufacture of pipe up to 16-inch diameter with
wall thickness varying from 1/8 inch to 1/2 inch in thickness.   Pipes
of larger diameter are usually fabricated by fusion welding or by
hammer welding, the latter being essentially a forge-welding process.

In the lap welding of pipe, the edges of the skelp are beveled as
it emerges from the furnace.   The skelp then is drawn through a
forming die, or between rolls, to give it cylindrical shape, with the
edges   overlapping.   After
being reheated, the bent skelp
is passed between two grooved
rolls as shown in Figure 19.
Between the rolls is a fixed
mandrel of a size to fit the
inside diameter of the pipe,
the edges being lap-welded
together by pressure between
the rolls and the mandrel.
Lap-weld is made in sizes
ranging from 2 to 16 inches
in diameter.

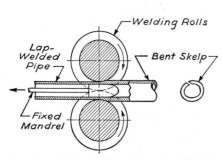

FIG. 19.   Method of Producing Lap-Welded
Pipe from Bent Skelp.

Piercing.   To produce seamless tubing, cylindrical billets of steel
are passed between two conical-shaped rolls operating in the same
direction.   Between these rolls is a fixed point or mandrel which
assists in the piercing, and controls the size of the hole as the billet
is forced over it.

The entire operation of making seamless tubing is shown diagrammatically in Figure 20.   The solid billet is first center-punched and

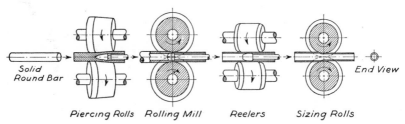

Courtesy National Tube Company.

FIG. 20.   Principal Steps in the Manufacture of Seamless Tubing.

then brought to forging heat in a furnace preparatory to piercing.
The billet is then pushed into the two piercing rolls which impart
to it both rotation and axial advance.   The alternate squeezing and
bulging of the billet opens up a seam in its center which is controlled

as to size and shape by the piercing mandrel. As the thick-walled tube emerges from the piercing mill, it next passes between grooved rolls over a plug held by a mandrel and is converted into a longer tube with specified wall thickness. While still at working temperatures, the tube passes through the reeling machine which further straightens and sizes it, and in addition gives the walls a smooth surface. Final sizing and finishing are accomplished in the same manner as that used for welded pipe.

This procedure applies to seamless tubes up to 6 inches in diameter. Larger tubes up to 14 inches in diameter are given a second operation on piercing rolls. To produce sizes up to 24 inches in diameter, reheated double-pierced tubes are processed on a rotary rolling mill as shown in Figure 21 and are finally completed by reelers and sizing rolls, as described in the single-piercing process.

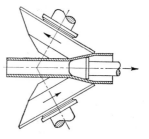

Rotary Rolling Mill

Courtesy National Tube Company.

Fig. 21. Rotary Seamless Process for Large Tubing.

In addition to the production of seamless pipe, piercing operations are frequently used on hydraulic presses in the production of short tubes, containers, and shells. Figure 10 illustrates the first operation in the manufacture of a 90-mm high-explosive shell. Many parts initiated this way are subsequently completed in a push bench by forcing the pierced blank through several dies of decreasing diameter.

Drawing or cupping. Some thick-walled tubes or cylinders are produced by drawing circular heated plates through a die, as illustrated diagrammatically in Figure 22. The entire procedure is made up of several drawing operations, between each two of which the cup-shaped cylinder must be reheated to provide the necessary plasticity for working. After about two drawing operations, the formed cup is pushed through a horizontal drawbench consisting of several dies of successively decreasing diameter, mounted in one frame. The hydraulically operated punch forces the heated cylinder through the full length of the drawbench. For long thin-walled cylinders or tubes, repeated heating and drawing may be necessary. If the final product is to be a tube, the closed end is cut off and the balance is sent through finishing and sizing rolls, similar to those used in the piercing process. To produce closed-end cylinders, similar to those for storing oxygen, the open end is swaged to form a neck. This process can be used also in the forming of heavy pots and short cylindrical objects not requiring the drawbench operations.

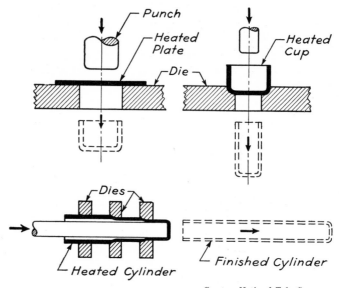

FIG. 22.   Drawing Thick-Walled Cylinders from Heated Plates.

Extrusion

Any plastic material can be extruded to uniform cross-sectional shape by the aid of pressure.   The principle of *extrusion*, similar to the simple act of squirting toothpaste from a tube, has long been utilized in processes ranging all the way from the production of brick, hollow tile, and soil pipe, to the manufacture of macaroni.   Some metals, notably lead, tin, and aluminum, may be extruded cold, whereas others require the application of heat to render them plastic or semisolid before extrusion.   In the actual operation of extrusion the processes differ slightly, depending on the metal and application, but in brief they consist of forcing metal (confined to a pressure chamber) out through specially formed dies.   Rods, tubes, molding trim, structural shapes, brass cartridges, and lead-covered cable are typical products of metal extrusion.   Typical samples of extruded products are shown in Figure 23.

The coating of wire cable with lead sheathing, as illustrated in Figure 24, is an important example of hot extrusion.   Molten lead is poured into the cylinder above the die and allowed to solidify under

Courtesy Schloemann Engineering Corporation

FIG. 23.   Samples of Extruded Products Formed in Large Extrusion Presses up
to 2500-Ton Capacity.

slight pressure of the cylinder plunger. When the correct extruding temperature is reached (around 500 F), the plunger, actuated by hydraulic means, forces the lead in two streams around the cable,

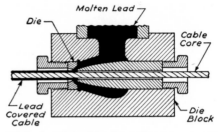

FIG. 24. Method of Extruding Lead Sheathing on Wire Cable.

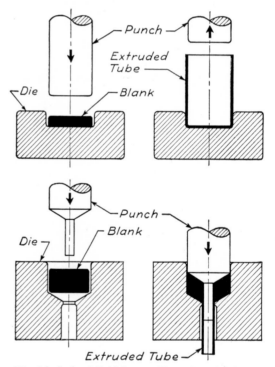

FIG. 25. Methods of Cold-Impact Extrusion of Soft Metals.

which weld together underneath. The lead is forced through the die, forming a uniform lead coating which grips the cable sufficiently to draw it through the die block. At the end of the stroke the plunger rises, more lead is added, and the cycle is repeated.

An interesting example of *extrusion by impact* is in the manufacture of collapsible tubes for shaving cream, toothpaste, and paint pigments. These extremely thin tubes are pressed out from slugs as illustrated in the upper half of Figure 25. The punch strikes a single blow of considerable force, causing the metal to squirt up around the punch.

The outside diameter of the tube is the same as the diameter of the die, and the thickness is controlled by the clearance between the punch and die. The tube shown in the figure has a flat end, but any desired shape can be made by properly forming the die cavity and the end of the punch. On the upstroke, the tube is blown from the ram with compressed air. The entire operation is automatic, with a production rate of 35 to 40 tubes per minute. The tubes are then inspected, trimmed, enameled, and printed. Zinc, lead, tin, and aluminum alloys are worked in this fashion.

In the lower half of the figure is illustrated a variation of what is known as the *Hooker process* for extruding small tubes or cartridge cases. Small slugs or blanks are used as in the impact-extrusion process, but in this case the metal is extruded downward through the die opening. The size and shape of the extruded tube are controlled by the space between the punch end and die cavity wall. Copper tubes having wall thicknesses from 0.004 to 0.010 inch can be produced in lengths of about 12 inches.

REVIEW QUESTIONS

1. Distinguish between hot- and cold-working of steel.
2. List the various methods of mechanically hot-working metals.
3. What defects are found in ingots, and how are they eliminated?
4. What are the following shapes used in connection with the rolling of steel: Ingot, bloom, slab, and billet?
5. Describe the process of drop forging.
6. What is a board hammer, and how does it operate?
7. What advantages does press forging have over drop forging?
8. What type of work is done by machine forging?
9. List the various methods used in producing pipe and tubular products.
10. Describe the process of producing seamless tubing.
11. Show by sketch how wire cable is coated with lead.
12. How are collapsible tubes produced?
13. Sketch and describe the Hooker process of extrusion.
14. How is steel improved by hot working?

BIBLIOGRAPHY

CAMP, J. M., and FRANCIS, C. B., *The Making, Shaping and Treating of Steel*, 5th edition, Carnegie-Illinois Steel Corporation.

FRIEDMAN, J. H., " Hot Press and Upset Forgings," *Trans. ASM*, Vol. 25, March 1937.

Impact Die Forging, Chambersburg Engineering Company, 1944.

KOENIG, PHIL, "Impact Extrusion and Cold Pressing of Airplane Parts," *Trans. S.A.E.,* Vol. 51, November 1943.

Metals Handbook, 1939 edition, American Society for Metals.

Metal Quality, Drop Forging Association, 1944.

PEARSON, C. E., and GENDERS, R., *The Extrusion of Metals,* John Wiley & Sons, 1944.

NANJOKS, W., and FABEL, D. C., *Forging Handbook,* American Society for Metals, 1939.

Pipes and Tubes in the Making, National Tube Company, 1945.

Steel in the Making, Bethlehem Steel Company, 1942.

CHAPTER 10

COLD FORMING OF METALS

Many of the same operations used in hot working may also be used in cold working, but the effect on the crystalline structure of the metal is different. Hot work, performed on the metal in a plastic state, actually refines the grain structure, whereas cold work merely distorts it and does little towards reducing its size. Cold working is normally done at room temperature; however, for steel, temperatures up to the critical range may be included.

Many products are cold-finished after hot rolling to make them commercially fit for their intended use. For example, hot-rolled strips and sheets are soft, have surface imperfections, and lack dimensional accuracy and certain desired physical properties. In the cold-rolling operation there is a slight reduction in size which permits accurate dimensional controls. A smooth surface is obtained, and both hardness and strength are increased. In general, the same results are obtained in the cold rolling of shafting, wire drawing, or other forms of cold work. Brittleness results if the metal is overworked, and an annealing operation is then necessary before further work can be done.

Cold-Working Processes

Press working may or may not accomplish the results just described. Operations involving bending, drawing, and squeezing metal result in grain distortion and changes in physical properties, while shearing or cutting operations change only form and size. The following classification lists the various methods of mechanically cold-working metals, including press operations.

COLD-WORKING OPERATIONS — METAL AT NEAR ROOM TEMPERATURE

1. Drawing
   (a) Blanks.
   (b) Tubes.
   (c) Embossing.
   (d) Wire.
   (e) Metal spinning.

2. Squeezing
   (a) Coining.
   (b) Cold rolling.
   (c) Sizing.
   (d) Swaging or cold forging.
   (e) Thread rolling and knurling.*

(*f*) Riveting.

(*g*) Embossing.

3. Bending

(*a*) Angle bending.

(*b*) Curling.

(*c*) Seaming.

4. Shearing

(*a*) Blanking.

(*b*) Punching.

(*c*) Cutting off.

(*d*) Trimming.

(*e*) Perforating.

(*f*) Notching.

(*g*) Slotting.

(*h*) Sprue cutting.

5. Extruding.†

(*a*) Cold.

(*b*) Impact.

6. Shot peening.

* See Chapter 13.

† See Chapter 9.

Tube drawing. Tubing which requires dimensional accuracy, smooth surface, and improved physical properties is finished by a cold-drawing operation. This method also produces tubes having smaller diameters or thinner walls than can be obtained by hot rolling. Special shapes and rounds up to 12 inches in diameter can be produced by this process.

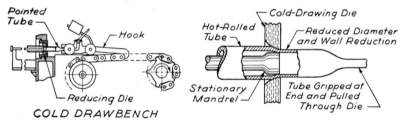

Courtesy National Tube Company.

Fig. 1.   Process of Cold-Drawing Tubing.

Hot-rolled tubing must first be treated by pickling and washing to remove all scale and then covered with a suitable lubricant. The drawing is done in a *drawbench,* shown diagrammatically in Figure 1. One end of the tube is reduced in diameter by a swaging operation to permit it to enter the die, and it is then gripped by tongs fastened to the chain of the drawbench. In this operation the tube is drawn through a die smaller than the outside diameter of the tube, the inside surface and diameter being controlled by a fixed mandrel over which the tube is drawn. This mandrel may be omitted for small sizes or for larger sizes if the accuracy of the inside diameter is not important. Drawbenches require a pulling power ranging from 50,000 to 300,000 pounds and may have a total length of 100 feet.

The operation of drawing a tube is very severe: the metal is stressed above its elastic limit to permit plastic flow through the die. The maximum reduction for one pass is around 40%. This operation increases the hardness of the tube so much that, if several reductions are desired, the material must be annealed after each pass.

Wire drawing. Wire is drawn by pulling a rod through several dies of decreasing diameter, as illustrated in Figure 2, until the final diameter is obtained. Rods from the mill, first cleaned in acid baths to remove scale and rust, are coated in various ways to prevent oxidation and to facilitate drawing through the die. The dies are usually made of carbide materials, although diamonds may be used for small diameters.

Both single-draft or continuous-drawing processes may be used. In the first method a coil is placed on a reel or frame and the end of the rod pointed so that it will enter the die. The end is grasped by tongs on a drawbench and pulled through to such length as may be wound around a drawing block or reel. From there on, the rotation of the draw block pulls the wire through the die and forms it into a coil. These operations are repeated with smaller dies and blocks until the wire is drawn to its final size.

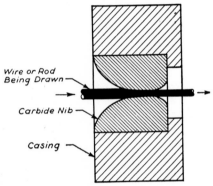

FIG. 2. Section Through Die for Drawing Wire.

In continuous drawing, as shown in Figure 3, the wire is fed through several dies and draw blocks arranged in series. This permits drawing the maximum amount in one operation before annealing is necessary. The number of dies in the series will depend on the kind of metal or alloy being processed and may vary from 4 to 12 successive drafts.

Wire thus manufactured is further processed into various wire products such as barbed wire, wire rope, woven fence, netting, nails, and bolts.

Swaging and cold forging. These terms refer to methods of cold working by pressure or impact which causes the metal to flow to some desired shape. *Swaging* is usually thought of as a means of reducing the ends of bars or tubes by rotating dies which open and close on the work. The rotating dies are rapidly opened and closed, and the end of the rod is tapered or reduced in size by a combination of pressure and impact. The swaging action, being rather severe,

Courtesy Wickwire Spencer Steel Company.

Fig. 3. View of Vaughn Continuous Drawing Bench Used to Reduce Round Stock to Around 0.072 Inch in Six Drafts. Dies Are All Tungsten Carbide and Are Water-Cooled. Roll Wire Pointer Is at Right Center.

hardens the metal and necessitates an annealing operation if much reduction is desired.

Cold forging of heads on bolts, rivets, and other similar parts is done in cold-header machines. Since the product of the cold-header is made from unheated material, the equipment must be able to withstand the high pressures that are developed; furthermore, there must be accuracy of alignment of the upsetting tool with the dies so that the work turned out will be accurate and free from defects. A large 5/8-inch capacity machine of this type is illustrated in Figure 4. The rod is fed by straightening rolls up to a stop and is then cut off and moved into the header die. The heading operation may be either single or double, and upon completion the part is ejected from the dies. Sixty-five 5/8-inch-by-3½-inch bolts per minute can be produced in this machine.

Courtesy The National Machinery Company.

FIG. 4.   National 5/8-Inch Double-Stroke Solid-Die Cold-Header.

Nails, rivets, and small bolts are made from coiled wire and forged cold, while large bolts require the end of the rod to be heated before the heading operation.

Shot peening. This method of cold working has recently been developed to improve the fatigue resistance of the metal by setting up compressive stresses in its surface. This is done by blasting or hurling a rain of small shot at high velocity against the surface to be peened. As the shot strikes, small indentations are produced, causing a slight plastic flow of the surface metal to a depth of a few thousandths of an inch. This stretching of the outer fibers is resisted by those underneath, which tend to return them to their original length, thus producing an outer layer having a compressive stress while those below are in tension. In addition, the surface is slightly hardened and strengthened by the cold-working operation.

Since fatigue failures result from tension stresses, having the surface in compression greatly offsets any tendency towards such a failure.

Shot peening is done by air blast or by some mechanical means. Figure 5 shows the unit from a machine which utilizes centrifugal force for hurling shot upon the work at a high velocity. This unit is similar to the one used in the machine shown in Figure 16, Chapter 3, for cleaning castings. Shot enters the funnel at A, which feeds it

Courtesy American Foundry Equipment Company.

FIG. 5.   Unit from Wheelapeening Machine Utilizing Centrifugal Force for Hurling Shot upon Work of High Velocity.

to the rotating wheel at G. The wheel then discharges the shot at a high velocity by its rotation. The surface obtained by this action is shown in Figure 6. Stress concentrations due to the roughened surface are offset for the reason that the indentations are close together and no sharp notches exist at the bottom of the pits.* Intense peening is not to be desired, for it may cause weakening of the steel. Shown in Figure 7 is a micrograph of the surface of SAE 1030 steel which has been shot-peened. The effect of the cold work may be seen by comparing the distorted crystalline grains near the surface with those underneath.

This process adds increased resistance to fatigue failures of working parts and can be done either on parts of irregular shape or on local

---

* H. F. Moore, *Shot Peening and The Fatigue of Metals,* American Foundry Equipment Company, 1944.

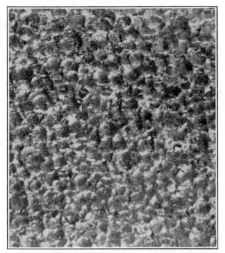

Courtesy American Foundry Equipment Company.

Fig. 6.   Surface of SAE 1030 Shot-Peened Steel.   Magnification ×7.

Courtesy American Foundry Equipment Company.

Fig. 7.   Micrograph of SAE 1030 Shot-Peened Steel.   Magnification ×200.

areas that may be subject to stress concentrations. Surface hardness and strength are also increased, and, in some cases, the process is used to produce a suitable commercial surface finish. However, it is not effective if there are cracks or seams in the base metal; nor is its effect as great as the sections increase in size.

Presses and Their Use

The machine used for most cold-working operations is known as a *press*. It consists of a machine frame supporting a bed and a ram, a source of power, and a mechanism to cause the ram to move in line with and at right angles to the bed. The illustrations in this chapter show numerous typical designs of press equipment.

A press in itself is not sufficient as a production machine but must be equipped with tools commonly called *dies* and *punches*, designed for certain specific operations. Although some presses are better adapted for certain types of work than others, most of the forming, punching, and shearing operations can be performed on any standard press if the proper dies and punches are used. This versatility makes it possible to use the same press for many different jobs and operations, which is a desirable feature for short-run production.

Presses are capable of rapid production, since the actual time of operation is only the time necessary for one stroke of the ram plus the time necessary to feed the stock. Accordingly, production costs may be kept very low. Any product that can be fabricated from thin metal and does not require extreme accuracy in dimensional tolerances can be economically made on this type of machine. Its special adaptation to mass-production methods is evidenced by its wide application in the manufacture of automotive and aircraft parts, hardware specialties, toys, and kitchen appliances.

Types of Presses

A classification of press machines is difficult to make, as most presses are capable of varied types of work. Hence it is not entirely correct to call one press a *bending press,* another an *embossing press,* and still another a *blanking press,* since all three types of operations can be done on one machine. However, some presses, especially designed for one type of operation, may be known by the operation name, as for example, a *punch press* or a *coining press.* The simplest classification would be according to source of power — either manually operated or power-operated. Many manually operated machines are used for thin sheet-metal work, particularly in jobbing work, but most production machines are power-operated. Other ways of grouping presses would be according to number of rams or method of operating the rams. Most manufacturers name them according to the general design of the frame, except in cases where the press is essentially a single-purpose machine. Using this method of classification, we can group most presses under the following headings:

POWER PRESSES

1. Bench.
2. Inclinable.
3. Open-back.
4. Gap or open-throat.
5. Arch.
6. Straight-side.
7. Horn or adjustable-table.
8. Punching.
9. Brake or bending press.
10. Squaring shears.
11. Circle shears.
12. Turret.
13. Percussion.
14. Extrusion.
15. Special-purpose—such as seaming, shell-drawing, spruecutting, straightening, veneering, coining, forging, and transfer feed.

Other descriptive terms may be added, as, for example, a *double-acting straight-side press*. Some terms frequently used, descriptive of the manner in which power is applied to the ram, are *single-crank, double-crank, cam, eccentric, toggle, knuckle-joint, screw, drop-hammer, geared,* and *hydraulic*.

In the selection of the type of press to use for a given job a number of factors must be considered. Among these are the kind of operation to be performed, size of the part being worked upon, power required, and speed of operation. For most punching, blanking and trimming operations, crank or eccentric-type presses are generally used. In these presses the energy of the flywheel may be transmitted to the main shaft either directly or through a gear train. For coining, squeezing, or forging operations, the knuckle-joint press is ideally suited. It has a short stroke and is capable of exerting a tremendous force. Presses for drawing operations normally operate at slower speeds than for operations such as blanking, and hydraulically operated presses are especially desirable for this work. The standard practice is not to exceed 65 feet per minute when working mild steel; however, aluminum and other nonferrous metals may be worked at speeds up to 125 and 150 feet per minute. Hydraulic presses may also be used for forging, straightening, sizing, and similar operations.

Inclinable Press

In Figure 8 is shown an *open-back inclinable press* engaged in trimming flash metal from aluminum die castings. Presses of this type are built in a wide variety of sizes and are adapted to many kinds of press operations. The fact that the frame of the press is

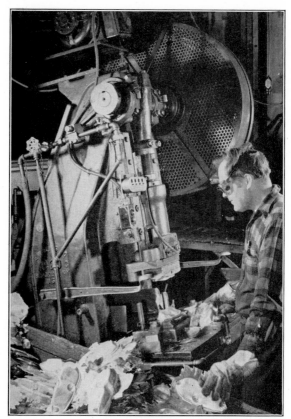

Courtesy Aluminum Company of America.

FIG. 8. Open-Back Inclinable Press Engaged in Trimming Flash Metal from Aluminum Die Castings.

easily inclinable is of great value where the work is discharged from the top of the die. When the press is in the inclined position, finished parts can slide by gravity into a box in the rear. If the parts are discharged through the die, it is best to have the frame in a vertical position. When the treadle is depressed, a large concentric spring slides the clutch sleeve into engagement, and the ram is forced down by the rotating crank. At the end of the stroke the clutch sleeve is positively withdrawn by a cylindrical cam machined in the sleeve. Only one stroke of the ram will take place even if the treadle is held down. For a second stroke the treadle must be raised and again depressed.

Inclinable presses are used in the production of an endless variety of metal stampings for automotive and aircraft parts, refrigeration equipment, business machines, hardware, and toys.

Straight-Side Double-Crank Press

As the size of the work increases, two cranks are necessary to distribute effectively the force to the ram or slide. A *straight-side double-crank press*, tooled for a forming operation on metal panels, is illustrated in Figure 9. The motor, mounted at the top of the press, drives through a belt to the flywheel clutch mounted on a drive shaft at the rear. The flywheel and driving disks of the clutch rotate continuously. When the press is operated, the brake is released, and the clutch is engaged by a shifting lever. Power is then transmitted to the crankshaft through two sets of gears. Double gearing is advisable on long-stroke presses where considerable power is required. The press shown in the figure is equipped with double push-button control for safety reasons. The push buttons are connected in series so that it is necessary for the operator to hold both push buttons depressed. If he should remove either hand before the slide reaches the bottom of the stroke, the clutch is automatically released and the brake applied. The air cylinder shown on the front of the press is used to counterbalance the slide. It restrains the slide on the down stroke and, at the end of the operation, helps to raise the slide on the up stroke and to keep it at top center when at rest.

These presses are used for blanking, embossing, forming, and drawing of sheet metal in the production of pieces having considerable area. They may also be used for perforating rows of holes and for straight-line bending operations.

Two-Point Double-Acting Eccentric Press

A large two-point suspension press equipped with dies for forming automobile hoods is shown in Figure 10. All-welded construction is used to give increased rigidity and strength over that obtained by the use of cast iron. Welded construction permits placing the metal where needed, and frame members need not be unnecessarily bulky to meet the usual casting requirements. The two plungers are driven by large-diameter eccentrics in place of the usual cranks. For this reason the press is known as a crankless steel press. The eccentrics are placed directly over the two points of power application to the

Courtesy Niagara Machine and Tool Works.

FIG. 9. Forming Operation on Double-Crank Press.

slide and are formed integrally with the steel driving gears. With this rigid construction, there is little chance for any deflection to occur. It is claimed by the manufacturer of this press that greater uniformity of load application is obtained with this design than with the conventional crankshaft drive. This press is built in three gen-

eral designs: One-, two-, or four-point suspension slides, each having
several possibilities so far as a plunger action is concerned. Capacities range from small presses of 100 tons to large units of 3000

Courtesy Clearing Machine Corporation.

Fig. 10.   Forming Automobile Hoods.

tons. This type of press has wide application in the forming of automobile body panels in addition to the usual press operations on
smaller parts.

Press Brake

The *press brake* is designed to handle a wide variety of press opera-tions and is especially adapted to process large sheets of relatively thin metal. Aside from the usual brake or forming operations, a press of this type may be used for corrugating, seaming, embossing, trimming, and multiple punching. These presses are made in lengths ranging from 4 to 20 feet, with metal thickness ranging from light-gage to 5/8 inch.

The pressure capacity of a press brake required for a given material is determined by the length of work it will take, the thickness of the metal, and the radius of the bend. The minimum inside radius of a bend is usually limited to a radius equal to the thickness of the material; however, the thickness rating of a press can be increased slightly if a larger radius is used. For bending operations the re-quired pressure varies directly as the tensile strength of the material. Press brakes have short strokes and are generally equipped with an eccentric-type drive mechanism.

A press brake prepared for corrugating light-gage metal is shown in Figure 11. The insert illustrates another die setup for forming stiffening rings in thin sheets. Products made with this type press include tubular track, steel channels, metal molding, sheet roofing, airplane-wing covering, and metal furniture.

Double-Acting Toggle Press

Figure 12 is a view of a large toggle press drawing steel tops for automobiles. Pressure is applied on the slide in four places, which is a distinct advantage in large-area presses, as such construction prevents tilting of the slide with unbalanced loads. This press is a double-acting toggle press, with built-in air-cushion drawing attach-ments. A *double-acting press* is one that has two motions in the slide, one for advancing and holding the work and the other for the drawing operation. Triple-acting presses have a third motion which is used for a second drawing operation. The third motion in this machine is from the air cushion in the bed which is synchronized to operate against the work and complete the drawing operation. This slide can be disconnected if desired and the press operated as a double-action machine.

The toggle mechanism in this machine is for the purpose of con-trolling the motion of the blank holder. " A toggle mechanism may be described as a grouping of cranks, levers, and slides with the

Fig. 11. (Right) Forming Stiffening Ribs in Thin Sheets. (Below) Corrugating Light-Gage Metal with Press Brake.

Courtesy E. W. Bliss Company.

FIG. 12. Toledo Double-Action Toggle Drawing Press Forming Steel Tops for Automobiles.

necessary connecting links, so arranged that the train of movement may contain several dead-center positions at approximately the same time. If the motion is so controlled that the several points pass through dead center a little way and back through it again in returning, the effective dwell period may be extended within certain limits."* The dwell period means no motion of the blank holder for an interval of time. It is necessary for blank holding on drawing operations, and it is frequently advisable to have a slight dwell on the punch to allow the metal to adjust itself properly under pressure.

The press shown in Figure 12 will produce four to five automobile tops per minute. The labor cost is insignificant compared to the cost of the machine and the expensive forming dies.

Knuckle-Joint Coining Press

Presses designed for coining, sizing, and heavy embossing must be quite massive to withstand the large concentrated loads imposed upon them. The press shown in Figure 13 is designed for this purpose and is equipped with a *knuckle-joint mechanism* for actuating the slide. The upper link or knuckle of this point is

Courtesy E. W. Bliss Company.

Fig. 13. Knuckle-Joint Coining and Embossing Press.

hinged at the upper part of the frame at one end and fastened to a wrist pin at the other. The lower link also is attached to the same wrist pin and the other end to the slide. A third link is fastened to the ends of the wrist pin and acts in a horizontal direction to move the joint. As the two knuckle links are brought into a straight-line

* E. V. Crane, *Plastic Working of Metals*, John Wiley & Sons, p. 302.

position, tremendous force is exerted by the slide. The press shown in the figure has a capacity of 150 tons.

This type of press has always been widely used in the striking of coins. A tube feed is provided for this operation. Mechanical fingers place the coin in position and remove it after the operation. According to tests made at the Philadelphia United States Mint, a pressure of 98 tons is required to bring out clear impressions on silver half dollars made in a closed die. This corresponds to a pressure of about 180,000 pounds per square inch.

Aside from striking coins, many other parts, such as medals, key blanks, car tokens, license plates, watch cases, and silverware, are cold-pressed in this type of machine. Sizing, cold-heading, straightening, heavy stamping, and similar operations can also be performed on this machine. As the stroke of this type press is short and slow, it is not adapted to some of the usual press operations.

Courtesy E. W. Bliss Company.

Fig. 14. Blanking Operation on an Arch Press.

Arch-Type Press

The *arch-type press* shown in Figure 14 is named from the peculiar shape of its frame. The lower part of the frame near the bed is wide to permit the working of large-area sheet metal; the upper part is narrow. The crankshafts are small in relation to the area of the slide and press bed, as these presses are not designed for heavy work. The press in the figure is shown set up for a blanking operation on light-gage metal, which is typical of the work this press is designed to do. It is also used for stamping, bending, and trimming in the manufacture of large-sized paint cans and numerous other tin products. Other applications are the blanking and forming of shovels, embossing letters on metal panels, and the manufacture of kitchenware.

Percussion Power Press

Figure 15 shows a percussion power press adapted for applications where a hard end pressure is required. The friction drive accelerates

Courtesy Zeh and Hahnemann Company.

Fig. 15. Percussion Power Press.

the flywheel gradually on the down stroke, and all its energy is utilized as it comes to a stop striking the work. Regulation of the blow is obtained by raising or lowering the position of the die. By raising the die the flywheel is stopped at a higher point and has less force, since its speed is less. For a given setting the blows are the same, and overloading is impossible. Since all the force of the blow is absorbed in the frame of the machine, expensive foundations are unnecessary.

Small machines of this type are used for striking medals and signet rings, stamping and embossing jewelry, and similar applications. Larger machines can be used for cold stamping and pressing of small metal parts as well as for hot pressing of brass and other forgeable metals. The largest of these machines, exerting 50,000 foot-pounds per stroke, can be used to replace drop-forging operations.

The hot pressing of brass (60% copper and 40% zinc) has proved very satisfactory with this type of press. Several types of dies are used in this work: open dies similar to drop-forging dies; extrusion dies, which confine the metal in the die and force it through an opening; and confined dies, in which the descending punch closes the die and compels the metal to flow into all cavities of the die. Parts made in this fashion are easily machined, are accurate as to size, homogeneous in structure, and have a smooth surface. Any metal of a forgeable nature such as aluminum, duralumin, or Monel metal can be forged in this manner.

Turret Press

A large *turret press*, with spacing table for handling sheets up to 50 inches wide by 100 inches long, is shown in Figure 16. This type of press is made in a wide variety of sizes from small hand-operated units to large power-operated machines having a pressure capacity of 160,000 pounds.

The principal features of this press are the upper and lower turrets designed for carrying the punches and dies. The turrets of the press shown are 50 inches in diameter and have 12 to 32 punching stations. The two turrets are geared together and, when operating, are securely locked in position for exact alignment. The turret can be operated by either hand wheel or motor drive. A large variety of sizes and shapes of dies are available at all times by simply indexing the turret.

The table shown in front of the machine offers an accurate and convenient means of locating the sheet under the punching station.

It is provided with both longitudinal and crosswise motion, controlled by the hand wheel in front of the operator. All controls for turret and table move with the table. Punching may be done directly from blueprints without previous layout, as the punch settings are made from full-length scales having large graduated dials for final accuracy. Only 15 seconds are required to change and align punches

Courtesy Wiedemann Machine Company.

Fig. 16.   Power-Operated Turret Punch.

and dies. The punch and die selected for use are located centrally under the ram and are the only ones that can operate. An eccentric drive is used to supply the power to the punch.

These punches are designed to handle short-run production and jobbing work in an efficient manner. Aside from ordinary metal punching, these machines can also be set up for slotting, embossing, notching, and louver operations.

Hydraulic Press for Aircraft Work

The hydraulic press shown in Figure 17, designed for aircraft metal-forming work, has a capacity of 2500 tons. It is a self-contained, down-acting type, the main rams being mounted on the top platen with all pumping equipment mounted on top of the press.

Fig. 17. Hydraulic Press and Loading Tables for Use in Aircraft Industry, Capacity 2500 Tons.

The top platen contains cylinders for one 38-inch and two 20-inch rams which act against the moving platen. Two additional 10-inch cylinders mounted on the top platen contain push-back rams and are connected to the moving platen by two steel rods. Although adapted to the cutting and forming of light sheet in aircraft work, this press has many other industrial applications.

Guerin Process

This process is a recent development which greatly reduces the cost of dies in the blanking and forming of thin sheet required for aircraft manufacture. In place of expensive steel-mating dies, this process employs a single die of low-cost material and a thick pad of rubber which adapts itself to the die while under pressure. Sheet metal placed between the resilient pad and the die can be cut readily or formed to the desired shape.

Rubber has proved satisfactory in this work because of its similarity to a fluid when properly restrained. Thick pads are mounted on the moving platen of the press and held in a container which extends about 1 inch past the pad. On the bed of the press is mounted a pressing block which fits into the container recess and upon which are mounted the cutting and forming dies. As the platen moves down and the rubber is confined, the force of the ram is exerted evenly in all directions, resulting in the sheet metal being pressed against the die block. Cutting die blocks are merely steel templates of the required parts and need not be over 3/8 inch thick. Forming dies may be made of Masonite, wood, aluminum, and magnesium, as well as steel.

In Figure 18 is shown a 2000-ton hydraulic press equipped with loading tables upon which is a miscellaneous assortment of cutting and forming dies. The loading tables facilitate loading and unloading the press and greatly reduce the labor involved. The adaptability of this process to group mounting of cutting blocks results in a considerable saving in materials. It is also possible to group both cutting and forming blocks all in a single press layout.

A forming die of Masonite is shown on the pressing block in Figure 19. Above the die is the blanked piece of metal, and below is the piece after the forming operation is complete. Another application, consisting of both blanking and forming, is illustrated in Figure 20. From left to right, the parts shown are: raw sheet, forming and cutting block, finished part before heat treatment, and scrap material remaining.

This process is limited in the cutting of soft aluminum to sheet thickness up to 0.051 inch. For bending and forming, the usual limit is around 3/16 inch thick. Thin gages of stainless steel may also be fabricated, and magnesium alloys may be hot-formed. In the latter application heater plates are mounted on the loading table, and the form blocks are maintained at the correct temperature.

The advantages of this process include simplicity of tooling, low

Courtesy Douglas Aircraft Company, Inc.

FIG. 18.   Hydraulic Press and Loading Table, Capacity 2000 Tons.

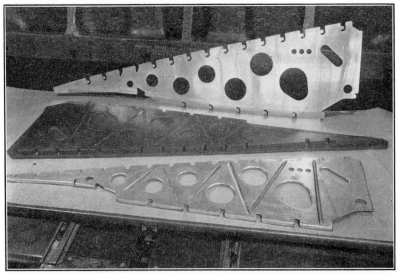

Courtesy Douglas Aircraft Company, Inc.

FIG. 19.   Masonite Forming Dies with Blank Before and After Forming.

tooling cost, use of gang setups, minimum material waste, uniform pressure on the metal, and adaptability to various press operations.

Courtesy Douglas Aircraft Company, Inc.

FIG. 20.   Method of Blanking and Forming Tray in One Operation by Guerin Process.

Fabrication of Light-Gage Metal

A large amount of work on light-gage steel, galvanized iron, tin, copper, and aluminum is done with small equipment, either power- or hand-operated.   Machines commonly known as sheet-metal equipment perform such simple operations as shearing, bending, folding, crimping, and rolling.

Thin metal is cut in *squaring shears,* a straight-shearing blade operating in conjunction with the edge of a table, with a suitable means for operating the blade. The squaring shears shown in Figure 21 is operated by foot pressure and has, in addition to the blade drive, a mechanism which exerts a hold-down pressure on the sheet being cut.   Gages are provided to hold the sheets square and to act as stops in cutting a number of sheets the same width.   Large shears are similar in construction, but are operated by power drive. A hand-operated *ring and circle shears* is shown in Figure 22.   The cutters in this machine are circular and must be ground carefully with proper cutting and clearance angles.   The sheet being cut is clamped at its center by the circle attachment.   This frame has a swivel base so that the clamping disks may be swung slightly off center to produce true circles and clean cuts.   This adjustment varies with the thickness of the metal and the type of cutters used.   Slitting and straight-shearing cuts can also be made on this machine, and irregular cuts can be made by guiding the sheet by hand.   Power-operated shears will cut steel plate up to 1/4 inch thick. A similar machine equipped with proper tools turns perpendicular flanges on the edges of circular disks. This operation is useful in making heads for tanks and containers of various sizes.

Simple straight-bending operations are performed in *folders* or *brakes*. Brakes have no limit to the width of fold or bend, whereas folders operate only to a limited width. The construction of these machines is simple: The upper of two steel straightedges, called the

Courtesy Niagara Machine and Tool Works.

FIG. 21. Squaring Shears Operated by Foot Pressure.

folding blade, is fixed; the lower, known as the *jaw,* is supported on pivots and may be moved upwards to any desired angle. A sheet to be bent is inserted in the space between the jaw and folding blade and held against the gage by one hand while the other hand raises the jaw with the operating handle, thus bending the sheet over the folding blade. Either sharp or rounded folds can be made, the latter being necessary if the edge has to be wired. Short single or double hems are often used to reinforce edges and to eliminate sharp corners.

For producing cylinders of sheet metal, a *slip roll-forming machine* is used. This machine has three rolls, two side by side in a horizontal position and one directly above, positioned between the two. A sheet to be rolled is fed between two of the rolls while the third deflects the material to produce the curvature. When a complete cylinder is produced, it encircles the upper roll and is removed by unfastening the roll and slipping it off one end. Although most sheet-metal machines are hand-operated, similar power machines of large capacity roll out heavy tank steel and armor plate.

Courtesy Niagara Machine and Tool Works.

FIG. 22. Ring and Circle Shears in Action.

A large number of rotary hand- and power-operation machines are used in the manufacture of cylindrical articles such as metal pipe, cans, pails, drums, and the usual type of work done by a tinsmith. Most of these machines are similar in construction and vary only in the circular tools they use. Typical operations performed on these machines are edging, wiring, crimping, flanging, seaming, and corrugating.

Seaming Machine

In the manufacture of metal drums, pails, cans, and numerous other products made of light-gage metal, several types of seams are used.

The most common of these are shown in Figure 23. The *lock seam* used on longitudinal seams is adapted for joints that do not have to be absolutely tight. After the container is formed, the edges are folded and pressed together. The *compound seam*, sometimes called

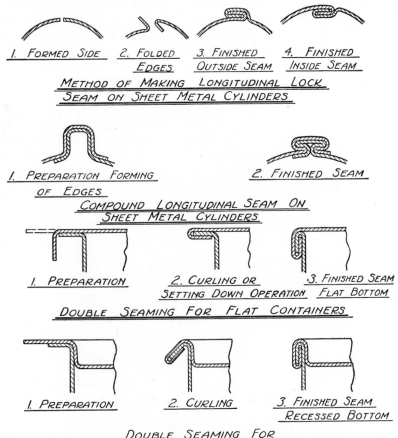

Courtesy Niagara Tool and Machine Company.

Fig. 23.   Typical Seams Used in Manufacture of Light-Gage Metal Containers.

the Gordon or box seam, is much stronger and tighter than the lock seam and is suitable for holding fine materials. Both these joints may be formed and closed on either hand- or power-seaming presses.

Bottom seams, which are somewhat similar to the longitudinal seams, are made in either *flat* or *recessed styles*. Flat-bottom recessing is limited to one end of a container, as the container must be open to make the joint. Double seaming with recessed bottoms can be done on both ends of a container. Edge flanging, curling, and flattening, the operations necessary to make a recessed double seam, are shown in the figure.

Double-seaming machines may be hand-operated, semiautomatic, or automatic. Semiautomatic machines must be loaded and unloaded by the operator, but the operation of the machine is automatic. In automatic machines the cans are brought to the machine by conveyor, and ends are supplied by magazine feed. The cans are fed from the conveyor to a star wheel, which transfers them to an automatic delivery turret. The delivery turret feeds them into position with the seaming heads, and the closing seam is made.

Courtesy E. W. Bliss Company.

Fig. 24. Metal Spinning Lathe.

Metal Spinning

Metal spinning is the operation of shaping thin metal by pressing it against a form while it is rotating. The nature of the process limits it to symmetrical articles of circular cross section. This type of work is done on a speed lathe similar to the one shown in Figure 24. A spinning lathe is like the ordinary wood lathe except that, in place of the usual tailstock, it is provided with some means of holding the work against the form. The forms are usually turned from hard wood and attached to the face plate of the lathe, although smooth steel chucks are recommended for production jobs. This type of chuck will not develop interior imperfection and is more economical where surface finish is to be considered.

Courtesy Aluminum Corporation of America.

FIG. 25.   Metal Spinning over an Off-Center Roll.

Practically all parts are formed by the aid of blunt hand tools which press the metal against the form. The cross slide has a hand- or compound-tool rest in the front for supporting the hand tools and some means for supporting a trimming cutter or forming roll in the rear. Parts may be formed either from flat disks of metal or from blanks that have previously been drawn in a press. The latter method is used as a finishing operation for many deep-drawn articles. Most spinning work is done on the outside diameter as shown in the figure, although inside work is also possible.

Bulging work on metal pitchers, vases, and similar parts is done by having a small roller, supported from the compound rest, operate on the inside and press the metal out against a form roller. Figure 25 shows a cutaway view of a part being spun over an off-center chuck or roll. In an operation such as this one, the part must first be drawn, and possibly given a bulging operation beforehand, as spinning cannot be done near the bottom. Contact of the spinning tools can

FIG. 26.   Sectional Chuck for Metal Spinning.

only take place next to the chuck. Sectional chucks, as illustrated in Figure 26, are used where the product is shaped so that it cannot be removed from a solid chuck after completion. The center part is first withdrawn, leaving sufficient clearance for removal of the outside sections. A close fit of the sections is necessary to avoid ridges in the final product.

Lubricants such as soap, beeswax, white lead, and linseed oil are used to reduce the tool friction. Of these, ordinary laundry soap proves very satisfactory, particularly for aluminum spinning. Since metal spinning is a cold-working operation, there is a limit to the amount of drawing or working the metal will stand, and one or more annealing operations may be necessary. This is frequently accomplished by applying a torch flame to the work while it is rotating. The percentage of elongation in 2 inches is a rough index to the amount of working that can be done before a heat treatment. The same applies to silver, copper, aluminum, and brass, as well as to soft steel.

The operation of metal spinning is very much like the drawing operation in press dies. If a large number of parts is to be produced, and the operation can be made in dies without too great expense, it is more economical. Spinning lends itself to short-run production jobs, although it has many applications in continuous production work. It is frequently used in the making of bells on musical instruments, light fixtures, reflectors, kitchenware, and large processing kettles.

Drive Mechanism Used on Presses

Figure 27 shows most of the drive mechanisms used in presses for transmitting power to the slide. The type of mechanism used in a given press is determined largely by the kind of work that is to be done.

The most common drive is the *single crank*, which gives a movement to the slide approaching simple harmonic motion. On a down stroke the slide is accelerated, reaching its maximum velocity at midstroke, and is then decelerated. Most press operations occur near the middle of the stroke, at maximum slide velocity. The *eccentric drive* gives a motion like that of a crank, and is often used where a shorter stroke is desired. Some proponents of this drive claim for it greater rigidity and less tendency for deflection than a crank drive might have. *Cams* are used where some special movement is desired, such as a dwell at the bottom of the stroke. This drive has some similarity to the eccentric drive, except that roll followers are used to transmit the motion to the slide.

Rack and gear presses are used only where a very long stroke is desired. The movement of the slide is much slower than in crank presses, and uniform motion is attained. Such presses are provided with stops to control the stroke length and may be equipped with some quick-return feature to raise the slide back to starting position. The common arbor press is a familiar example of this type.

Hydraulic drive is used in many presses for a wide variety of work. It is especially adapted to large pressures and slow speeds in form-

ing, pressing, and drawing operations.   The length of stroke is readily
adjustable, pressure is applied gradually, and a smooth even application of fluid power follows.   Pressure can be adjusted to suit conditions, and, if desired, a dwell period at the end of the stroke can be
obtained.

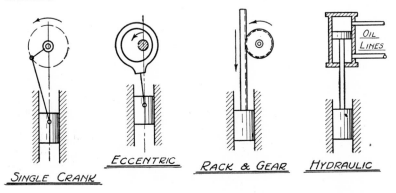

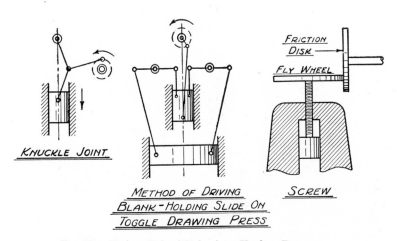

FIG. 27.   Various Drive Mechanisms Used on Presses.

In the *screw* drive, the slide is accelerated by means of the friction
disk engaging the flywheel, and as the flywheel moves down greater
speed is applied to it.   From beginning to end of the stroke, the
slide motion is an accelerated one.   At the bottom of the stroke
the entire amount of stored energy is absorbed by the work.   The
action resembles that of a drop hammer, but it is slower and there
is less impact.

Several link mechanisms are used in press drives, either because of the type of motion they have or because of the mechanical advantage they develop. The *knuckle joint* is very commonly used, because it

Courtesy Zeh and Hahnemann Company.

FIG. 28.   Reclinable Press with Double-Roll Feed and Stock Reel.

has a high mechanical advantage near the bottom of the stroke when the two links approach a straight line. Because of the high load capacity of this mechanism, it is used for coining and sizing operations. The drive may be eccentric or hydraulic in place of the crank shown in the figure. *Toggle mechanisms,* used primarily as a means of holding the blank on a drawing operation, are made in a variety of designs. The auxiliary slide in the figure is actuated by a crank, but eccentrics or cams may likewise be used. The principal aim of this mechanism is to obtain a motion having a suitable dwell so that the blank may be held effectively.

Feed Mechanisms

Safety is a paramount consideration in press operation, and every precaution must be taken to protect the operator. Wherever possible, material should be fed to the dies by some means which eliminates any chance of the operator having his hands near the dies. In long-run production jobs such features can be economically worked out in various ways. Feeding devices are best applied to small- and medium-sized presses, and have the advantage of rapid uniform machine feeding in addition to the safety features.

One of the most common types of feeding mechanisms is the double-roll feed, as shown on the reclinable press in Figure 28. Strip steel for this machine comes in rolls which are mounted on the stock reel at the side of the machine. The material is fed intermittently by the rolls of the dies, and the remaining scrap is fed through the second set of rolls at the side of the machine.

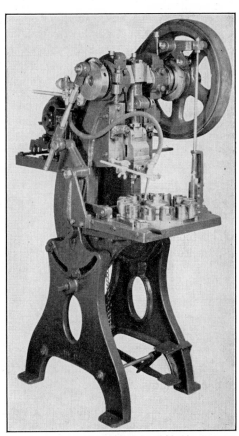

Courtesy F. J. Littell Machine Company.

FIG. 29. Littell 12-Station Dial Feed.

The operation of the rolls is controlled by an eccentric on the crankshaft through a linkage to a ratched wheel. Each time the ram moves up, the rolls turns and feed the proper amount of material for the next stroke. By providing the machine with a variable eccentric, the amount of stock fed through the rolls can easily be varied. The rolls are relieved before the stroke to permit proper alignment of the stock. For heavy material, straightening rolls can be used which also act as feeding rolls.

Another type of feeding device is the dial-station feed shown in Figure 29. This method is designed to take care of single parts previously blanked or formed in some other press. Again the indexing is controlled by an eccentric on the crankshaft through a suitable link mechanism to the dial. Each time a stroke is made, the dial indexes one station. All feeding by the operator takes place at the front of the machine away from the dies. This method of feeding is widely used in the drawing presses for the production of 75-mm shells.

Light parts can be stacked in a magazine and successfully placed in position by a suction device. A blank is lifted off the top of the stack by suction fingers and placed against a stop gage on the die. Magazine feeds may also be used with a reciprocating mechanism which feeds blanks from the bottom of the stack. Gravity feed is sometimes used on inclined presses, the blank sliding into a recess at the top of the die.

Punches and Dies

The tools used in most presses come under the general heading of punches and dies. The *punch* refers to that part of the assembly which is attached to the ram of the press and is forced into the die cavity; the *die* is usually stationary and rests on the press bed. It has an opening to receive the punch, and the two must be in perfect alignment for proper operation. Punches and dies are not interchangeable, but must work together as a unit. A single press may do a large variety of operations, depending on the type of dies used.

Dies are classified according to the type of work they can do. A simple classification is this:

TYPES OF DIES

1. Bending — angle bending, curling, folding, and seaming.
2. Drawing — forming tubes, cupping, bulging, embossing, and reducing.
3. Squeezing — coining, sizing, flattening, swaging, cold forging, riveting, upsetting, hot pressing, and extruding.

4. Shearing — blanking, trimming, cutting off, punching, perforating, notching, slitting, and shaving.

This classification does not include all the various arrangements and combinations that are possible. Many dies combine two or more operations, such as blanking and drawing, and accomplish them both in one stroke of the press. Multiple dies may punch or blank several holes simultaneously. With the use of strip stock, several operations may be arranged progressively in proper sequence as the stock moves through the die. Each time the stock is indexed one station, all operations are performed and one piece is completed. Such tools are known as progressive or multistage dies.

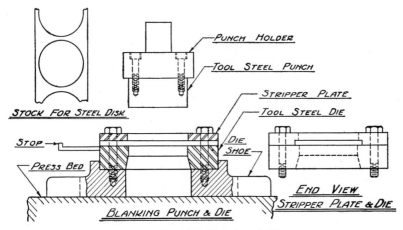

FIG. 30.   Blanking Punch and Die.

Misalignment of punch and die causes excessive pressures, shearing or chipping of die edges, or actual breaking of the tools. Such action may occur through shifting, even though the setup is originally correct. To prevent such occurrences, proper alignment is insured by providing guide rods at two or four corners of the die which fit into holes provided in the punch holder. Such dies are known as *pillar dies*. This arrangement of having the punch and die always held in proper alignment greatly facilitates the setting up of the tools. A similar arrangement, known as a *subpress die* (occasionally used on small work), employs a punch and die mounted in a small frame so that accurate alignment is always maintained. Pressure is applied by a plunger which extends out of the top of the assembly.

A simple blanking punch and die are shown in Figure 30. In making up such tools the parts that do the cutting are made of tool steel and are built in as inserts.   The punch is made up of a holder having a shank and the tool-steel punch.   The shank of the punch fits into the press slide or a punch plate attached to the slide.   The die is supported on a cast-steel die shoe which, in turn, is fastened to the bolster plate on the press bed.   The die shown in the figure is designed for blanking disks from strip metal.   Steel is fed in the opening at one end of the die up to a stop provided at the other end.   The steel

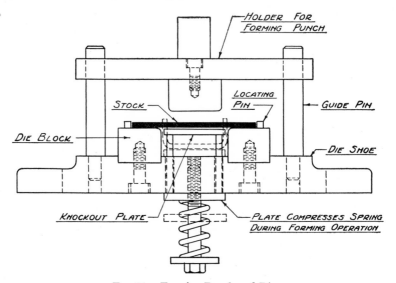

FIG. 31.   Forming Punch and Die.

plate over the stock is called a stripper plate, since it holds the blanked strip in place as the punch moves up to its starting position.   Blanks which have been sheared by the punch drop through it to a container underneath the press.

The clearance between punch and die will vary according to the type of material being processed.   It is expressed in percentage of stock thickness and ranges from 5 to 12%.*   The smaller percentages are for small work and soft materials; larger jobs using a fairly hard steel require more clearance.

A typical forming die, designed to bend a flat strip of steel to a U shape, is shown in Figure 31.   The punch holder is made large and is provided with accurate holes to fit with the guide rods set in the die shoe.   As the punch descends and forms the piece, the knockout plate is pressed down, compressing the spring at the bottom of the

* F. A. Stanley, *Punches and Dies*, McGraw-Hill Book Company.

press. When the punch moves up, the plate forces the work out of
the die with the aid of the spring. Such an arrangement is necessary
in most forming operations, as the formed metal presses against the
walls of the die, making its removal difficult. Parts that tend to
stick to the punch are removed by a knockout pin which is engaged
on the up stroke.

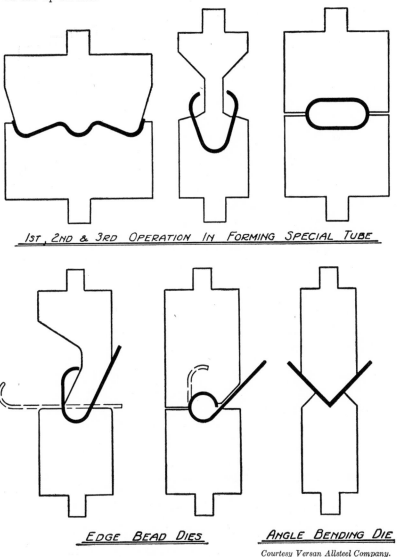

1ST, 2ND & 3RD OPERATION IN FORMING SPECIAL TUBE

EDGE BEAD DIES   ANGLE BENDING DIE

Courtesy Versan Allsteel Company.

FIG. 32. Typical Forming Dies.

Other forming dies are shown in Figure 32. These are of the type used in brake presses for long work. The first three figures in the illustration show the steps necessary to form a special tube for

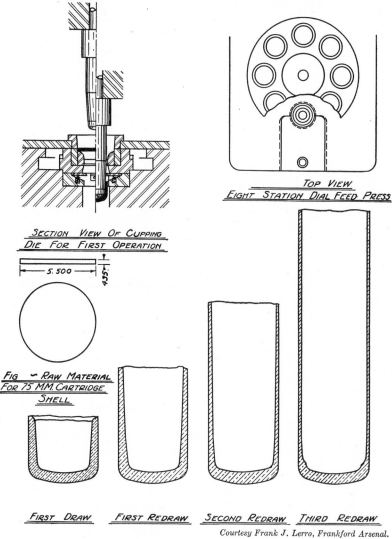

SECTION VIEW OF CUPPING
DIE FOR FIRST OPERATION

TOP VIEW
EIGHT STATION DIAL FEED PRESS

5.500

FIG — RAW MATERIAL
FOR 75 MM. CARTRIDGE
SHELL

FIRST DRAW   FIRST REDRAW   SECOND REDRAW   THIRD REDRAW

Courtesy Frank J. Lerro, Frankford Arsenal.

FIG. 33.   Various Stages in Drawing 75-Mm Shells.

use in metal-furniture manufacture. Although not shown in the figure, stop pins or gage bars must be used to locate the stock centrally with the die. Two of the figures below show the last two operations in making an edge bead. The operations shown here are similar to those used in making hinges. The last figure in the illustration is a common angle-bending operation. Parts up to 14 feet in length can be formed in presses of this type.

Figure 33 illustrates the preliminary drawing operations in the manufacture of 75-mm shell cartridges. The material comes to the first drawing press as a disk $5\frac{1}{2}$ inches in diameter and 0.435 inch thick. A sectional view of the press and dies for the first cupping operation is shown at the upper part of the illustration. Since there is a tendency for the cupped blank to stick to the end of the punch, it is removed by the two lugs shown near the bottom of the die

Courtesy E. W. Bliss Company.

FIG. 34. Typical Punches and Dies for Blanking, Drawing, Forming, and Similar Press Operations.

as the punch moves up to its starting position. For feeding the blanks to the press an eight-station dial feed is provided. When the dial is indexed, one station is aligned with the die and the blank is placed in proper position. The redrawing operations and dies are similar to the one just described.

The aforementioned dies are typical of the average ones used for small and medium work. Another group of punches and dies for blanking, drawing, and similar operations is shown in Fig. 34. So many variations in die design and arrangement are possible that it is not within the scope of this book to discuss them.

REVIEW QUESTIONS

1. What is the purpose of cold-finishing steel strips and shafting?
2. Describe the process of putting heads on rivets and small bolts.
3. What is an inclinable press, and what advantages does it have over rigid-frame presses?

4. For what type of work is an arch press recommended?

5. What is a subpress unit?

6. Briefly describe two methods of automatic feed for punch presses.

7. What advantage does hydraulic power have over other drive arrangements, and for what type of work is it recommended?

8. What type of equipment is used in metal spinning?

9. Explain how steel tubes are cold-drawn.

10. Briefly describe a screw press, and state the type of work for which it is adapted.

11. What various arrangements do presses have to protect the worker from injury?

12. Show by sketches the operations necessary to make a bottom seam on a metal drum.

13. What is the purpose of stripping plates and knockout pins on dies? Illustrate their use by sketch.

14. What is a progressive die?

15. What type of a press do you recommend for forming steel tops for automobiles? For stamping coins?

16. Describe the Guerin process for forming and stamping airplane parts.

17. Name and sketch five drive mechanisms used on presses.

18. Describe the method of producing 75-mm shells.

19. Sketch a punch and die for bending U-shaped pieces from strip steel.

20. Classify presses for cold-working operations.

21. Assume you are to manufacture steel butt hinges. Describe the operations necessary to make this article.

22. How is wire manufactured?

23. What is the purpose of shot peening, and how is it done?

BIBLIOGRAPHY

BROOTZKOOS, S. D., *The Selection of Presses*, Dryden Press, 1941.

CAMP, J. M., and FRANCIS, C. B., *The Making, Shaping and Treating of Steel*, 5th edition, Carnegie-Illinois Steel Corporation.

"Carbon Steel Wire," Section 16 of *Steel Products Manual*, American Iron and Steel Institute, 1943.

CHARNOCK, G. F., and PARTINGTON, F. W., *Mechanical Technology*, 2d edition, Constable & Company, London, 1934.

CORY, C. R., *Die Design Manual*, Parts I and II, General Motors Corporation, 1939.

CRANE, E. V., *Plastic Working of Metals and Power-Press Operations*, 2d edition, John Wiley & Sons, 1939.

HINMAN, C. W., *Die Engineering Layouts and Formulas*, McGraw-Hill Book Company, 1943.

HINMAN, C. W., *Press Working of Metals*, McGraw-Hill Book Company, 1941.

JEVONS, J. D., and SWIFT, W. W., *The Metallurgy of Deep Drawing and Pressing*, John Wiley & Sons, 1942.

JONES, FRANKLIN D., *Die Design and Die-Making Practice*, Industrial Press, 1930.

MARSHALL, E. G., *Practical Die Design and Die Making*, McGraw-Hill Book Company.

Shot Peening, American Foundry Equipment Company.

STANLEY, FRANK A., *Punches and Dies*, McGraw-Hill Book Company, 1937.

TURNBULL, D. C., "Fatigue Life of Stressed Parts Increased by Shot Peening," *American Machinist*, August 31, 1944.

WOODWORTH, JOSEPH V., *Punches, Dies, and Tools*, Norman W. Henley Publishing Company, 1935.

CHAPTER 11

INSPECTION — MEASURING INSTRUMENTS AND GAGES

Mass production requires that all parts be made according to rigid specifications and working drawings. No matter how carefully these drawings and specifications are prepared, they lose their value unless they are adhered to by the production department. It is the function of inspectors to see that the standards established by the engineering department are maintained in the shop.

Inspection

Inspection departments in companies demanding close quality control of their product are separate from the production department. Since it is the aim of the production department to produce goods as fast as possible, there is some tendency to lower quality standards if maintaining quality means lowering the output. For this reason it is advisable to place the inspection department in a position on the organization chart such as will insure sufficient authority to act independently and for the best interest of quality control. Frequently it is directly responsible to the engineering department, since it is from this department that the drawings originate.

The inspector occupies a very important place in an organization. He should have personal qualities which warrant placing him in an authoritative position: ability, tact, impartiality, and thoroughness are all essential qualities for an inspector. To understand the problems of the operator, he must have a knowledge of materials, manufacturing processes, and tools. His acceptance or rejection of work must be based entirely on merit and on established specifications. He must avoid arbitrary methods of inspection at all times.

The amount of inspection given to the product will vary according to the nature of the product, degree of accuracy required, and type of equipment used. As greater accuracy is demanded in the product, more inspection is necessary. A watch factory may use one inspector for seven to ten workmen, while a foundry requires only one inspector for 30 to 40 workmen. After being set up, certain types of machines, particularly presses for blanking, punching, and forming, require very little attention from the inspector. In such cases the tools and first parts produced are carefully inspected at the start, and, from then on, periodic inspection (aside from the attention given the machines by the operator) is sufficient. Automatic screw machines and other similar automatic equipment can be handled the same way. Once the machine is set up, the change in the product due to wear of the tools is so slight that the periodic inspection given by a roving inspector is sufficient.

A system of inspection known as *sampling* is used on most bulk materials such as coal, batch materials, and foundry sand. It is used also in the dimensional inspection given parts in machine shops. If a proper sample can be determined readily, this method offers a means of reducing inspection costs. The frequency or method of sampling must be such that there is no possibility of producing a large number of defective parts before an error is discovered.

Many parts that require accurate machine work should be given 100% inspection to eliminate any possibility of performing expensive operations on defective parts. Crankshafts, bearing races, and gears are typical parts that should be treated in this manner. In many cases 100% inspection is necessary at several points in the manufacturing cycle as well as at completion. In any event, a final inspection should always be given prior to assembly operations.

Types of Fits

The term " interchangeable manufacture " implies that the parts which go into the assembly of the machine can be selected at random from a large number of parts. In such a system of manufacture, selective fitting is unnecessary except possibly for special close-working parts. To make this possible, manufacturing methods must be standardized and limits of accuracy specified on details. Extreme accuracy is not always necessary or desirable, since manufacturing costs increase greatly as working limits become closer. In many cases, on modern production machines, it is possible to maintain a limit of accuracy in excess of that required by the part with no added expense; however no part should be made with any greater degree of accuracy than is required by its use in a given mechanism or machine. A balance must be established between cost of manufacture and ease of assembly.

The fact that there is a need for various types of fits in manufacturing work is clearly evident. A given industry may require only a few; others will maintain that a large number is necessary. In general there are but three types of fits: a *clearance fit,* a *tight fit,* and an *interference fit.* It is quite obvious that these three conditions will not satisfy all needs, as the amount of clearance or interference of the mating parts is also an important factor. Hence it becomes necessary to subdivide these classifications further to include those fits most commonly used in manufacturing work. Any such classification or standard will probably not satisfy all manufacturers, but it should include the general needs of all industry.*

According to the American Standards Association, fits are classified as follows:

Loose fit. This fit provides for a large allowance giving considerable freedom and is used where accuracy is not essential.

Free fit. For running fits with speeds of 600 rpm or over, and journal pressures of 600 pounds per square inch or over. Closer allowance than for a loose fit.

Medium fit. Used for parts revolving easily. For speeds under 600 rpm and with journal pressures less than 600 pounds per square inch. This is also applied to sliding parts and is the largest allowance for freedom consistent with accuracy,

Snug fit. This calls for zero allowance and is the closest fit that can be assembled by hand without appreciable pressure. It will not rotate easily, and no shake is permissible. A snug fit is not intended to move freely under a load.

Wringing fit. This is a metal-to-metal contact with no negative allowance. It allows for no movement and is assembled with slight pressure. Wringing fits are not usually interchangeable.

Tight fit. This is a wringing fit with slight negative allowance. This fit is for parts permanently assembled or subject to pressure. It is much used in ordnance work.

Medium-force fit. For permanently assembled parts, but subject to disassembly without severe pressure. The fit is the tightest possible for cast iron or parts where internal stress will be detrimental.

Heavy-force fit. Used for steel holes where the metal can be highly stressed without exceeding its elastic limit. Parts united by force fit form one unit without other means of holding.

Shrink fit. This is the heavy-force fit applied to larger parts where a force fit is impractical, such as for locomotive wheel tires. A definite negative allowance is given, and the outer part is expanded by heat before assembly.†

* John Gaillard, *Tolerances for Cylindrical Fits,* American Standards Association.

† *Tolerances, Allowances, and Gages for Metal Fits,* American Standards Association, B4a — 1925.

Tolerance and Allowance

In dimensioning a drawing, the figures placed in the dimension lines represent *nominal sizes*. Nominal sizes are only approximate; they do not represent any degree of accuracy unless it is so stated by the designer. To specify a degree of accuracy, it is necessary to add *tolerance* figures to the dimension. Tolerance is the amount of variation permitted in the part or the total variation allowed in a given dimension. A shaft might have a nominal size of $2\frac{1}{2}$ inches, but for practical reasons this figure could not be maintained in manufacture without great cost. Hence a certain tolerance would be added, and, if a variation of ±0.003 inch could be permitted, the dimension would be stated 2.500 ± 0.003. Where dimensions are given close tolerances, the reason is that the part must fit properly with some other part. Both must be given tolerances in keeping with the type of fit and allowance desired.

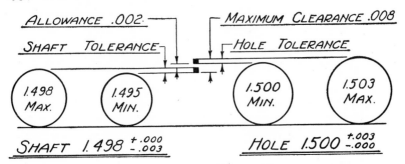

Fig. 1.   Loose or Clearance Fit.

Allowance, which is sometimes confused with tolerance, has an altogether different meaning. It is the minimum clearance space intended between mating parts and represents the condition of tightest permissible fit. Figures 1 and 2 illustrate exaggerated conditions for clearance and interference fits. The tolerances for shaft and hole are indicated by the black bars. In the figure showing a clearance fit, the allowance is the difference between the largest shaft size and the smallest hole size indicated as 0.002 inch. This value represents the minimum allowable clearance space, and 0.008 inch represents the maximum. All shafts and mating parts have tolerances, which, if maintained, will give clearances between these two extremes. Figure 2, representing an interference fit, has tolerances limiting the interference to values between 0.001 and 0.005 inch. In both cases there is probably one clearance or interference value that is best, but for manufacturing reasons a variation is necessary. To obtain the best value, selective or assembly fitting would have to be resorted to.

Tolerances may be either *unilateral* or *bilateral*. Unilateral tolerance means that any variation is made in only one direction from the nominal or basic dimension. Referring again to Figure 1, we see that the hole is dimensioned 1.500 $+0.003 \atop -0.000$, which represents a unilateral

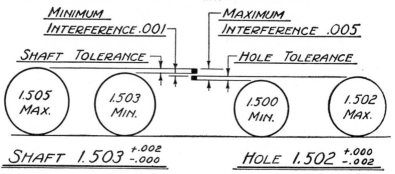

FIG. 2.   Tight or Interference Fit.

tolerance. If the dimensions were given as **1.500 ± 0.003**, the tolerance would be bilateral; that is, it varies both over and under the nominal dimension. The majority of manufacturing concerns in this country use the unilateral system. This system permits changing tolerance values without altering the esssential characteristics of the respective fits. Tolerances and allowances are specified by the designers when parts to be made are detailed. It is the function of the mechanic to produce according to these dimensional specifications, and it is the duty of the inspector to see that the specifications are maintained. Both mechanic and inspector should have a clear understanding of the various type fits and the necessity for maintaining standards of accuracy.

MEASURING INSTRUMENTS

Standard of Measurement

The standard of measurement in the United States is the *meter*. This standard, adopted by Congress in 1866, has reference to the international meter at the International Bureau of Weights and Measures at Sèvres, France. Our legal *yard* is defined as 3600/3937 of the length of the meter at a temperature of 68 F, from which 1 meter is equal to 39.37 inches. The British standard yard is slightly different, 1 meter being equal to 39.370113 inches. An inch under this system is equal to about 25.39998 mm, whereas the United States inch is equal to about 25.40005 mm. Although this difference does not seem to be of any great importance in ordinary shopwork, it is noticeable in accurate measurements. In 1933 these values were changed in both countries to permit the use of a uniform ratio: 1 inch equals 25.4 mm.* This change makes it possible to convert readily from one system to the other and eliminates any possible confusion.

The standard of angular measurement is the *degree* which is obtained by dividing a circle into 360 parts. A degree is further divided into 60 minutes, and each minute is divided into 60 seconds. This standard of measurement is universal.

Classification of Measuring Instruments

A measuring instrument is any device which may be used to obtain a dimensional or angular measurement. Some instruments, such as a steel rule, may be read directly; others, like the caliper, are used for transferring or comparing dimensions. Also, various principles are employed in obtaining measurements. A micrometer, for example, utilizes a different principle from a steel rule or a vernier caliper. Here are a number of the common measuring instruments listed according to use:

MEASURING INSTRUMENTS

1. Linear measurement
   - (a) Steel rule.
   - (b) Micrometer.
   - (c) Vernier caliper.
   - (d) Depth gage.
   - (e) Vernier height gage.
   - (f) Calipers.
   - (g) Dividers.
   - (h) Telescopic gages.
   - (i) Combination square.
   - (j) Measuring machine.
     - 1. Mechanical.
     - 2. Optical.

2. Angular measurement
   - (a) Adjustable bevel.
   - (b) Bevel protractor.
   - (c) Sine bar.
   - (d) Square.
   - (e) Optical protractor.
   - (f) Dividing head.

3. Plane surface measurement
   - (a) Level.
   - (b) Straight edge.
   - (c) Surface gage.
   - (d) Profilometer.
   - (e) Optical flat.

Linear Measuring Instruments

Rule. The most common measuring device in the shop is the steel *rule* — made of tempered steel, carefully ground, and accurately graduated on both sides. Usually one side is graduated in eighths

* American Standard Practice for Inch-Millimeter Conversion for Industrial Use, B48.1 — 1933, and British Standard Conversion Tables BS350 — 1930.

and sixteenths and the other in thirty-seconds and sixty-fourths, although numerous other graduations in both metric and English systems are used. This tool is very satisfactory for rough machine

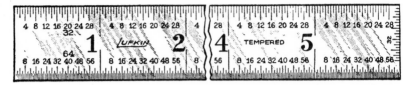

Showing "Readable" Graduations

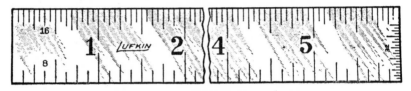

Showing End Graduations

Courtesy The Lufkin Rule Company.

Fig. 3. Steel Scale.

work, layout work, checking dimensions, and many other shop applications. Mechanics with considerable skill and experience can attain a high degree of accuracy in measuring with a rule and calipers. A steel rule with standard graduations is shown in Figure 3.

Courtesy The Lufkin Rule Company.

Fig. 4. Combination Set Including Square, Center Head, Protractor, and Scale.

Combination set. A combination set (see Figure 4) consists of a steel rule or blade on which is mounted a *square head*, a *center head*, and a *bevel protractor*. Although a set includes all three accessories, only one is used at a time. With the square head mounted on the blade, it serves as both try and miter squares; and it can be adjusted to be used as a marking gage. Placing it on the end converts the tool into a height gage. The head alone may be used as a level. When the center head is mounted on the blade, centers of all cylinderical work can be determined. The bevel protractor, used in connection with the blade, permits the measurement, layout, and checking of angles.

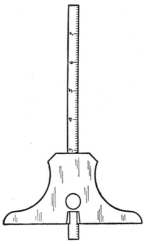

Depth gage. As shown in Figure 5, narrow steel scales are frequently mounted in a head which has a straight edge at right angles to the scale. This forms a depth gage, with a scale that can be adjusted and clamped so as to extend a given

FIG. 5. Rule Depth Gage.

amount below the straight edge. Similar gages are made with micrometer adjustments.

Calipers. Calipers are used for approximate measurements, both external and internal. They do not measure direct, but must be

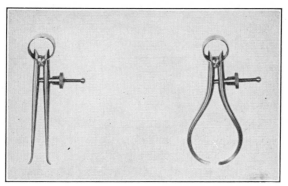

Courtesy The Lufkin Rule Company.

FIG. 6. Inside and Outside Calipers.

set to size, with a steel rule or some form of gage being used. Most shop calipers, known as *spring calipers*, consist of two legs with a flat spring head plus a nut and screw to hold them in position. Some calipers are provided with a " quick nut " for making rapid adjustments, as shown in Figure 6.   On release of the pressure, this nut slides freely along the screw, but with the slightest leg pressure it grips the threads of the screw firmly.   *Hermaphrodite* calipers are used principally for locating centers and layout work.   They have one leg similar to the leg on an outside caliper while the other is a straight point.   In layout work the curved leg rests against the edge of the work while the other leg is used as a scriber.

Dividers.   Dividers are similar in construction to calipers except that both legs are straight with sharp hardened points at the end. This tool is used for transferring dimensions, scribing circles, and general layout work.

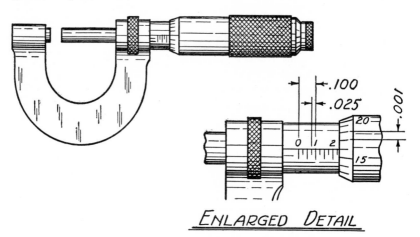

ENLARGED   DETAIL

FIG. 7.   Micrometer with Enlarged View Showing Graduations.

Micrometer calipers.   The *micrometer* is used for quick, accurate measurements to the thousandth part of an inch.   It is direct reading and eliminates the personal element.   This tool illustrates the use of an accurate screw thread as a means of obtaining a measurement. The screw is attached to a spindle and is turned by movement of a thimble at the end.   The barrel, which is attached to the frame, acts as a nut to engage the screw threads, which are very accurately made with a pitch of 40 threads per inch.   Each revolution of the thimble advances the screw 1/40 of an inch, or 0.025 inch.   The outside of the barrel is graduated in 40 divisions, and any movement of the thimble down the barrel can be read next to its beveled end.   When the spindle is in contact with the anvil on a 1-inch micrometer, the zero readings on barrel and thimble should coincide.

The scale on the barrel and thimble edge can best be understood by reference to the enlarged view of Figure 7. On the beveled edge of the thimble are 25 divisions, each division representing 0.001 inch. To read the micrometer, the division on the thimble coinciding with the line on the barrel is added to the number of exposed divisions on the barrel converted into thousandths. Thus, the reading shown in Figure 7 is made up of 0.200 plus 0.025 on the barrel, or 0.225 inch, to which is added 0.016 on the thimble to give a total reading of 0.241 inch.

Since a micrometer reads only over a 1-inch range, in order to cover a wide range of dimensions, several micrometers are necessary.

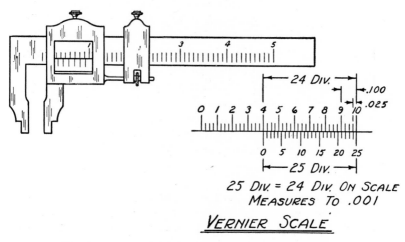

FIG. 8.   Vernier Caliper and Enlarged View Showing Scale.

A 2-inch micrometer reads 1 to 2 inches; a 3-inch, 2 to 3 inches, and so on. Micrometers equipped with a vernier scale can be read to ten-thousandths of an inch. The micrometer principle of measurement is also applied to inside measurements, depth reading, and to the measurements of screw threads.

Vernier calipers. In Figure 8 is a *vernier caliper* which may be used for taking both inside and outside measurements over a wide range of dimensions. It consists of a main scale graduated in inches and an auxiliary scale having 25 divisions. Each inch on the main scale is divided into tenths and each tenth into four divisions, so that in all there are 40 divisions (each 0.025 inch) to the inch. The 25 divisions on the auxiliary or sliding scale correspond to the length of 24 divisions on the main scale and are equal to 24/40 of an inch. One division would be equal to 1/25 times 24/40, or 24/1000 inch, which is 1/1000 inch less than a division on the main scale. Hence, if the two scales were on zero readings, the first two lines would be 0.001 inch apart, the tenth lines 0.010 inch apart, and so on.

In actual use, the reading on the main scale is first observed and converted into thousandths, and to this figure is added the reading on the vernier. The vernier reading is obtained by noting which line coincides with a line on the main scale. If it is the 15th line, then 0.015 inch is added to the main scale. These scales are shown in some detail in the enlarged view of the vernier scale. As shown, the vernier reads exactly 0.400 inch.

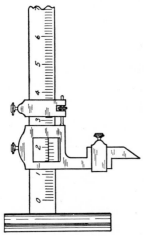

FIG. 9. Vernier Height Gage.

This scale is also used extensively on both height and depth gages, as readings can be taken up to a maximum of 12 or more inches. Outside measurements are taken with the work between the jaws; inside measurements, with the work over the ends of the two jaws. This method of measurement is not so rapid as a microm- eter; but has the advantage of having a wider range with equal accuracy. It also has some use on protractors for angular measurement.

Vernier height gage. An application of the vernier scale to a *height gage* is shown in Figure 9. This tool differs from a vernier caliper in that it rests on a heavy base and has a beveled pointer on the movable jaw. In using this instrument the work is placed on a surface plate, and distances are measured above this reference elevation. The reading of the scale is identical with that of a vernier caliper. This measuring tool is used principally for accurate height measurements and scribing lines in layout work.

Telescopic gage. The *telescopic gage*, used for measuring the inside size of slots or holes, is much quicker than other methods. The gage shown in Figure 10 consists of a handle and two plungers, one telescoping into the other and both under spring tension. The plungers may be locked in any position by the knurled screw at the end of the handle. In using the telescoping gage, the plungers are first compressed and locked in position. Next, the plunger end is inserted into the hole and the screw released, allowing the plungers to expand to the hole size. Finally, the plungers are locked in place and removed, after which the over-all length of the plungers is measured with an outside micrometer. To cover a range from 1/2 to 6 inches, a set of five gages is required.

Toolmakers' microscope. Because of their extreme accuracy and their ability to measure parts without pressure or contact, numerous optical instruments have been devised for inspecting and measuring. A typical microscope for toolroom work is shown in Figure 11. An object viewed is greatly enlarged, and the image is not reversed as in the ordinary microscope. To be measured, a part is first clamped in

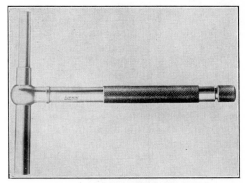

Courtesy The Lufkin Rule Company.

Fig. 10. Telescoping Gage for Internal Measurements.

proper position on the cross-slide stage. The microscope is focused and the part to be measured brought under the crossline seen in the microscope. The micrometer screw is then turned until the other extremity is under the crossline, the dimension being obtained from the difference in the two readings. The micrometer screws operate in either direction and read to an accuracy of 0.0001 inch. In the illustration the microscope is shown with a screw-measuring attachment in position for checking the lead of the screw.

Angular Measurements

As previously stated, the basic unit of angular measurement is the degree. This unit is defined as the angle formed by two radii subtending an arc 1/360 of the circumference of a circle. Although degrees may be further subdivided by fractions, the usual smaller subdivisions are minutes and seconds. Common angular measuring instruments read the degrees directly from a circular scale scribed on the dial or circumference. There are also devices that require the aid of other measuring instruments and calculations to obtain the result. Some few, such as the ordinary square, measure only a single angle and are not adjustable. Fixed instruments of this type are more correctly classified as gages.

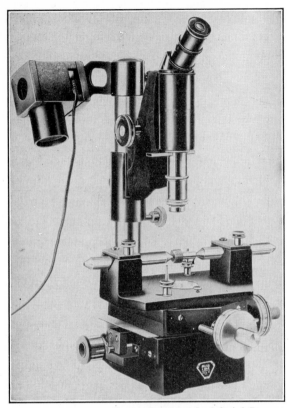

FIG. 11. Toolmakers' Microscope with Lead Measuring Attachment in Position.

Bevel protractor. The plain or universal *bevel protractor* measures directly in degrees and is adapted to all classes of work where angles are to be laid out or established. The universal protractor, shown in Figure 12, is graduated in degrees and, in addition, is provided with a vernier scale for fine measurements. Most such protractors read to 5 minutes or 1/12 of a degree.

Adjustable bevel. An instrument known as an *adjustable bevel* or a bevel gage is widely used for checking or transferring angles. This tool consists of two blades, which can be set and locked in relation to each other. No direct reading is obtained, and the angle must be set or checked from some other angular measuring device.

Sine bar. A *sine* bar is a simple device used either for accurately measuring angles or for locating work to a given angle. Mounted

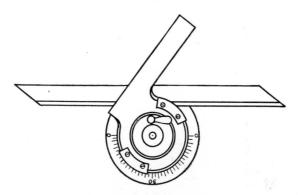

Fig. 12. Vernier Bevel Protractor.

on the center line are two buttons of the same diameter and at a known distance apart, the distance on most sine bars being either

Courtesy The Taft Peirce Manufacturing Company.

Fig. 13. Checking the Included Angle of a Taper Plug Gage With a 10-Inch Sine-Bar Fixture.

5 or 10 inches. For purposes of accurate measurement the bar must be used in connection with a true surface.

The operation of the sine bar is based on the trigonometric relationship that the sine of an angle is equal to the opposite side, divided by the hypotenuse. Hence, if the hypotenuse is known, the angle may be determined by measuring the height of the opposite side, dividing it by the known figure, and referring to trigonometric tables. Measurement of the unknown side is accomplished with either a height gage or by means of precision blocks.

Courtesy The DoALL Service Company.

FIG. 14. Checking the Angle of a Bevel Gear Using a 5-Inch Sine Bar.

Figure 13 shows the setup for checking the included angle of a taper plug gage, using a 10-inch sine bar. The 10-inch dimension is the center distance between the two buttons shown on the side of the bar. The difference in elevation of the two buttons represents a value ten times the sine of the included angle. Using a five-place trigonometric table, we may obtain the angle in degrees and minutes without further calculation.

An interesting setup is shown in Figure 14 illustrating how the accurate checking of angle can be accomplished by the use of a sine bar, gage blocks, and a master flat. The face angle of the bevel gear is checked by clamping the gear to the sine bar and sliding the sine bar and gage blocks across the master flat and under the dial indicator gage. A sine bar should always be used on a precision surface plate or a master flat.

When work is set up to be machined at a given angle, the operation is reversed. The bar is then set at the proper angle, which in turn acts as a gage to position the work correctly. Various designs of the sine bar have been worked out, but the method of measuring the angle is the same in all cases.

Optical protractor. An optical protractor for accurately measuring angles is shown in Figure 15. This instrument is used for measuring and checking the exact angular relationship between surfaces, edges, or holes. It has the advantage over the sine bar in that readings to 1 minute of the arc can be read directly and no trigonometric calculations are required. The protractor is built upon an adjustable base, and the ring center revolves, carrying with it a level vial and protractor scale which are viewed through an eyepiece In using this instrument the base is first adjusted to compensate for any errors in the bed of the machine. The center ring is then adjusted by hand or by the micrometer screw until

Courtesy Bausch and Lomb Optical Company.

FIG. 15. Optical Protractor for Measuring Angle Accurately.

the required angle appears on the scale. The protractor is then placed upon the work, and the work is turned until the level bubble becomes central. It is used both for setting up work on machines and for checking parts already machined.

Dividing heads. Index or *dividing heads* were originally developed for use on milling machines, but their use has been extended to inspection work for checking angles about a common center. The head is made up of a worm and worm-gear set having a ratio of 40 to 1. Hence, one turn of the crank will turn the spindle 1/40 of a revolution or 9 degrees. By using index plates with the head, we may obtain any desired angle with great accuracy. The operation of this device is discussed in the chapter on milling machines, Chapter 16. For inspectional work it is used in connection with a surface plate for checking parts already machined.

Surface Measurements

Surface checking instruments are for the purpose of obtaining some measure of the accuracy of a surface or the condition of a finish. Much of this type of work is done on a flat accurately machined casting known as a *surface plate*. It is the base upon which parts are laid out and checked with the aid of other measuring tools. These plates are very carefully made and should be accurate to within 0.001 inch from the mean plane, any place on the surface. Small plates, known as toolmakers' flats, are lapped to a much greater degree of accuracy. Their field of application is limited to small parts, and in most cases they are used with precision gage blocks.

Square. The *square*, previously mentioned in connection with angular measurements, is used a great deal in surface work for checking the squareness of two surfaces. Hardened-steel squares with rigid mounting are made in a large number of sizes and find wide application in toolroom work.

Straightedge. This tool is a bar of steel having either one or two straight edges. It is used for the inspection of surfaces for straightness, checking flat surfaces before straightening, and for accurately scribing straight lines. The ordinary shape is rectangular, but for accurate work one edge is beveled or formed into a knife edge. Minute variations in curvature can be detected by using the latter type in front of a good light. When one is working to a close tolerance, a feeler gage should be used to check the surface variations.

Surface gage. The *surface gage,* shown in Figure 16, is used to check the accuracy or parallelism of surfaces and, in addition, finds much use in transferring measurements by scribing them on a vertical surface. The instrument consists of a base, an adjustable spindle, and a scriber. When in use it is set in approximate position and locked. Fine adjustment of the spindle can be made by turning the knurled nut which controls the rocking bracket. When used with the scriber, it is a line-measuring or locating instrument. If the scriber is replaced by a dial indicator, it then becomes a precision instrument for checking surfaces.

Optical flat. Measurements to the millionth part of an inch are made by interferometry, the science of measuring with light waves. Measurement by this principle is made with a small instrument known as an *optical flat*. Optical flats are usually made from natural quartz because of its hardness, low coefficient of expansion, and resistance to corrosion. They are flat lenses having very accurately polished surfaces with light-transmitting quality. The usual optical-flat set consists of two lenses 2 inches in diameter and 5/8 inch thick, although they may be obtained in various sizes and shapes up to 5 inches. It is not necessary that the two surfaces of a flat be absolutely parallel. A light having a single color (monochromatic light) is used, because it gives interference bands that are complete and dark in color.

One of the common uses for optical flats is the testing of plane surfaces. The optical flat is placed on the flat surface to be tested, and light is reflected both from the optical flat and the surface being tested. The interference between the rays reflected from the bottom of the flat and from the top of the work causes dark bands (Newton's Rings) or areas to appear. If the surface is irregular, the appearance is similar to a contour map. The position and number of lines show the location and extent of the irregularities. When bands are straight, evenly spaced, and parallel to the line of contact, the surface is perfectly flat. Since we know the wave length of the light source, any deviation from this pattern indicates an error in the surface, the amount of which can be measured. For ordinary daylight, the difference between bands is 10 millionths of an inch (one-half a wave length). Sodium light has a separation in one-half wave length equal to about 12 millionths of an inch. In Figure 17 is shown a 6-inch optical flat used in examining

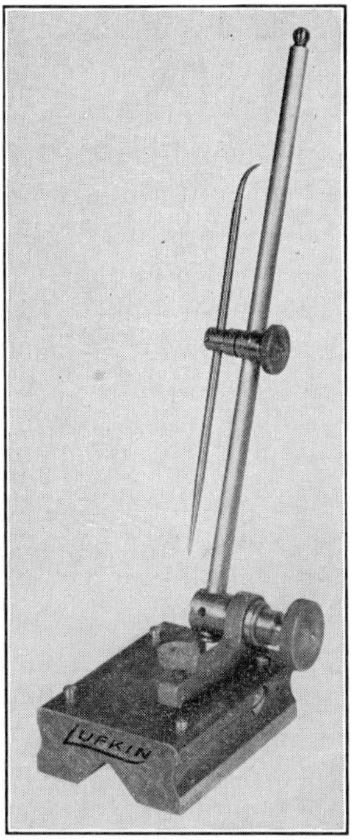

Courtesy The Lufkin Rule Company.

FIG. 16. Surface Gage.

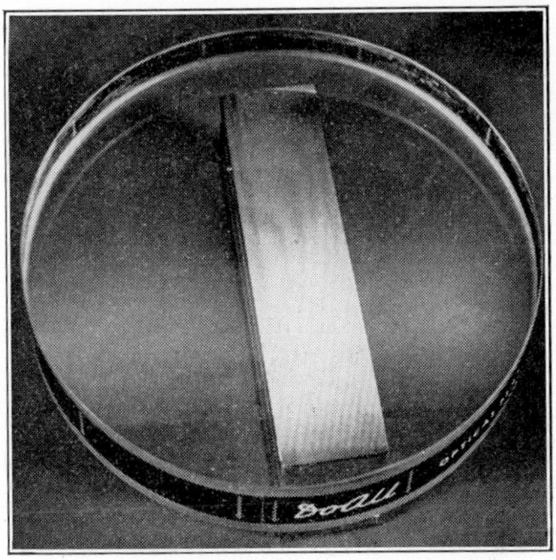

Courtesy The DoALL Service Company.

FIG. 17. Examining the Surface Condition of a Rejected Comparator Anvil.

the surface condition of a rejected comparator anvil. The center left edge is 174 millionths higher than the lower right corner.

In Figure 18 are shown two optical flats being used in connection

Courtesy Bausch and Lomb Optical Comapny.

Fɪɢ. 18. Optical Flat Showing Newton's Rings.

with a gage block to check the height of an object. The part and gage block are wrung to the lower flat, and the upper flat is placed securely on the two top surfaces. If the parts are not the same size, the flat will be tilted slightly, and parallel interference lines will appear on the top surface of the gage block. The difference in height can be determined from the position and number of bands that appear. Each band indicates a difference of 12 millionths of an inch. To facilitate calculations the the contact lines of the flat on the part and the gage blocks are usually placed apart a distance equal to the width of the gage block.

Profilometer. Several devices have been developed recently which have as their purpose the measurement of surface roughness. The need for some measure of surface finish for various machine operations is brought about by the necessity for smooth bearing surfaces in high-speed machinery. The unit of measurement in this work is the " microinch " (0.000001 inch).

Several methods of roughness measurement have been established, the simplest being a visual comparison with some established standard. Direct measurement of scratch depth can be made by light interference in a manner similar to that described in the discussion of optical flats. This method requires surfaces having good reflecting qualities. Another optical method employs a microscope which magnifies a shadow cast by the scratches of a surface. Light is directed at an angle to a straight edge, and the shadows from it are projected from the scratches. This method is not suitable for very fine finishes and its use is further limited in that only a small part of the entire area can be inspected.

The *profilometer* is a direct-reading instrument which measures roughness by passing a fine tracing point over a surface. The vertical movement of the tracer is magnified electrically so that a readable profile curve can be obtained. Some of the important features of this instrument are that it has variable magnification, is adapted to quick reading of large areas, can be used for both plane and curved surfaces, and is capable of measuring any surface roughness up to 1000 microinches. This instrument is used in the production inspection of parts where surface finish is important.

GAGES

Production work requires speed as well as accuracy in measuring parts. Measuring instruments, such as those just described, can be used for this type of work, but many of them are more elaborate than necessary and require adjustments for each individual reading. Consequently, they are used only on short-run jobs where the expense of fixed gaging equipment is unwarranted. To attain the quick measurements required in production work, a measuring device that has a fixed shape or size is used. It represents a standard with which the manufactured parts are compared. Since its use is limited to one dimension and no adjustments are required, the operation of inspecting a part requires a minimum of time. Much gaging is done by operators in the shop while their equipment is in operation, with no loss of production time.

Classification of Gages

Numerous kinds of gages are used, varying widely in shape and size. Gages are classified, by the Ordnance Department of the Army and many industries as well, as *inspection gages* and *manufacturing gages*. Inspection gages are those used by inspectors in the final acceptance of the product. They are to insure that the product is made in accordance with the tolerance specifications on the working blueprints. Manufacturing or working gages are those used by the machine operators in the actual production of parts. These gages are frequently made to slightly smaller tolerances than the inspection gages, the idea being to keep the size near the center of the limit tolerance. Parts, then, made around limit sizes may still pass the inspector's gage. In addition to these, a third type of gage, known as a *master gage,* is sometimes used. Such a gage is merely a reference gage with which inspection gages are periodically compared. Most industries have abandoned this type of gage and have adopted precision measuring instruments and methods for this purpose. Inspection gages are checked at intervals determined by the rate of production to see if they are still within the allowable variation.

The equipment used for this purpose consists of such tools as precision-gage blocks, measuring machines, optical flats, microscopes, and projecting equipment.

The following gages represent those most commonly used in production work. The classification is principally according to the shape or purpose for which each is used. A complete classification would require more subdivisions under the various headings.

GAGES

1. Plug.
2. Ring.
3. Snap.
4. Length.
5. Form
   (a) Screw thread.
   (b) Fillet.
   (c) Center.
   (d) Drill point.
   (e) Angle.
   (f) Gear tooth.
   (g) Special contour, etc.

6. Thickness
   (a) Precision gage blocks.
   (b) Sheet metal.
   (c) Feeler.
   (d) Wire, etc.
7. Indicating.
8. Projecting.
9. Special
   (a) Planer.
   (b) Sine bar.
   (c) Taper parallels, etc.

Gage materials. The true value of a gage is measured by its accuracy and service life, which, in turn, depend upon the workmanship and materials used in its manufacture. Since all gages are continually subject to abrasive wear while in use, the selection of the proper material is of great importance. High-carbon and alloy tool steels have been the principal materials used for many years. These materials can be accurately machined to shape, and they respond readily to heat-treating operations which increase their hardness and abrasive resistance. Objections to steel gages are that they are subject to some distortion because of the heat-treating operation and that their surface hardness is limited.

These objections are largely overcome by the use of chrome plating or cemented carbides as the surface material. Chrome plating permits the use of steels having inert qualities, as wear resistance is obtained by the hard chromium surface. This process also is widely used in the reclaiming of worn gages. Cemented carbides applied on metal shanks by powder-metallurgy technique provide the hardest wearing surface obtainable. Although their cost is several times that of a steel gage, their life is much greater, and the additional cost is more than justified.

A recent developement in gage materials is the use of flint and borosilicate glass for many types of gages of small size.* This material has proved successful, because it has a low thermal conductivity, and the body heat of inspector has little effect on the functioning of the gage. Other advantages are that it is not subject to corrosion, has high abrasive-resisting qualities, has no tendency to burr at the edges, and affords visibility in inspection. The use of glass gages also has the desirable effect of causing inspectors to handle gages carefully.

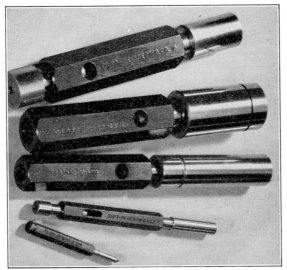

Courtesy Taft Peirce Manufacturing Company.

FIG. 19.   Plug Type "Go — Not-Go" Gages.

Description of Gages

Plug gage. A plain plug gage is an accurate cylinder used as an internal gage for the size control of holes. It is provided with a suitable handle for holding and is made in a variety of styles. These gages may be either single- or double-ended, as shown in Figure 19. Double-ended plain gages have " go " and " not-go " members assembled on opposite ends, whereas progressive gages have both gaging sections combined on one end.

Three separate designs have recently been recommended for adop-

* Stone, Col. J. A., " Glass for Precision Gages," *American Machinist,* February 4, 1943.

tion by the American Standards Association.* One is known as the taper-lock design and is the same as those shown in Figure 19. This design applies to all gages from 0.059 inch to and including 1.510 inches. For diameters varying from 1.510 to 8.010 inches the trilock design with reversible gaging members is used. This gage has three wedge-shaped prongs on the handle which are forced into three locking grooves in the gaging cylinder by means of a through screw. The cylinder may be reversed when it becomes worn and thus prolong the life of the gage. Large gages ranging from 8.101 to 12.010 inches are made in the form of a rim with reinforcing web. Holes are drilled and tapped in the web to receive two ball handles. The annular design has proved very satisfactory for large gages, as they are light in weight and easily handled. These same designs are recommended for thread, taper, and special-form plug gages.

Courtesy Taft Peirce Manufacturing Company.

Fig. 20. Micrometer Plug Gage.

Micrometer plug gage. This tool (see Figure 20) is a variation from the usual form of internal micrometer and is adapted for accurate checking of holes in tool manufacture as well as for inspection work where the expense of permanent fixed gages is not justified. It is made in a variety of sizes ranging from 3/4 to 4 inches, each having a range of around 1/16 inch. The blades, seated on a hardened and ground cone, are uniformly expanded and provide a means for checking hole roundness and straightness, as well as size. Readings may be made to 0.0001 inch.

Ring gages. Plain ring gages are also standardized in construction and general proportion. The three shown in Figure 21 are typical of gages up to 1.510 inches. A plain knurled surface indicates a " go " gage, whereas a " not-go " gage is identified by an annular groove on the periphery. Above 1.510 inches all gages are flanged to reduce weight and facilitate handling. Large gages are provided with ball handles. Details of construction, with dimensions of all sizes, have been worked out by the Standards Committee.

* *Commercial Standard — CS8-41*, American Gage Design Committee, National Bureau of Standards.

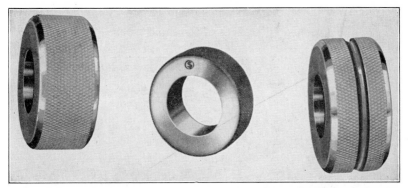

Courtesy Sheffield Gage Corporation.

Fig. 21. Plain Ring Gages.

Taper gages. Taper gages are made in both the plug and ring styles and, in general, follow the same standard construction adopted for plain and ring gages. Two taper plug gages are shown in Figure 22. Taper gages are not dimensional gages but rather a means of checking in terms of degrees or inches per foot. Their use in testing

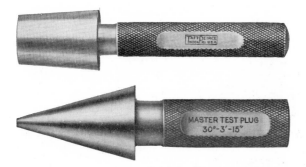

Courtesy Taft Peirce Manufacturing Company.

Fig. 22. Special Taper Plug Gages.

is a matter of fitting rather than. measuring. If size is involved, " go " and " not-go " tolerances are indicated on the end of the gage by grooves or by milling off a portion of the gage. This type of gage is widely used in checking the standard tapers and sockets used on tools and production machines.

Spline gages. For inspectional work on splined hubs and shafts, spline gages similar to those shown in Figure 23 are used. These gages are in effect special forms of the ordinary plug and ring gages. Standardization of these gages is impractical because of the wide variations in the design of splined parts.   Numerous other special shapes are inspected with gages of the same type.

FIG. 23.   Spline Plug and Ring Gage.

Thread gages.   In Figure 24 are shown standard plug and ring gages for inspecting threads.   Both these gages are standardized with size variations in the same manner as provided for plain plug and ring gages.   Double-ended plug gages have the " go " and " not-go " feature according to the tolerances desired.   All ring gages are of

FIG. 24.   Thread Plug and Ring Gage.

the adjustable type, as shown in the figure, with effective means for locking the adjusting screw in position.   This feature is desirable, as it is very difficult to measure internal threads accurately in a blank during its manufacture.   After completion they are set to correct size by means of a threaded plug of exact dimensions.

Screw-pitch gages, as shown in Figure 25, are only for the purpose of checking the form and pitch of the threads. The consecutive arrangement of the blades on the case enables the desired blade to be selected quickly. The form of the blades is such that they can be used for both internal and external threads.

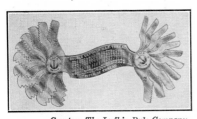

Courtesy The Lufkin Rule Company.

FIG. 25. Thread-Pitch Gage.

Snap gages. A snap gage, used in the measurement of plain external dimensions, consists of a U-shaped frame having jaws equipped with suitable gaging surfaces. A plain gage has two parallel jaws or anvils which are made to some standard size and cannot be adjusted. This type of gage is largely being replaced by adjustable gages to provide means of changing tolerance settings or adjusting for wear. Most gages are provided with the " go " and " not go " feature in a single jaw, and this design is both satisfactory and rapid. A wide variety of snap gages has been developed by various companies, although it is recognized that some disadvantage results from this lack of standardization. Four types, standardized by the American Gage Design Committee, are:

Model A — Employing 4 gaging pins.
Model B — Employing 4 gaging buttons, either square or round.
Model C — Employing 2 gaging buttons, either square or round, and a single block anvil.
Model MC — A miniature snap gage with 2 gaging buttons, either square or round, and a single block anvil.*

The model-B gage is shown in Figure 26, and the model-C is shown in Figure 27. Model-A gages are similar to the model-B design, except that straight pins are used in place of buttons as the gaging members. The model-MC is similar to the model-C gage, except that it is in miniature and is used only in gaging diameters up to 0.760 inch.

The general design shown in Figures 26 and 27 has been selected, because it incorporates most of the advantages of all similar gages now in use. It is light in weight, sufficiently rigid, easy to adjust and provided with suitable locking means, and is designed to permit interchangeability of as many of the parts as possible.

Precision gage blocks. These blocks, of hardened steel, are square or rectangular in shape, having two parallel sides very accurately

* Commercial Standard CS8-41.

lapped to some size.   Blocks up to one inch are made to size within 5 to 8 millionths of an inch per block and per inch of length on larger blocks.   Laboratory sets may be obtained with a guaranteed accuracy

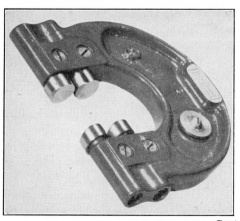

Courtesy Taft Peirce Manufacturing Company.

FIG. 26.   Adjustable-Type Snap Gage (AGD Model B).

FIG. 27.   Snap Gage Equipped with Square Buttons (AGD Model C).

within 2 millionths of an inch per block.   The principal uses for these blocks are for reference in setting gages and for accurate measurements in tool, gage, and die manufacture.

In Figure 28 is shown a set of 81 gage blocks which includes the following:

One ten-thousandth series	0.1001–0.1009 inch, inclusive	9 blocks
One thousandth series	0.101 –0.149 inch, inclusive	49 blocks
Fifty thousandths series	0.050 –0.950 inch, inclusive	19 blocks
Inch series	1.000 –4.000 inches, inclusive	4 blocks

With this set it is possible to obtain practically any dimension in increments of 0.0001 inch from 0.100 to over 10 inches by combining blocks of the proper size.

Precision gage blocks are assembled by a wringing process.   The blocks must first be thoroughly cleaned.   One is placed on the other centrally and oscillated slightly.   It is then slid partially out of engagement, and they are wrung together under slight pressure.   A slight liquid film between the surfaces of the gages causes them to adhere firmly.   Gage blocks put together in this fashion should not be so assembled for more than a few hours.

FIG. 28. Set of 81 Precision Gage Blocks.

Many interesting applications of gage blocks can be made with the aid of special holders and base blocks. The holders secure the blocks in one rigid accurate unit as illustrated in Figure 29. This height gage combination can be set by means of a vernier gage block to accuracies of 0.000010 of an inch. When the graduated block is slid to the right one graduation, the height of the vernier gage is increased 10 microinches. Snap gages can also be assembled quickly for accurate inspection on short-run jobs. The casting shown in Figure 30 is being inspected with a " go " and " not-go " snap gage combination.

Thickness or feeler gage. The thickness gage, shown in Figure 31, consists of a number of thin blades which are held together in a metal case.

FIG. 29. Height-Gage Combination Using Block and Vernier Gage.

The blades in this gage vary in thickness from 0.0015 to 0.015 inch, but similar gages are available in any thickness range desired. They are used for checking clearances and for gaging in narrow places.

Dial indicator. A dial indicator is composed of a graduated dial having a hand connected to a test point with suitable means for

Courtesy The DoALL Service Company.

Fig. 30.   Snap-Gage Combination Checking a Machined Casting.

supporting or clamping it firmly. The dial is graduated in thousandths of an inch, the number depending on the accuracy desired. Most indicators have a spindle travel equal to $2\frac{1}{2}$ revolutions of the hand. Between the test point and the hand is interposed an accurate multiplying mechanism which greatly magnifies on the dial any movement of the point. This tool may be considered either a measuring device or a gage. As a measuring device it is used to measure inaccuracies in alignment, eccentricity, and deviations on surfaces supposed to be parallel. In gaging work it gives a direct reading of tolerance variations from the exact size.

A dial test indicator equipped with a permanent-magnet base is shown in Figure 32. This base operates in the same fashion as the permanent-magnet chuck described on page 602. With the handle in the base turned to the " on " position, the indicator is held securely on the horizontal, vertical, or overhead flat surface of any machine. Other methods of support are a suitable clamp or a heavy base as used on a surface gage.

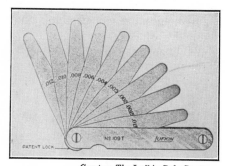

Courtesy The Lufkin Rule Company.

Fig. 31.   Thickness Gage Having Nine Blades of Different Thicknesses.

A special gage for checking the diameter of four crankshaft bearings, simultaneously at both ends and in the center of each bearing, is shown in Figure 33. As the crankshaft is rotated, the 12 dial indicators give tolerance readings at the various points. This gage is typical of many gaging devices developed for special applications.

Comparator or visual gage. A visual gage giving direct plus or minus readings to close tolerance is shown in Figure 34. This gage employs a feed mechanism in connection with a light beam to obtain high magnification of any movement of the gaging point. The

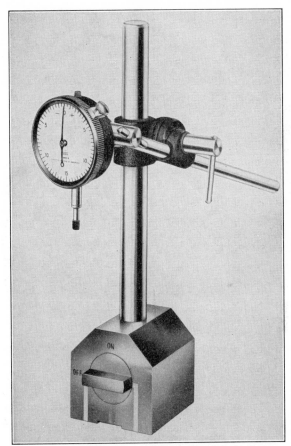

Courtesy Brown and Sharpe Manufacturing Company.

FIG. 32. Dial Indicating Gage with Permanent-Magnet Base.

magnification in this instrument is 5000 to 1, and the range on the 5-inch scale is only 0.001 inch. Each graduation on the scale represents 0.000025 inch. This gage is used for checking inspection gages and for toolroom work, as well as for routine production work requiring close size control.

An electric comparator having four ranges of magnification is shown in Figure 35. A common model of this type comparator has magnification of 10,000, 5000, 1000, and 500 to one. For the highest magnification each graduation is 0.000010 of an inch. An

Courtesy Sheffield Gage Corporation.

FIG. 33. Gage for Checking Crankshaft Bearings.

electric current is arranged so that a minute motion of the gage spindle produces the height indication on the scale. In the illustration shown a thread gage is being checked for pitch diameter by the three-wire method. The gage is set by means of gage blocks.

Projecting comparators. Projecting comparators are designed on the same principle as the ordinary projection lantern. An image is placed before a light source, and the shadow of the profile is projected on the screen at some enlarged scale. Usual magnifications are ×20 and ×50, although others up to ×100 can be used.

A pedestal-type comparator showing the threads on an airplane cylinder barrel is illustrated in Figure 36. Frequently, in checking

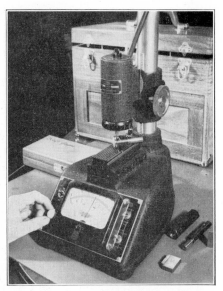

Courtesy The DoALL Service Company.

Courtesy Sheffield Gage Corporation.

FIG. 34. Model 5000 Visual Gage.

FIG. 35. Checking the Pitch Diameter of a Thread Gage on an Electric Comparator Gage.

threads, a chart showing tolerance variations is prepared on a transparent sheet and placed against the glass screen. This permits the inspector to check the thread for dimensional accuracy as well as for contour and lead. The attachment holding the cylinder is arranged so that it can be moved across the path of the light beam, thus allowing the entire thread to be inspected. Micrometer adjustment to one ten-thousandth of an inch is provided on the work table of some of these machines for accurate checking purposes.

The path of light in a different type projector is shown diagrammatically in Figure 37. The object to be inspected is supported at 6 in the figure, and, as the light beam passes by the contour of the object, it enters the projection lens at 7 and is reflected to the screen. Contour inspection is of great value in many tools, dies, gages, and formed products. It is employed in the inspection of many small

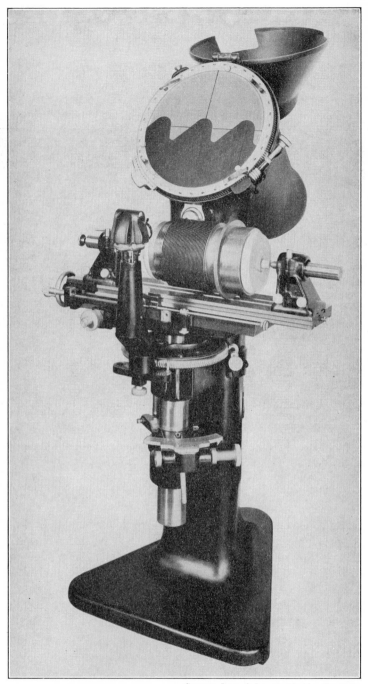

FIG. 36.   Pedestal-Type Comparator Showing Threads on Airplane Cylinder Barrels.

parts such as needles, saw teeth, threads, forming tools, taps, and gear teeth. Since it checks work to definite tolerances, it can be used for studying wear on tools or distortion caused by heat treatment.

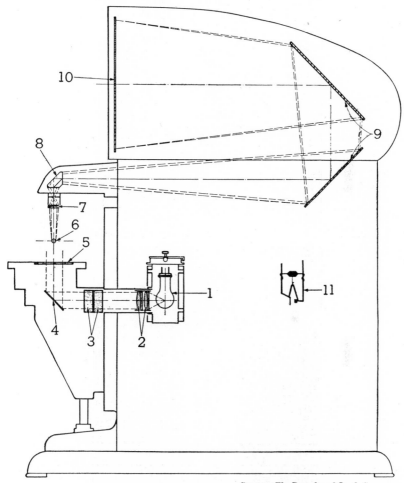

Fig. 37. Path of Light in Bausch and Lomb Contour Measuring Projector.

Its limitations are based on the size of the object and the number of magnifications desired.

Planer and shaper gage. This gage, shown in Figures **38** and **39**, is used primarily as a means of setting cutting tools accurately so that the first cut may give correct dimension. It consists of a

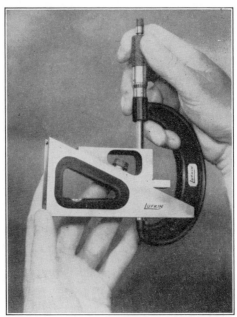

FIG. 38.   Method of Setting Planer and Shaper Gage to Correct Height.

Both Courtesy The Lufkin Rule Company.

FIG. 39.   Use of Planer and Shaper Gage in Setting Cutting Tool.

triangular-shaped base and slide which, accurately fitted, slide upon one another, always maintaining a parallel relationship between the extreme surfaces. It may be used on either its base or its end; and, by the use of an extension pin, dimensions may be gaged ranging from 1/4 to 9 inches. For ordinary dimensions, the gage is set with a micrometer, but a surface or height gage can be used if more convenient. Aside from its use in setting cutting tools, it is also used in connection with layout work and in the measuring of parts on a surface plate.

REVIEW QUESTIONS

1. (*a*) What is the standard of length in the United States? (*b*) How is a yard defined? *36/39.37 at 68°F*

2. (*a*) What is a marking gage? (*b*) What is the angle between the measuring surfaces of a center head? *90°*

3. What is a combination set? State the use of each part.

4. (*a*) How would you set a divider to $2\frac{15}{32}$ inches? (*b*) Give two uses for a surface gage.

5. (*a*) How is the ordinary 6-inch steel rule graduated? (*b*) What tool is used for checking lead on screw threads? *tool makers microscope*

6. (*a*) What are gage blocks used for? (*b*) When several are used together, how should they be assembled?

7. Distinguish between tolerance and allowance.

8. A bearing is dimensioned 1.750 plus 0.003 minus 0.000. The shaft which runs in the bearing is 1.748 plus 0.0005 minus 0.0015. What is the allowance? Are these dimensions unilateral or bilateral?

9. If the diameter of a hole is specified as 1.375 plus 0.002 inch minus 0.005 inch, what should be the diameters of the gaging members of a " go " and " not-go " plug gage?

10. State how you would gage a hole with a telescopic gage.

11. Show by sketch the micrometer reading for 0.762 inch.

12. Does a taper plug gage measure both the size and taper of a tapered hole? Explain.

13. How is a " go " ring gage distinguished from a " not-go " ring gage? *not grooved*

14. Name and describe three standard types of snap gages.

15. Name three or four methods of gaging threads.

16. What is the purpose of a dial indicator and how is it used?

17. Explain how measurements on a vernier caliper are read.

18. What advantages does a vernier caliper have over the ordinary micrometer? *larger range*

19. (*a*) Explain the use of a surface plate. (*b*) What is a toolmaker's flat?

20. Show by sketch how an angle would be checked using a sine bar.

21. A 1-inch gage block is set up under one end of a 5-inch sine bar. What height would have to be used on other end to check an angle of 12° 30″?

22. (*a*) What is the purpose of a projecting comparator? (*b*) Is it possible to check for dimensional tolerance with this instrument?

23. (a) Why is monochromatic light recommended for use with optical flats? (b) Of what material is an optical flat made?

24. Explain how you would check the height of a 1-inch gage block using an optical flat.

25. Prepare a classification of measuring instruments and gages.

BIBLIOGRAPHY

American Standards Association, *Tolerances, Allowances, and Gages for Metal Fits*, B4a, 1925.

COLE, C. B., *Tool Making*, American Technical Society, 1939.

Dimensional Control, Sheffield Corporation, 1942.

FULLMER, IRVIN H., "Fundamentals of Mechanical Dimensional Control," *Mechanical Engineering*, Vol. 57, no. 12, 1935.

Gages, Pratt & Whitney.

Gages, Jigs, and Fixtures, International Textbook Company, 1936.

KURTZ, H. F., "Optical Projection," *Mechanical Engineering*, Vol. 60, no. 6, 1938.

Measuring Instruments, International Textbook Company, 1928.

Production Handbook, Ronald Press Company, 1944.

Quality Control, DoALL Service Company, 1945.

Taft-Peirce Handbook, Taft-Peirce Manufacturing Company.

CHAPTER 12

LATHES AND LATHE TOOLS

The lathe is a machine which removes material by rotating the work against a cutter. Parts to be machined can be held between centers, attached to a face plate, supported in a jaw chuck, or held in a draw-in chuck or collet. Though this machine is particularly adapted to cylindrical work, it may also be used for many other purposes. Plain surfaces can be obtained by supporting the work on a face plate or in a chuck. Work held in this manner can likewise be centered, drilled, bored, or reamed. In addition, the lathe can be used for cutting threads and turning tapers; with the proper attachment, it can be adapted to simple milling or grinding operations. It is probably the oldest of all the machine tools as well as the most important machine in modern production.

A suitable classification of lathes is difficult to make, as there are so many variables in the size, design, method of drive, arrangement of gears, and purpose. In general, the following classification covers most of the lathes used today. The turret and automatic lathes listed in this classification are discussed in Chapter 18.

CLASSIFICATION OF LATHES

1. Speed lathe
   - (a) Wood working.
   - (b) Centering.
   - (c) Metal spinning.
   - (d) Polishing, etc.
2. Engine lathe
   - (a) Step-cone pulley drive from line shaft.
   - (b) Step-cone pulley drive from individual motor.
   - (c) Gear-head drive.
   - (d) Variable-speed drive.
3. Bench lathe.
4. Toolroom lathe.
5. Special-purpose lathes
   - (a) Crankshaft.
   - (b) Car wheel.
   - (c) Gap.
   - (d) Multicut, etc.

347

6. Turret lathe.
7. Vertical turret lathe.
8. Vertical automatic multistation lathe.
9. Automatic screw machine.
10. Horizontal multispindle automatic.
11. Automatic lathe.

Speed Lathe

The *speed lathe* shown in Figure 10 of Chapter 2 is the simplest of all lathes. It consists of a bed, headstock, tailstock, and an adjustable slide for supporting the tool. Usually this lathe is driven by a variable-speed motor built into the headstock, although in some cases the drive may be a belt to a step-cone pulley. Because hand tools are used and the cuts are small, these lathes are driven at high speeds. The work is either held between centers or attached to a face plate on the headstock.

This lathe is used principally in the turning of wood for small cabinet work or for patterns, and for the centering of metal cylinders prior to further work on the engine lathe. In the latter operation, the center drill is held in a small chuck fastened to the headstock, and the work is guided to the center drill either by a fixed center rest or a movable center in the tailstock. Metal spinning is done on lathes of this type by rapidly revolving a stamped or deep-drawn piece of thin ductile metal and pressing it against a form by means of blunt hand tools. In all these applications, the work is revolved at high speeds, and hand tools are used.

Engine Lathe

The *engine lathe* derives its name from the early lathes which obtained their power from engines. This lathe differs from a speed lathe in that it has additional features for controlling the spindle speed and for supporting and controlling the feed of the fixed cutting tool. There are several variations in the design of the headstock through which the power is supplied to the machine. Lathes which receive their power from an overhead line shaft are belt-driven and equipped with a step-cone pulley, usually a four-step pulley. This gives a range of four spindle speeds driven directly from the line shaft. In addition, these lathes are equipped with back gears which, when connected with the cone pulley, provide four additional speeds. These speeds, which are slower, are used in cutting hard materials and parts of large diameter. This type of lathe requires an overhead

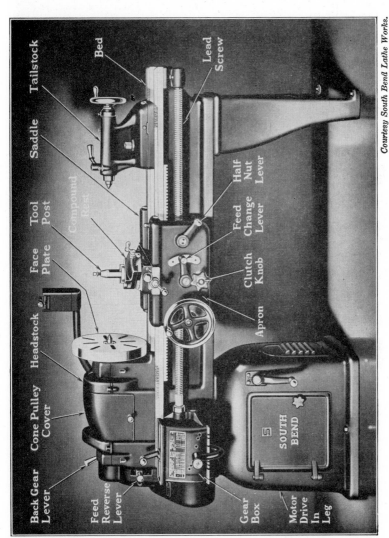

Courtesy South Bend Lathe Works.

FIG. 1.   Engine Lathe.

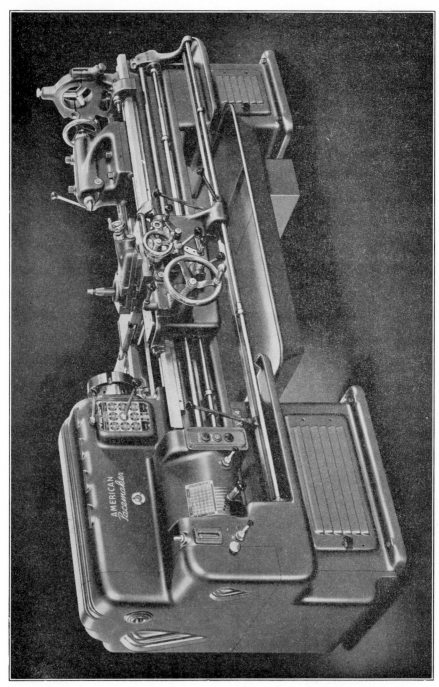

Courtesy American Tool Works.

countershaft carrying a cone pulley which matches the one on the lathe, plus two additional pulleys equipped with clutches to cause forward and reverse rotation of the work.

Another type of engine lathe receives its power from an individual motor mounted either on the side or beneath the lathe. In this case the general design of the headstock is the same, but the power is supplied through a short belt from the motor or from a small cone-pulley countershaft driven by the motor. This machine has the advantage of an individual motor drive. A machine of this type with the principal parts labeled is shown in Figure 1.

A *geared-head lathe* (see Figure 2) varies the spindle speeds by means of a gear transmission. Various speeds are obtained by setting certain speed levers in the head. Such lathes are usually driven by a constant-speed motor mounted on the lathe, but in a few cases variable-speed motors are used. This type of lathe has the advantage of a positive drive and has a greater number of spindle speeds available than are usually found on a step-cone-driven lathe. A few geared-head lathes are equipped with a single pulley and driven from an overhead line shaft. Recently some lathes have been developed with a variable-speed mechanism built into the headstock, thus permitting close regulation of the spindle rotation.

Bench Lathe

The name *bench lathe* is given to small lathes which are mounted on a work bench. In design they have the same features found in ordinary speed or engine lathes, and differ from these lathes only in their size and mounting. They are adapted to small work having a maximum swing capacity of 9 inches at the face plate. The spindle speeds are increased so as to permit the efficient machining of small parts. Many lathes of this type are used for precision work on small parts: jeweler's lathes, for example, are included in this group. A bench lathe adapted for precision work is shown in Figure 3.

Toolroom Lathe

The *toolroom lathe*, the most modern type of engine lathe, is equipped with all the accessories necessary for accurate tool work. Such lathes (see Figure 4) are usually individually driven geared-head lathes with a considerable range in spindle speeds. They are equipped with center steady rest, quick change gears, lead screw, feed rod, taper attachment, thread dial, chuck, indicator, draw-in collet attachment, pump for coolant, and frequently a relieving attachment

Courtesy Rivett Lathe and Grinders, Inc.

FIG. 3. Cabinet-Mounted Bench Lathe.

Courtesy Lodge and Shipley Machine Tool Company.

FIG. 4. A 14-Inch Selective-Head Toolroom Lathe.

to control the motion of the tool. In some cases a " live " or anti-friction center is used in the tailstock which eliminates any scoring of the center, preserving its accuracy. All toolroom lathes are carefully tested for accuracy and, as the name implies, are especially adapted for making small tools, test gages, dies, and other precision parts.

Lathe Size

The size of a lathe is determined by the diameter of the work it will swing and the length of the lathe bed. Thus a 9-inch-by-4-foot lathe is one having sufficient clearance over the bed to take work 9 inches in diameter and having a lathe bed 4 feet long. It should be noted that the clearance over the tool rest is somewhat less than the swing over the bed, so that work supported between centers must be less in diameter than the capacity indicates. Also the length of the work that may be turned is less than the bed length by an amount equal to the space taken up by the headstock and the tailstock.

Lathe Construction

In studying the construction of a lathe, reference to Figure 1 will be of assistance, as all the principal parts are labeled. All lathes receive their power through the *headstock*, which may be equipped either with a step-cone-pulley drive or a geared-head drive. Figure 5 shows a typical back-geared headstock equipped for a belt drive through a step-cone pulley. Four spindle speeds are available when the pulley is directly conected to the lathe spindle. If slower speeds are desired, the back gears are thrown in mesh with the two gears on either end of the cone pulley, and the large gear to the right of the cone pulley is disengaged. Power then is transmitted through the train of four gears to the main spindle, providing four additional speeds.

Figure 2 shows the construction of a typical geared-head lathe where spindle-speed variations are obtained by regulating the levers on the side of the transmission.

The lathe *tailstock*, shown in Figure 6, can be adjusted along the bed of the lathe to accommodate different lengths of stock being turned. It is provided with a hardened center which may be moved in and out by the wheel adjustment, and is also equipped with set-over screws at its base to be used for adjusting the alignment of the centers and for taper turning.

The *lead screw* is a long carefully threaded shaft, located slightly

FIG. 5.   Back-Geared Headstock with Gear Guards Removed.

FIG. 6.   Lathe Tailstock.

below and parallel to the bed and extending all the way from the headstock to the tailstock. It is geared to the headstock in such a way that it may be reversed, and is fitted to the carriage assembly so that it may either engage or be released from the carriage during cutting operations. Power may be brought to the lead screw by two different methods. Figure 7 shows an end view of a lathe equipped with standard change gears. The top four gears shown are the *spindle gear*, two *reverse gears*, and the *stud gear*. These gears are

FIG. 7. End View of Standard-Change Gear Lathe.

fixed and are not changed to obtain various lead-screw speeds. When it is necessary to change the speed of the lead screw, the driving gear, idlers, and lead-screw gears are removed and the proper gears put in place. Most lathes today do not rely on manual gear changing but are equipped with a quick-change gearbox, located at the left end of the lead screw. (See Figures 1 and 2.) This box contains a number of gears so arranged that the lead screw can be driven at the proper speed to cut any of the standard threads. To change the feed it is necessary only to move the extending levers.

The carriage assembly includes the *compound rest, tool post, saddle,*

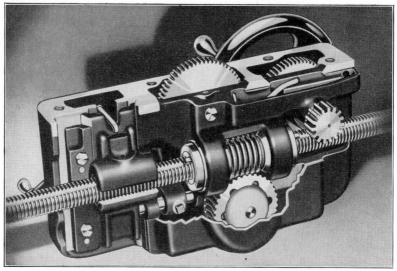

FIG. 8.   Interior View of Double-Wall Apron.

FIG. 9:   Relieving Attachment on Lathe:

and *apron*. Since it supports and guides the cutting tool, it must be constructed with great accuracy and rigidity. Two hand feeds are provided to guide the tool on a crosswise motion. The upper hand wheel controls the motion of the compound rest, and, as this rest is provided with a swiveling adjustment protractor, it can be placed in various angle positions for short taper turning. A third hand wheel is used to move the carriage along the bed, usually to pull it back to starting position after the lead screw has carried it along the cut. The portion of the carriage which extends in front of the lathe is called an apron. On the face of the apron are mounted the various control wheels and levers. Figure 8 shows an interior view of an apron and the method of drive.

Lathes which are designed for toolroom use are provided with a *relieving attachment*. The function of this attachment is to relieve or back off the flutes of rotary cutters, taps, reamers, end mills, dies, and such. Figure 9 shows a lathe setup for relieving a hob cutter. Power for the attachment is taken from the headstock end of the lathe and transmitted through a shaft, provided with universal joints, to a cam located in front of the tool slide. The rotation of the shaft is synchronized with the spindle rotation and controls the movement of the cam. The cam in turn moves the tool slide back and forth and causes the tool to give proper relief to the cutter teeth.

Lathe Operation

The most common way to support work on a lathe is to mount it between centers, as shown in Figure 10. This method has the advantage of being able to resist heavy cuts and is convenient for long parts. Since the work is mounted between two tapered centers, it will not turn uniformly with the spindle unless it is attached in some further fashion. Such attachment is made through a pear-shaped forging known as a *dog*, which consists of a main body with an opening to accommodate the stock being turned, a setscrew at the lower end to fasten the work securely, and an elongated portion at the top (known as the tail) which is bent at a right angle — parallel with the stock — so that it may engage a slot in the face plate. After mounting, the tail of the dog is fitted into the face plate, the setscrew is tightened, and the work is ready to be turned, as shown in Figure 10. The center in the headstock turns with the work; consequently, no lubrication of that center is necessary. The tailstock center, or dead center, acts as a conical bearing and for this reason must be kept clean and well lubricated. It should not be too tight against the work, nor should it be so loose that there is end play. A check should be made at frequent intervals, for if it is too tight the oil film will be broken down, and either the work or the center will be ruined.

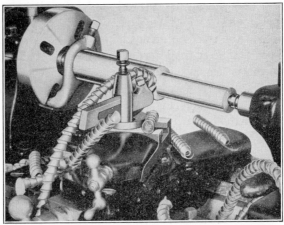

Courtesy South Bend Lathe Works.

Fig. 10. Turning a Steel Shaft Mounted between Centers.

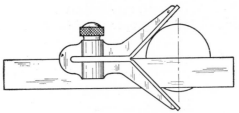

USE OF CENTER HEAD TO LOCATE CENTERS

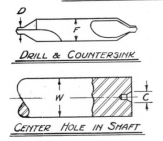

DRILL & COUNTERSINK

CENTER HOLE IN SHAFT

SIZE OF CENTER HOLES			
DIA. OF WORK W	DIA. OF HOLE C	DIA. OF DRILL D	DIA. OF BODY F
$\frac{3}{16}$″ TO $\frac{5}{16}$″	$\frac{1}{8}$″	$\frac{1}{16}$″	$\frac{13}{64}$″
$\frac{3}{8}$″ TO 1″	$\frac{3}{16}$″	$\frac{3}{32}$″	$\frac{3}{10}$″
$1\frac{1}{4}$″ TO 2″	$\frac{1}{4}$″	$\frac{1}{8}$″	$\frac{3}{10}$″
$2\frac{1}{4}$″ TO 4″	$\frac{5}{16}$″	$\frac{5}{32}$″	$\frac{7}{16}$″

Fig. 11. Size of Center Holes and Method of Locating.

In preparing cylindrical work, care must be exercised in locating the center holes before they are drilled. Probably the most convenient method is with a combination square and scriber. The center head of the combination square should be held firmly against the shaft, as shown in Figure 11, and two intersecting lines scribed close to the blade. The intersection of the lines represents the center of the shaft, and this point should be center-punched to facilitate proper starting of the center drill. The sizes of center holes for various-size shafts are given in the table at the lower part of the illustration.

In turning long slender shafts, or boring and threading the ends of spindles, a *center rest* is used to give additional support to the work.

Courtesy Lodge and Shipley Machine Tool Company.

Fig. 12. Lathe Center Rest.

A standard center steady rest is shown in Figure 12. It is attached to the bed of the lathe and supports the work by means of the three jaws shown. Another rest, somewhat similar, is known as a follower rest. It is attached to the saddle of the carriage and supports work of small diameter that is likely to spring away from the cutting tool. This rest moves with the tool, whereas the center rest is stationary.

Cylindrical work that has been bored and reamed to size may be pressed on a steel *mandrel* and supported between centers for further machining, as shown in Figure 9. Lathe mandrels have hardened, ground surfaces and are are available in all standard sizes. The surface is ground with a taper of about 0.0006 inch per inch in length, the small end being 0.0005 inch undersize to facilitate starting the work. The work should be pressed on the mandrel in an arbor press, as considerable pressure is needed in this operation. After the work is mounted on the mandrel, it is placed between the centers, and any further machining done is similar to other cylindrical turning.

Supporting Work in Chuck or on Face Plate

In addition to being held between centers, the work can also be held by being bolted to the *face plate*, by a jaw *chuck*, or by a draw-in *collet*. Figure 13 illustrates work supported by being bolted to a

Courtesy South Bend Lathe Works.

Fig. 13. Boring an Eccentric Hole on the Face Plate of a Lathe.

face plate. Such mounting is suitable for flat plates and parts of irregular shape. The figure illustrates the use of a boring bar for internal turning.

Lathe chucks are made in several designs, as shown in Figures 14, 15, and 16 and may be classified as follows:

1. *Universal chuck.* The jaws all maintain a concentric relationship when the chuck wrench is turned.
2. *Independent chuck.* Each jaw has an independent adjustment.
3. *Combination chuck.* Each jaw has an independent adjustment and, in addition, has a separate wrench connection which controls all jaws simultaneously.
4. *Drill chuck.* A small universal screw chuck used principally on drill presses, but frequently used on lathes for drilling and centering.
5. *Draw-in collect chuck.* Holds standard-shape bar stock in a centered position. A separate collet is necessary for each work size.

Lathe chucks are used for holding short pieces of stock of irregular shape that cannot be held between centers. For example, in making a small gear blank from solid stock, the piece would first be

Courtesy Warner & Swasey Manufacturing Company.

FIG. 14.   Four-Jaw Independent Chuck.

FIG. 15.   Three-Jaw Geared Scroll Chuck.

mounted in the chuck and one side of it faced true. To produce a hole through the blank, the dead center should be removed from the tailstock and a drill chuck mounted in its place. Successive operations of centering, drilling, boring, and reaming can then be performed. The blank is then removed from the chuck and mounted on a mandrel for further machining.

Draw-in collet chucks are the most accurate of all types of chucks and are especially adapted for holding bar stock. In mounting a collet chuck attachment to a lathe, the live center is removed and a tapered sleeve bushing put in its place. The proper-sized col-

Courtesy Skinner Chuck Company.

FIG. 16.   Combination Geared Scroll Chuck.

let is placed in this sleeve and screwed to the draw bar extending through the spindle, as shown in Figure 17. Work can then be placed in the collet and held by turning the hand wheel on the end of the draw bar. This forces the collet back against the tapered surface of the sleeve and causes the collet jaws to grip the work. Collets are made for round, square, and other shapes, as shown in Figure 18.

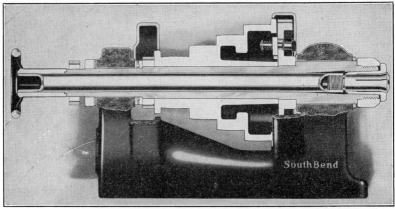

Courtesy South Bend Lathe Works.

FIG. 17.   Cross Section of Headstock Showing the Construction of Draw-in Collet Chuck Attachment.

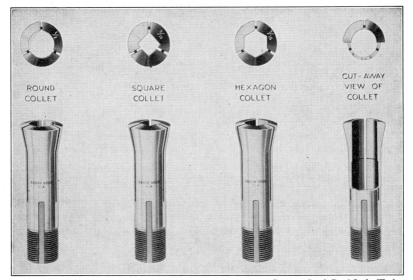

Courtesy South Bend Lathe Works.

FIG. 18.   Side of End View of Various Types of Spring Collets.

This means of holding stock is used in precision work for such parts as small tools, spindles, bushings, and screws. The work held by the collet must be the same size as the collet, or its accuracy will be impaired.

Taper Turning

Many parts and tools made in lathes have tapered surfaces, varying from the short steep tapers found on bevel gears and lathe center ends to the long gradual tapers found on lathe mandrels. The shanks of twist drills, end mills, reamers, arbors, and other tools are examples of taper work. Such tools, supported by taper shanks, are held in true position and are easily removed.

There are several taper standards found in commercial practice. The following classification lists the standards commonly used:

1. *Morse taper.* Largely used for drill shanks, collets, and lathe centers. The taper is approximately 5/8 inch per foot.

2. *Brown & Sharpe taper.* Used principally in milling machine spindles: 1/2 inch per foot.

3. *Jarno and Reed tapers.* Used by some manufacturers of lathe and small drilling equipment. Both systems have a taper of 0.6 inch per foot, but the diameters are different.

4. *Sellers taper.* Used principally in equipment manufactured by William Sellers and Co. The taper is 3/4 inch per foot.

5. *Taper pins.* Used as fasteners. The taper is 1/4 inch per foot.

Each of these standards is made in a variety of diameters and designated by a number. In addition to the taper standards listed, there are many other tapered surfaces found in lathe work which vary considerably from these standards.

External tapers may be cut on a lathe in several ways:

1. By using a taper-turning attachment on the lathe, as illustrated in Figure 19. This attachment is bolted on the back of the lathe and has a guide bar which may be set at the desired angle or taper. As the carriage moves along the lathe bed, a slide over the bar causes the tool to move in or out, according to the setting of the bar. In other words, the taper setting of the bar is duplicated on the work. The advantages of this system are that the lathe centers are kept in alignment, and the same taper may be turned on various pieces, even though they vary in length.

2. By using the compound rest on the lathe carriage. This rest, which has a circular base, may be swiveled to any desired angle with the work. The tool is then fed into the work by hand. This method is especially adapted for short tapers such as truing up a lathe center.

3. By setting over the tailstock center. If the tailstock is moved horizontally out of alignment 1/4 inch, and a cylinder 12 inches long is placed between centers, the taper will be 1/4 inch per foot. However, a cylinder 6 inches long will have a taper of 1/2 inch per foot. Hence the amount of taper obtained on a given piece depends on the length of the stock, as well as on the amount the center is set over.

4. By manually operating both hand feeds.   This method is **not recom**
mended, as it is impossible to cut an accurate taper.

Internal tapers can be machined on a lathe only by using the compound rest or the taper-turning attachment.   Small holes for taper
'pins are first drilled and then reamed to size with a taper reamer.

Courtesy The Hendey Machine Company.

FIG. 19.   Lathe Taper-Turning Attachment.

Lathe Tools

Single-pointed metal-cutting lathe tools consist of small pieces of
rectangular tool steel rigidly supported in suitable holders.   The
tool holder is held on the tool post of the lathe, and the work revolves against point of the tool.

In order to cut metal efficiently and accurately, the cutting tool
must be properly ground to provide a keen cutting edge with correct
angles for the kind of metal to be cut.   The shapes and angles of
the tool vary considerably, depending on the type of cutting operation, kind of material being cut, tool material, and finish desired.

The machinability,* or ease with which a given material can be cut, is influenced greatly by the cutting tool. It must be recognized, however, that machinability is a relative term and is expressed only in such terms as length of tool life, power required to make the cut, and cost of removing a given amount of metal. Other factors influencing the machinability include cutting speed and feed and type of coolant used, a proper selection of which must be made to obtain optimum results for a given material.

Tool materials. Present-day production practices make rather severe demands on machine-tool materials. To accommodate the many conditions imposed upon them, a wide variety of tool materials have been developed. The principal materials used in cutting tools are as follows:

1. HIGH-CARBON STEELS. For many years, prior to the development of high speed and various special alloys, carbon steels were used entirely for all cutting tools. With the carbon content ranging from 0.9 to 1.2%, these steels have great hardening ability and, with proper heat treatment, attain as great a hardness as any of the high-speed alloys. However, because of a tendency to lose hardness at around 400 F, they are not suitable for high speeds and heavy-duty work, and their use is confined to light work in metals and to work in soft materials such as wood.

2. HIGH-SPEED STEELS. High-speed tool steels will retain a keen cutting edge at temperatures up to around 900 F. The first of these steels was developed by Robert Mushet about 1868. Later, in 1900, Frederick W. Taylor and M. White developed the first tool steel that would hold its cutting edge to almost a red heat. This was accomplished by adding 18% tungsten and 5.5% chromium as the principal alloying elements. Present-day practice in the manufacture of high-speed steels still uses these two elements in nearly the same percentages. Other common alloying elements are vanadium, molybdenum, and cobalt. The present high-speed steel containing 18% tungsten, 4% chromium, 1% vanadium, and 0.6% carbon (known as 18-4-1) is considered to be one of the best all-purpose tool steels. Some high-speed steels of similar composition, but having 2 to 3½% vanadium, give slightly better results and are generally used for heavy-duty work. Many high-speed steels use molybdenum as the principal alloying element, as one part will replace two parts of

---

* For discussion of this subject see report of Committee on Machinability of Steel, *ASM Metals Handbook*, 1939.

tungsten. Molybdenum steels, such as 8-4-1 and 8-4-2, are somewhat tougher than the tungsten steels, but not quite so resistant to high temperatures. Many high-speed steels have cobalt added in amounts ranging from 5 to 12%, as this element increases the cutting efficiency, especially at high temperatures.

3. NONFERROUS ALLOYS. Several nonferrous alloys containing principally cobalt, tungsten, and chromium have been developed which give excellent results in metal cutting. Among these are the Stellite alloys developed by the Haynes Stellite Company. These alloys are cast to shape and have the property of "red hardness," which is so important to cutting tools. All are quite tough at a dull red heat, and some will maintain a cutting edge at temperatures up to 1500 F. The principal elements in these alloys are 12 to 19% tungsten, 40 to 48% cobalt, 30 to 35% chromium, with smaller amounts of several other elements. Stellite can be used at cutting speeds ranging from 50 to 100% greater than for high-speed steel.

4. CEMENTED CARBIDES. Carbides of tungsten, tantalum, and titanium have proved to be excellent tool materials. Inserts of these materials are usually made by powder metallurgy: the metal powders are hydraulically pressed to shape, semisintered to facilitate handling and forming to final shape, then sintered at a high temperature, and finally finished with a grinding operation. A typical analysis of tungsten carbide is 5.3% carbon, 12.7% cobalt, and 81.4% tungsten. Tungsten with cobalt as a binder is the usual combination, but in some cases tantalum or titanium is used in combination with the tungsten. Nickel may also be used as a binder. Tantalum carbide has about the same properties as tungsten carbide and, in addition, has a low coefficient of friction which reduces the heat generated by contact with the chips. Since there is little tendency for the steel to adhere to the tool, there is no cratering or top wear. Carbide tools are made by brazing the formed inserts on the ends of commercial steel pieces. Carbides are very hard and will withstand great compressive strength. When they are being mounted, they must be rigidly supported to prevent cracking. In addition to being used for cutting tools, carbides are successful as a die material for wear- and corrosion-resisting purposes.

5. DIAMONDS. Diamonds used as single-point tools for light cuts and high speeds must be rigidly supported because of their high hardness and brittleness. They are used either for hard materials, difficult to cut with other tool materials, or for light high-speed cuts on softer materials where accuracy and surface finish are im-

portant. Diamonds are commonly used in the machining of plastics, hard rubber, pressed carbon, and aluminum with cutting speeds from 1000 to 5000 feet per minute. In addition to being used for cutting tools, diamonds are also used for dressing grind wheels, small wire-drawing dies, and in certain grinding and lapping operations.

Methods of supporting lathe tools. In most cases, the tools used in lathe work are small rectangular pieces of high-speed tool steel

(A)

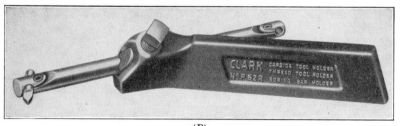

(B)

Both Photos Courtesy Robert H. Clark Company.

FIG. 20. A: Lathe Tool Holder for High-Speed Steel and Carbide Bits. B. Tool Holder Used for Holding Boring Bar.

which are held in a tool holder. Because of the high cost of tool materials, it is much more economical to use these small inserts held in special holders than to use solid forged tools. In Figure 20A is shown an adjustable tool holder capable of accommodating several different sizes of tool bits. The tool bit is securely clamped as near the cutting edge as possible and held by the setscrew and vise jaw with extreme rigidity. Internal lathe cuts can be made by clamping a boring bar in the holder as shown in B of the figure. The tool post on the lathe carriage supports the tool holder in proper position for cutting. A variety of forged tool holders designed to hold the tool in correct position for directional cuts is available.

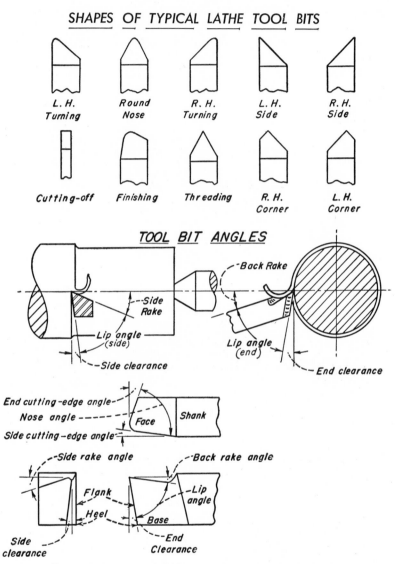

SHAPES OF TYPICAL LATHE TOOL BITS

L. H.
Turning

Round
Nose

R. H.
Turning

L. H.
Side

R. H.
Side

Cutting-off

Finishing

Threading

R. H.
Corner

L. H.
Corner

TOOL BIT ANGLES

Back Rake

Side
Rake

Lip angle
(side)

Side clearance

Lip angle
(end)

End clearance

End cutting-edge angle

Nose angle

Side cutting-edge angle

Face

Shank

Side rake angle

Back rake angle

Flank

Lip
angle

Heel

Base

Side
clearance

End
Clearance

FIG. 21. Recommended Shapes and Angles for Lathe Tools.

Tool shapes and angles. In order to understand the cutting action in metal turning on a lathe, refer to Figure 21. The view to the left center shows a cross section of the tool near its cutting end. The tool has been ground wedge-shaped, the included angle being called the *lip* or *cutting angle*. The *side-clearance angle* between the side of the tool and the work is to prevent the tool from rubbing. The angle is small, usually 6 to 8 degrees for most materials. The top angle, known as the *side-rake angle*, varies with the lip angle, which in turn depends on the type of material being cut. The view to the right center is a side view of the tool with similar angles indicated. If the tool is supported in a horizontal position, the *back-rake angle* is obtained by grinding. However, most tool holders are designed to hold the tool in approximate position for correct back rake. *End clearance* is also necessary to prevent a rubbing action on the flank of the tool. The complete nomenclature of the various parts of a tool is labeled on the finishing tool at the lower part of the figure.

In grinding tools it should be noted that the lip or cutting angle varies with the kind of material being cut. A properly selected cutting angle will have the cutting edge supported well enough to withstand heavy cuts and be capable of carrying away the heat generated, yet keen enough to cut well without requiring too much power. A compromise is necessary, and in general it is based on the hardness of the material. Hard materials require a cutting edge of great strength with a capacity for carrying away heat. Soft materials permit the use of smaller cutting angles, around 22 degrees for wood tools. Soft and ductile metals, such as copper and aluminum, require larger angles, ranging from 37 to 57 degrees, whereas brittle materials, where chips crumble or break easily, require still larger angles. An interesting variation in tool angles is that recommended for brass and duralumin. These materials work best with practically zero rakes, the cutting action being a scraping one. Because of the high ductility, the tool will dig in and tear the metal if a small cutting angle is used. Figure 22 illustrates suggested high-speed steel tool angles for various materials.

Carbide tools require slightly greater cutting angles than those shown in the figure because of the brittleness of the material. *Side-cutting-edge angles* of from 5 to 20 degrees are recommended for these cutters. As the tool starts a cut, the load is not on the end but at a point back of the tip where the tool is stronger. Also, at the end of the cut there is a gradual reduction of the load. With the cutting edge at an angle, the length of the cutting edge is increased and the pressure per unit length decreased. An *end-cutting-edge angle* of from 8 to 15 degrees with a small nose radius is recommended. This shape is a compromise between a point contact, which is apt to break, and a large nose radius which results in a thin chip, excessive tool wear, and chatter.

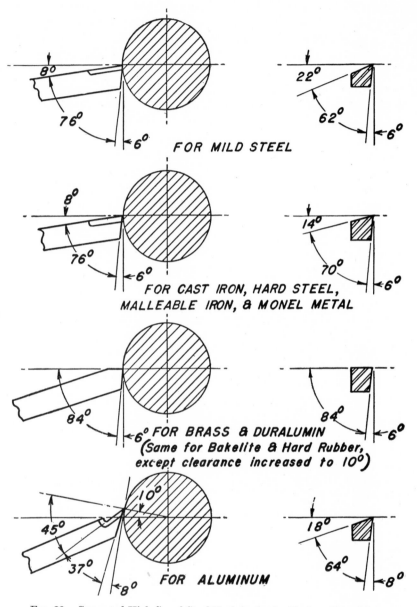

FOR MILD STEEL

FOR CAST IRON, HARD STEEL,
MALLEABLE IRON, & MONEL METAL

FOR BRASS & DURALUMIN
(Same for Bakelite & Hard Rubber,
except clearance increased to 10°)

FOR ALUMINUM

FIG. 22.   Suggested High-Speed Steel-Tool Angles for Various Materials.

Recent experience in the rough turning of artillery shells has demonstrated that carbide-tipped turning tools having a 5-degree negative side and back rake provide longer tool life before sharpening is necessary.* This is particularly true on roughing cuts, where materials are rough, scaly, or slightly eccentric, because with negative rake the surface irregularities strike the cutting edge of the tool back from the tip where the tool is much stronger. Naturally more heat is generated than by conventional turning; so an adequate coolant supply must be provided.

Chip control. In high-speed production turning, the control and disposal of chips is important to protect both the operator and the tools. This is best accomplished by breaking the chip into small lengths, which also facilitates its easy removal from the machine. Various means can be provided to accomplish this end:†

1. Grinding on the face of the tool along the cutting edge a small flat to a depth of about 0.020 inch. This is known as a *chip breaker*, as it stresses the chips to their breaking point as they are formed. The width varies according to the feed and depth of cut used.

2. Mechanical chip breaking: secured by brazing or screwing a thin carbide-faced plate on the face of the tool. As the chip is formed, it hits the edge of the plate and is curled to the extent that it breaks in short pieces.

3. Proper selection of tool angles, which controls the direction of the curled chip and forces the chip into some obstruction which stresses the chip to its breaking point.

Grinding and setting tools. Experienced personnel with adequate grinding equipment are needed to obtain proper cutting angles on tools. Most uniform results are realized with special tool grinders which can be set to grind accurately any angle desired. If off-hand grinding is necessary, gages and templets should be used by the operator.

Tool bits should be removed from the holder before attempting to grind them. The procedure for grinding the various faces is not of great importance, but it is suggested that the side-rake and side-clearance surfaces be prepared first. After these two angles are ground, the tool end may be given any shape desired. Shapes of typical lathe tool bits are shown in the upper part of Figure 21.

Aluminum oxide wheels are best for rough and finish grinding of

* C. Edgar, "Negative Rake-Turning Tools Improve Roughing Cuts in Steel," *American Machinist*, August 16, 1945.

† "Selection and Application of Single-Point Tools," *American Machinist*, November 23, 1944.

high-speed steel tools, whereas silicon carbide and diamond wheels are necessary for cemented carbide tools. A final honing of the cutting edge is recommended to increase tool life.

In most cases it is assumed that the tip of the tool is in line with the center of the work as shown in the various figures; however, some authorities recommend that the cutting edge of the tool point should be about 5 degrees above center, or 3/64 inch per inch in diameter of the work. The position of the cutter bit must be taken into account when the various angles are being ground, as the height has considerable influence on the front clearance. For example, if the point is 1/10 inch above the center lines in turning a 1-inch diameter, the front clearance practically disappears and the back rake is increased materially. All lathes are provided with spherical seats or rockers to assist in setting the tool properly with the work.

Tool life. The life of a tool is an important factor in production work, since considerable time is lost each time a tool is ground and reset. It is well to consider the following reasons for tool failure so that preventive measures can be taken where possible.

1. IMPROPER GRINDING OF TOOL ANGLES. Cutting angles depend on the material to be cut; and their values are well established in handbooks, manufacturers' literature, and other sources.

2. LOSS OF TOOL HARDNESS. This is brought about by excessive heat generated at the cutting edge. This situation is relieved by the use of coolants or by reducing the cutting speed.

3. BREAKING OR SPALLING OF TOOL EDGE. This may be caused by taking too heavy a cut or by having too small a lip angle.

4. NATURAL WEAR AND ABRASION. All tools will gradually become dull by abrasion. In some cases this is accelerated by the development of a crater just back of the cutting edge. As the crater increases in size, the cutting edge becomes weaker and breaks off. This can be avoided by the proper selection of tool material.

5. FRACTURE OF TOOL BY HEAVY LOAD. This condition will be reduced materially if the cutting tools are rigidly supported with a minimum of overhang.

Tool life can be prolonged by careful selection of tool angles, feed, cutting speed, depth of cut, and by the use of a proper coolant when the job is originally set up. Frequently a tool failure is caused by disregard of one or more of these factors. What has been said about lathe tools applies equally well to tools used by other machines.

Coolants. Coolants are various liquids, emulsions, or gases applied to the material being cut to improve the cutting operation. In addition to the fact that they cool the tool and the work, there are several other reasons for their use. Chips are washed away, and the finish of the machined surface is improved. Wear on the cutting edges is reduced and tool life increased because of the lubrication of the tool. Also, less power is required. Finally, coolants reduce possible corrosion on both the work and the machines.

Coolants used should be capable of absorbing and carrying away heat. In addition, they must be free from odor, should not injure the skin of the operator, and should not corrode either the work or the machine. In most cases the coolant is recirculated by a small pump on the machine unless it is used in such small quantities that this arrangement is unnecessary. Best results are generally obtained when a generous supply of cutting fluid is directed on the tool and work.

Various kinds of coolants are used, depending primarily on the kind of material being machined. Typical coolants used for various materials are:

Cast iron. Compressed air or worked dry. The use of compressed air necessitates an exhaust system to remove the dust caused by blowing the fine particles of iron.

Aluminum. Kerosene lubricant or soda water. Soda water consists of water with a small percentage of some alkali which acts as a rust preventive.

Malleable iron. Water-soluble oil lubricant. These coolants consist of a light mineral oil held in suspension by caustic soda, sulphurized oil, soap and other ingredients which forms an emulsion when mixed with water.

Brass. Worked dry or paraffin oil.

Steel. Water-soluble oil, sulphurized oil, or mineral oil.

Wrought iron. Lard oil, or water-soluble oil.

In addition to these listed, may other coolants are used, the selection depending upon the type of work being done. For example, lard oil would be very satisfactory for a tapping operation on steel, whereas, for high-speed cutting of the same steel on a turret lathe, a water-soluble oil emulsion would be better. Grinding operations use a water solution with some alkali such as soda or sodium carbonate to act as a rust preventive, a little oil being added to keep the soda in suspension.

Grinding coolants should not only cool efficiently but also keep the wheel clean. A poor coolant often contributes to wheel loading or glazing, resulting in the ineffective cutting action of abrasive particles. Coolants for grinding should be used in considerable volume and should not contain too much oil or gumming material.

The water-soluble oil emulsions, resembling milk in appearance, are widely used for all forms of operations. They are inexpensive, have high cooling properties, and have low viscosity, which permits the oil to separate readily from the chips. In addition to the coolants mentioned, there is a wide variety of compounded mineral, fixed, and sulphurized oils that are used as cutting fluids.

Cutting Speeds and Feeds

Cutting speed is expressed in feet per minute and on a lathe is the surface speed or rate at which the work passes the cutter. It may be expressed by the simple formula: $CS = \pi DN$, where D is the diameter of the work in feet and N is the revolutions of the work per minute. The cutting speed in this expression is seldom unknown, since cutting

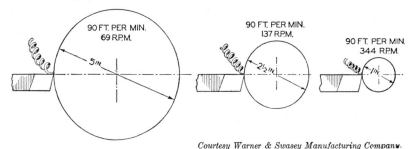

Courtesy Warner & Swasey Manufacturing Company.

Fig. 23.   Relation of Rpm to Surface Speeds on Different Diameters.

speeds for various materials may be found in many textbooks and handbooks. In lathe work the unknown factor is the speed of the work, or the term N. Referring to Figure 23, we may note that, to maintain a recommended cutting speed of 90 feet per minute, it is necessary to increase the work revolution materially, as the diameter is decreased from 5 inches to 1 inch.

The term " feed " is used to express the distance that the tool moves for each revolution of the work, and is sometimes expressed as chip thickness. Since so many factors must be given consideration, it is difficult to state definitely what the speed and feeds for a given material should be. In cutting steel, the color of the chip and the general feel of the machine are important indications, a blue chip indicating too much speed or feed. Vibration of the machine or rapid dulling of the tool also indicates that the tool is overloaded. In general, the speed and feed are determined by the following factors:

1. Kind of material being cut. Materials vary greatly in hardness and other physical properties which affect the cutting speed and life of the tool.

2. Kind of material in the cutter. Carbon-steel cutters can take about one-half the cutting speed of a high-speed tool-steel cutter. Stellite and carbide cutters will stand still greater speeds.

3. Types of finish desired. In general, high speeds with fine feeds give the best finish.

4. Rigidity of the machine. No work should be done at speeds and feeds that cause vibration in the machine.

5. Kind of tool being used. Forming tools, taps, and other tools that are expensive and difficult to sharpen should be operated at speeds and feeds that insure long life.

6. Type of coolant used.

Table **7**, prepared by the Warner & Swasey Company, gives recommended cutting speeds of various materials for turret lathes.

TABLE 7

CUTTING SPEEDS OF VARIOUS MATERIALS

Material	H. S. Steel		Carbide	
	Rough	Finish	Rough	Finish
Cast iron	50– 60	80–110	120–200	350–400
Semisteel*	40– 50	65– 90	140–160	250–300
Malleable iron*	80–110	110–130	250–300	400–300
Steel casting* (0.35C)	45– 60	70– 90	150–180	200–250
Brass (85–5–5)	200–300	200–300	600–1000	600–1000
Bronze (80–10–10)*	110–150	150–180	600	1000
Aluminum†	400	700	800	1000
SAE 1020*	80–100	100–120	300–400	300–400
SAE 1050*	60– 80	100	200	200
Stainless steel*	100–120	100–120	240–300	240–300

* Water-soluble oil lubricant.
† Kerosene lubricant.

REVIEW QUESTIONS

1. Prepare a classification of lathes.

2. What is a speed lathe, and for what type of work is it used?

3. How is the size of a lathe determined?

4. Name three methods of turning accurate tapers on a lathe.

5. What is an engine lathe? A geared-head lathe?

6. What is the purpose of the following lathe parts: Face plate, center rest, compound rest, lead screw, and back gears?

7. How may the direction of feed on a lathe be reversed?

8. What is wrong with a lathe that turns a slightly tapered surface instead of a cylinder? Assume the small diameter to be on the tailstock end.

9. How does a toolroom lathe differ from an ordinary engine lathe?

10. What is a relieving attachment, and for what purpose is it used?

11. Name four types of chucks used in lathe work.

12. Prepare an operation sheet for machining a shaft collar, listing all operations and tools in sequence.

13. How does a Morse taper differ from a Brown & Sharpe taper?

14. Sketch a tool bit, and indicate the lip angle, top rake, and side rake.

15. Define the term " cutting speed." How is it determined on a lathe?

16. Sketch and label six typical lathe tool bits.

17. How does a tool bit for cutting brass differ from one for cutting mild steel?

18. What is a collet, and for what type of work is it used?

19. What desirable properties should a coolant possess?

20. Prepare a classification of cutting fluids.

21. What factors determine the cutting speed and feed to use in lathe work?

22. What is the correct rpm to be used in cutting SAE 1020 steel with a high-speed steel tool? Assume the diameter of the work is 2½ inches.

BIBLIOGRAPHY

Boston, O. W., *Metal Processing*, John Wiley & Sons, 1941.

Class, George M., " Tool Life in Metal Turning," *American Machinist*, December 25, 1940.

Colvin, Fred H., *Running an Engine Lathe*, McGraw-Hill Book Company, 1941.

Edgar, C., " Negative Rake Turning Tools for Roughing Steel," *American Machinist*, August 16, 1945.

How to Operate a Lathe, South Bend Lathe Company, 1940.

Jones, F. D., *Machine Shop Training Course*, Industrial Press, 1940.

Judkins, Malcolm F., " Metal Cutting," *Mechanical Engineering*, May 1937.

Metals Handbook, American Society for Metals, 1939.

Moir, H. L., and Boston, O. W., " A New Study in Cutting Fluid Recommendations," *SAE Journal*, 1940.

Williams, W. J., " How Carbide Tools Reduce the Cost of Machining Steel," *Machinery*, February 1941.

CHAPTER 13

THREADS AND THREAD CUTTING

A *screw thread* is a ridge of uniform section, in the form of a helix, on the surface of a cylinder. The terminology relating to screw threads is clearly illustrated in Figure 1. Screw *sizes* are expressed by the outside or major diameter and the number of threads per

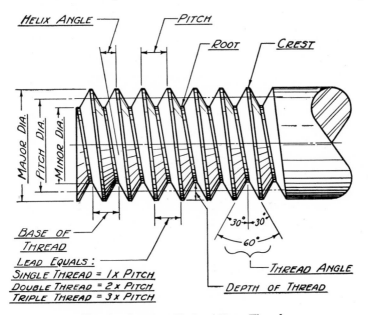

Fig. 1.   American National Form Thread.

inch.   Thus a 1/2-inch 13-thread stud indicates a screw 1/2 inch in diameter and having 13 threads per inch. *Pitch* is expressed by a fraction with 1 as the numerator and the number of threads per inch as the denominator. Thus, a screw having 16 single threads per inch has a pitch of 1/16. It should be kept in mind that only on single-threaded screws does the pitch equal the lead. By definition the *lead* is the amount a screw advances axially in one revolution. Hence on a double-threaded screw the lead is twice the pitch; on a triple-threaded screw the lead is three times the pitch; and so on.

377

Screw threads are used principally on fasteners such as machine bolts, stove bolts, and wood screws. Threads of this nature are simple in design and easy to produce. The usual form is a V, although there are several slight variations of this form.

Another use for screw threads is to transmit power. The mechanical advantage obtained in the ordinary screw jack illustrates this application. Closely associated with this is the use of threads for transmitting motion, such as the lead screw on a lathe.

Finally, there are screw threads used for such measuring devices as micrometers. Screw threads sometimes fulfill several of these uses. For example, the screws controlling the work table of a milling machine may be used either for accurate measuring or for controlling the table movement. The form in which the screw is made is naturally influenced by the function it has to fulfill.

Screw threads have been standardized according to their cross-sectional form. Figure 2 shows the common forms in use and the relationships that exist between the pitch and the principal dimensions. All bolts and similar fasteners have a *V-shaped thread*, as shown in *A* of the figure. There are two standards in the United States which utilize this form of thread, namely, the *National Coarse Screw Thread* and the *National Fine Screw Thread*. The National Fine series differs from the National Coarse series only in having more threads per inch for a given size than the National Coarse series. Such threads have been adopted by the automotive and aeronautical industries, since there is less tendency for them to work loose because of the vibrations they may be subjected to. This type of thread is characterized by a small flat on top and at the root of the thread, which adds to its strength. V-type threads without these flats, shown at *B*, have a greater tendency to fail at the sharp root corners when subjected to loading conditions. The *International Standard Metric thread* shown in *D* is essentially the same as the National Standard except for a smaller flat at the root and a different number of threads per unit distance. From a standpoint of design, the *Whitworth Standard* used in England is perhaps better than any of those already mentioned, as the filleted top and root add strength to the thread by eliminating sharp corners where fatigue cracks may start. This thread is shown at *C* in the figure. The *Acme thread* shown at *F* is principally used for the transmission of power and motion. It has an advantage over other similar screws in that wear may be compensated for by adjusting the half nuts in contact with the screw. This would be impossible with square threads shown at *E*.

Another advantage over the square thread is that these threads may be cut with suitable taps and dies. *Square threads* are more suitable for transmitting power where there is a large thrust on one side of

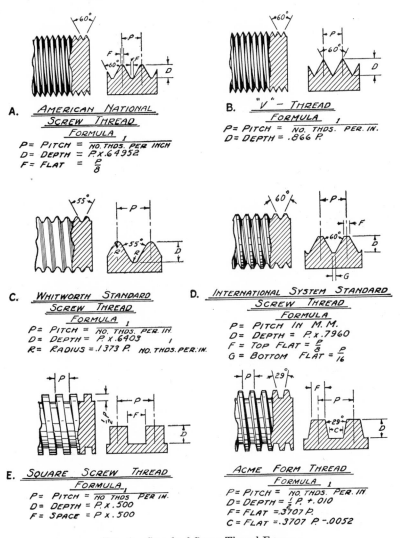

A. AMERICAN NATIONAL SCREW THREAD
FORMULA
P = PITCH = $\frac{1}{NO. THDS. PER INCH}$
D = DEPTH = P x .64952
F = FLAT = $\frac{P}{8}$

B. "V" - THREAD
FORMULA
P = PITCH = $\frac{1}{NO. THDS. PER. IN.}$
D = DEPTH = .866 P.

C. WHITWORTH STANDARD SCREW THREAD
FORMULA
P = PITCH = $\frac{1}{NO. THDS. PER. IN.}$
D = DEPTH = P x .6403
R = RADIUS = .1373 P. $\frac{1}{NO. THDS. PER. IN.}$

D. INTERNATIONAL SYSTEM STANDARD SCREW THREAD
FORMULA
P = PITCH IN M.M.
D = DEPTH = P x .7960
F = TOP FLAT = $\frac{P}{8}$
G = BOTTOM FLAT = $\frac{P}{16}$

E. SQUARE SCREW THREAD
FORMULA
P = PITCH = $\frac{1}{NO. THDS. PER IN.}$
D = DEPTH = P x .500
F = SPACE = P x .500

ACME FORM THREAD
FORMULA
P = PITCH = $\frac{1}{NO. THDS. PER. IN.}$
D = DEPTH = $\frac{1}{2}$ P. + .010
F = FLAT = .3707 P.
C = FLAT = .3707 P - .0052

Fig. 2. Standard Screw Thread Forms.

the thread. These threads, however, cannot be cut with taps and dies and must be machined on a lathe. Another type, similar to the square thread, is known as a *buttress thread*. It has one side that is sloping 45 degrees while the other is perpendicular. The principal application of this thread is for the transmission of power, although it has the disadvantage that the thrust can be in only one direction. *Worm threads* are similar to the Acme Standard except that they have a greater depth. This form of thread is used exclusively for worm-gear drives.

Pipe threads have been standardized according to the American National Standard shown in Figure 3. To insure tight joints, the

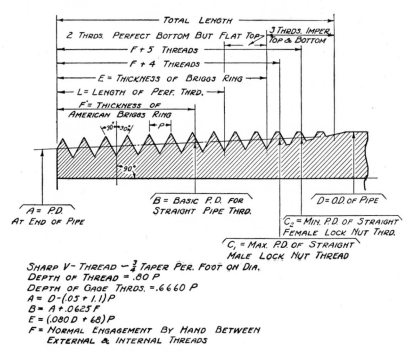

FIG. 3. American (Briggs) Standard Pipe Thread.

thread has a taper of 3/4 inch per foot. The threads have the conventional V shape, except for the last four or five, which have flat tops. These last threads all have imperfect bottoms, as shown in the figure. The usual method of cutting these threads is with suitable taps and dies, although they may also be cut on a lathe using the taper attachment.

Methods of Making Threads

External threads may be produced by the following manufacturing processes:

1. Cutting to shape on an engine lathe.
2. Using die and stock (manual).
3. Automatic die head (turret lathe).
4. Milling machine.
5. Threading machine (plain or automatic).
6. Rolling between dies (flat or circular).
7. Die casting.

Internal threads may be produced by:

1. Cutting to shape on an engine lathe.
2. Using tap and holder.
3. Automatic collapsible tap.
4. Milling machine.

In addition to these processes, threads are sometimes produced by grinding. This method is usually a finishing operation for screws that have been heat-treated or for those requiring a high degree of accuracy. However, complete threads can be produced by this process.

Cutting Threads on a Lathe

The lathe is the most versatile of all machine tools for cutting threads, since on this machine it is possible to cut all forms of threads; however, the lathe is usually selected when only a few threads are to be cut or when special forms are desired. The form of the thread is obtained by grinding the tool to the proper shape. To insure getting the proper shape, a suitable gage or templet should be used. Figure 4 shows a cutter bit ground for cutting 60-degree V threads and the gage which is used for checking the angle of the tool. This gage is known as a *center gage*, as it is also used for gaging lathe centers. Special form cutters as shown in Figure 16, may also be used for cutting these threads. These cutters are previously shaped to the correct form and are sharpened by grinding only on the top face.

In setting up the tool for V threads, there are two methods of feeding the tool. First, the tool may be fed straight into the work and the threads formed by taking a series of light cuts, as shown at *A* in Figure 4. With this method there is cutting action on both sides of the tool bit. The disadvantage of this method is that it is impossible to provide any side rake on the cutting tool, although some back rake may be obtained. On materials such as cast iron or brass, where little or no side rake is recommended, this method is satisfactory. However, in cutting steel threads it is advisable to use a side rake on the tool. This necessitates feeding the tool in at an angle

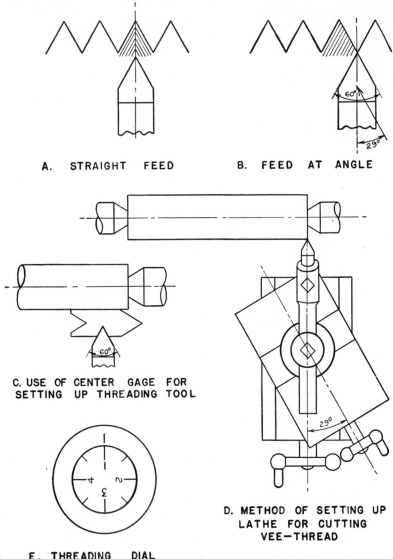

A. STRAIGHT FEED

B. FEED AT ANGLE

C. USE OF CENTER GAGE FOR SETTING UP THREADING TOOL

D. METHOD OF SETTING UP LATHE FOR CUTTING VEE-THREAD

E. THREADING DIAL

FIG. 4. Method of Setting Tool for Thread Cutting on Lathe.

as shown at B and D. To do this, the compound rest is turned to an angle of 29 degrees, and, by using the cross-feed on the compound rest, the tool is fed into the work so that all cutting is done on the left-hand side of the tool. The tool bit, being ground to an angle of 60 degrees, allows 1 degree of the right-hand side of the tool to smooth off that side of the thread.

It is necessary that the tool be given a positive feed along the work at the proper rate to cut the desired number of threads per inch. This is accomplished by a train of gears located on the end of the lathe, which drive the lead screw at the required speed with relation to the headstock spindle. This gearing may be changed to cut any desired pitch of screw. The lead screw, in turn, engages the half nuts on the apron of the lathe, which provides a positive **drive for** the tool.

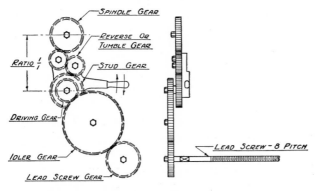

SIMPLE GEARING AT END OF LATHE

FIG. 5.

The older-type standard change-gear lathes require that the gears be changed manually. Referring to Figure 5, one may see the usual arrangement of gears on such lathes. The *spindle gear* drives through the small *reversing gears* to the *stud gear*. As shown in the figure, both the reversing or tumble gears are in mesh. By raising the hand lever, one of the gears is thrown out of mesh, and, by eliminating one gear from the train, the rotation of the lead screw is reversed. The speed ratio from spindle gear to stud gear is 1 to 1, or, in other words, there is no increase or decrease in the rotational speed of the stud gear. Keyed to the stud gear is another gear called the *driving gear*. It connects with the *lead-screw gear* by means of an *idler*. In simple gearing the only two gears to be changed are the driving and lead-screw gears. The correct selection of these gears depends on the number of threads to be cut and the pitch of the lead screw.

Assume, as an example, that it is desired to cut a thread with 13 threads per inch on a lathe having a lead screw with 8 threads per inch. If the ratio from driving gear to lead-screw gear is 1 to 1, the tool will advance 1/8 inch for each revolution of lead screw and will cut 8 threads per inch. In order to cut 13 threads per inch, the lead-screw speed must be slowed down somewhat in relation to the spindle speed. The ratio that must exist in this case is 8 to 13. Hence, a driving gear and a lead-screw gear having this ratio will cut the desired thread. The rule for determining the proper gear ratio is:

$$\frac{\text{Number of threads on lead screw}}{\text{Number of threads to be cut}}$$

In this case the ratio is 8 to 13.

By multiplying both the numerator and denominator by a number, the ratio may be expressed as numbers of teeth in each gear. Hence,

$$\frac{8 \times 3}{13 \times 3} = \frac{24T}{39T} \quad \begin{matrix} \text{(driving gear)} \\ \text{(lead-screw gear)} \end{matrix}$$

If no gears are available with these numbers of teeth, some other multiplier must be used, or, if the ratio is too large, compound gearing is necessary.

All the newer-type lathes are provided with quick-change gearboxes, as shown in Figures 1 and 2 of Chapter 12. No computation is necessary, as the chart on the cover of the gearbox states the correct position of levers needed to obtain the number of threads per inch desired.

After the lathe is set up, the cross-feed screw is set at some mark on the micrometer dial, and a light cut is taken to check the pitch of the thread. At the end of each successive cut, the tool is removed from the thread by turning back the cross-feed screw. This is necessary, as any back play in the lead screw would prevent the tool from returning in its previous cut. The tool is returned to original position, the cross-feed screw set at the same reference mark, the tool is fed the desired amount for the next cut, and another cut taken. These operations are repeated until the thread is cut to a proper depth. To check the work a ring thread gage or a standard nut is used. Figure 6 illustrates the tool setup for thread cutting on a lathe.

Most lathes are equipped with a *thread dial indicator* as shown in Figures 4 and 7. Close by the dial is a lever (shown in Figure 7) which is used to engage and disengage the lead screw with a matching

Courtesy South Bend Lathe Works.

FIG. 6. Thread Cutting on a Lathe.

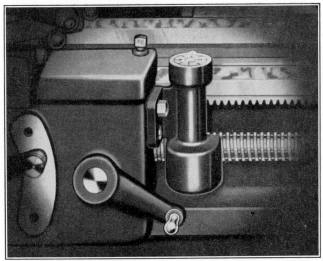

Courtesy South Bend Lathe Works.

FIG. 7. Threading Dial on Lathe.

set of half nuts in the carriage. At the end of each cut, the half nuts are disengaged and then re-engaged at the correct time so that the tool always follows in the same cut. The indicator is connected to the lead screw by means of a small worm gear, and the face of the dial which revolves is numbered to indicate positions at which the half nuts may be engaged. The position at which the half nuts should be closed depends upon the size of thread, as follows:

1. For even-number threads: any line on dial.
2. For odd-number threads: any *numbered* line.
3. For threads involving half threads: any odd-numbered line.
4. For threads involving one-quarter threads: return to original starting point each time.

Taps and Dies

Taps are used principally for the manual production of internal threads, although with proper mounting they may also be used in machine threading. Figure 8 is a graphic illustration of a *tap* with the various parts of the tool labeled. The tool itself is a hardened

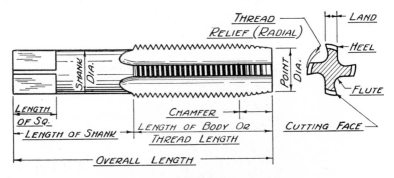

Fig. 8. Tap Nomenclature.

piece of carbon or alloy steel resembling a bolt, with flutes cut along the side to provide the cutting edges. For hand tapping they are furnished in sets of three for each size, as shown in Figure 9. In starting the thread, the *taper tap* should be used, since it insures straighter starting and more gradual cutting action on the threads. If it is a through hole, no other tap need be used. In the case of closed or blind holes where it is desired to have threads to the very bottom, the *taper plug*, and *bottoming taps* should all be used in the order named.

Many other taps are available and are usually named according

to the kind of thread they are to cut. Several of these special taps
are shown in Figure 10. The regular taper *pipe tap* is used for
tapping pipe fittings. The diameter of the tap increases from bottom to top of the threaded
portion at the rate of 3/4
inch per foot. The *interrupted taper pipe tap* shown
is used for tapping tough
metals in which the threads
tear. The removal of alternate teeth also provides a
groove which assists in the
removal of the chips. The
combined tap and drill
shown is designed for drilling and tapping in one operation. This style of tap is
well adapted for tapping
cored holes, as the sizing of

Taper Hand Tap

Plug Hand Tap

Bottoming Hand Tap

Courtesy Greenfield Tap and Die.

FIG. 9. Hand Taps.

the hole prior to tapping requires the removal of a certain amount
of material which can be taken care of with the drill section of the
tap. It is also used for tapping into water and gas mains that are
under pressure. Other styles are known as machine-screw, stove-

Regular Taper Pipe Tap

Interrupted Thread Taper Pipe Tap

Combination Drill and Tap

Courtesy Greenfield Tap and Die.

FIG. 10. Pipe Taps.

bolt, nut-bent-shank, pulley, stay-bolt, acme, boiler, and mud or washout taps.

In all cases, where a hole is to be tapped, the hole that is drilled prior to the tapping operation must be of such size as to provide the necessary metal for the threads. Such a hole is said to be a " *tap-size* " hole. Referring to Table **8**, note that if a 3/8-inch 16-thread hole is to be cut, the tap-size drill is listed as being 5/16 inch in diameter. This is equivalent to the root diameter of a 3/8-inch **16** standard screw and allows sufficient metal in the hole for the threads.

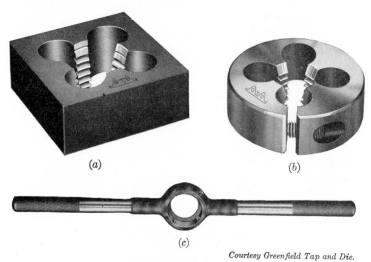

(a)   (b)

(c)

Courtesy Greenfield Tap and Die.

Fig. 11.   (a) Square Die, (b) Round Die, and (c) Stock.

In order to cut external threads, *dies* similar to those shown in Figure 11 are used. The most common type is the adjustable die, as it can be made to cut slightly undersize or oversize. When used for hand cutting, the die is held in a die stock which provides the necessary leverage to turn the die in making a cut.

For successful operation of either taps or dies, some consideration must be given to the nature of the material to be threaded. The shape and angle of the cutting face influence the performance, since no tool can be made to work successfully for all materials. Another important factor is proper lubrication of the tool during the cutting operation. This insures longer life of the cutting edges and results in smoother threads. Since no one lubricant can be recommended for all cases, it is advisable to consult specialists or handbooks in making a selection.

TABLE 8

Tables of Standard Screw–Thread Pitches and Recommended
Tap–Drill Sizes

American National Coarse-Thread Series Standard Thread (NC) *Formerly US Standard*					American National Fine-Thread Series Standard Thread (NF) *Formerly SAE Thread*				
No. or Diam.	Thr'ds Per Inch	Outside Diam. of Screw	Tap Drill Sizes	Decimal Equivalent of Drill	No. or Diam.	Thr'ds Per Inch	Outside Diam. of Screw	Tap Drill Sizes	Decimal Equivalent of Drill
1	64	0.073	53	0.0595	0	80	0.060	$\frac{3}{64}$	0.0469
2	56	0.086	50	0.0700	1	72	0.073	53	0.0595
3	48	0.099	47	0.0785	2	64	0.086	50	0.0700
					3	56	0.099	45	0.0820
4	40	0.112	43	0.0890					
5	40	0.125	38	0.1015	4	48	0.112	42	0.0935
6	32	0.138	36	0.1065	5	44	0.125	36	0.1040
					6	40	0.138	33	1.1130
8	32	0.164	29	0.1360					
10	24	0.190	25	0.1495	8	36	0.164	29	0.1360
12	24	0.216	16	0.1770	10	32	0.190	21	0.1590
					12	28	0.216	14	0.1820
$\frac{1}{4}$	20	0.250	7	0.2010					
$\frac{5}{16}$	18	0.3125	F	0.2570	$\frac{1}{4}$	28	0.250	3	0.2130
$\frac{3}{8}$	16	0.375	$\frac{5}{16}$	0.3125	$\frac{5}{16}$	24	0.3125	I	0.2720
					$\frac{3}{8}$	24	0.375	Q	0.3320
$\frac{7}{16}$	14	0.4375	U	0.3680					
$\frac{1}{2}$	13	0.500	$\frac{27}{64}$	0.4219	$\frac{7}{16}$	20	0.4375	$\frac{25}{64}$	0.3906
$\frac{9}{16}$	12	0.5625	$\frac{31}{64}$	0.4843	$\frac{1}{2}$	20	0.500	$\frac{29}{64}$	0.4531
					$\frac{9}{16}$	18	0.5625	0.5062	0.5062
$\frac{5}{8}$	11	0.625	$\frac{17}{32}$	0.5312	$\frac{5}{8}$	18	0.625	0.5687	0.5687
$\frac{3}{4}$	10	0.750	$\frac{21}{32}$	0.6562	$\frac{3}{4}$	16	0.750	$\frac{11}{16}$	0.6875
$\frac{7}{8}$	9	0.875	$\frac{49}{64}$	0.7656	$\frac{7}{8}$	14	0.875	0.8020	0.8020
1	8	1.000	$\frac{7}{8}$	0.875	1	14	1.000	0.9274	0.9274
$1\frac{1}{8}$	7	1.125	$\frac{63}{64}$	0.9843	$1\frac{1}{8}$	12	1.125	$1\frac{3}{64}$	1.0468
$1\frac{1}{4}$	7	1.250	$1\frac{7}{64}$	1.1093	$1\frac{1}{4}$	12	1.250	$1\frac{11}{64}$	1.1718

Both the taps and dies described may also be used in the machine
cutting of threads. Because of the nature of the cutting operation,
they must be held in a special holder so designed that the tap or die
can be withdrawn from the work without injury to the threads. This
is frequently accomplished by reversing the rotation of the tool or
work after the cut has been made. Numerous tapping attachments
are available for internal thread cutting on a drill press. These

attachments are usually provided with two spindles which operate in opposite directions. The tap is rotated into the work at the proper cutting speed until the threads are made. As soon as the tap is

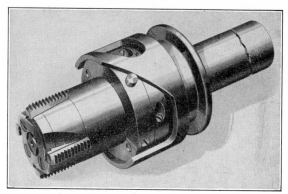

Courtesy The National Acme Company.

Fig. 12.   Revolving Tap — Collapsing Type.

Courtesy Landis Machine Company, Inc.

Fig. 13.   Receding Circular Chaser Collapsible Tap.

raised upward, the other spindle is engaged by means of a ball or friction clutch, and the rotation of the tap is reversed, thus removing it from the work. The withdrawing speed is usually much faster than the cutting speed. The same procedure is used in cutting external threads with nonopening dies.

In small production work on a turret lathe, the tap is held by a special holder, which prevents the tap from turning as the threads are cut. Near the end of the cut the turret holding the tool is stopped, and the tap holder continues to advance until it pulls away from a stop pin a sufficient amount to allow the tap to rotate with the work. The rotation of the work is then reversed, and, when the tap holder is withdrawn, it is again engaged with the stop and held until the work is rotated from the tap. External threads may also be cut with a die utilizing this same procedure, although in most cases such threads are cut with self-opening dies.

For large-diameter internal threads it is common practice to use a *collapsing tap*. In Figure 12 is shown such a tap, widely used on automatic screw machines, drill presses, and other machines which have facilities for rotating the threading tool. This tap revolves while the threads are cut. Collapsing is automatic when the proper length of thread has been cut, and resetting is by means of an outside yoke. Similar taps are devised for stationary spindles. The collapsing in this type is also automatic, but the resetting is by means of a handle which must be moved by the operator.

In Figure 13 is shown a receding circular chaser collapsible tap which was specially designed for production tapping of line pipe, casing, and drill-pipe couplings in the tube mills. It produces a high-quality thread well within all API tolerance specifications on seamless, welded steel, or wrought-iron pipe. The head of the tap is a self-contained unit that is detachable from the body, which permits using heads covering a thread range of 4 to 12 inches. Any taper thread from 0 to 3/4 inch per foot can be obtained by a cam adjustment. For manual operation, the tap is expanded by a short movement of the crank handle, to bring the tripping ring into contact with the face of the coupling, thus bringing into operation the receding mechanism as the tap is fed into the coupling. At the end of the thread the tap is collapsed, which permits it to be removed without interference with the threads.

Two types of *automatic die heads* are shown in Figures 14 and 15. In one type the cutters or chasers are mounted tangentially; in the other, they are in a radial position. The die head shown in Figure 14 is for use on machines having revolving spindles. Those commonly used on turret lathes and screw machines are of the stationary type (see Figure 15), requiring the work to rotate, but in general appearance there is not much difference. In all cases the chasers open automatically at the end of the cut and may be withdrawn from the

Courtesy The National Acme Company.

FIG. 14.   Revolving Die Head.

Courtesy Warner & Swasey Manufacturing Company.

FIG. 15.   Stationary Tangent Die Head.

work without damage to the threads. Dies of this type may be adjusted to various diameters within practical limits; and, in addition, they have micrometer adjustments to control the thread size.

Courtesy The Bodine Corporation.

FIG. 16. Four-Spindle Automatic Nut-Tapping Machine.

Tapping Machines

Production tapping machines for threading nuts and similar parts are made in several types, depending on the nature of the product and number to be threaded.   A fully automatic nut-tapping machine equipped with four spindles is shown in Figure 16.   Nut blanks are automatically fed to working plates from two motor-driven oscillating hoppers.   Each hopper feeds two blanks to working position through adjustable chutes.   The blanks are then clamped in working position, and the four tapping spindles are fed to the work by individual lead screws.   When the operation is finished, the spindles are reversed at approximately double the tapping speed, and the nuts are discharged to individual containers.   The capacity of the machine depends on such factors as the tap size, nut thickness, type of material; however, for small-size nuts a production close to 9000 per hour can be attained.

Another type of threading machine is equipped with a dial feed arrangement and is provided with several vertical spindles for performing successive operations.   Drilling, milling, tapping, and screw inserting can all be done if desired, as the part is indexed from one station to another.   These machines are semiautomatic; the operator has only to place the work in position.

A common type of tapping machine, usually a multispindle arrangement, is provided with taps having extra long shanks.   The tap is advanced through the nut by the lead screw and, upon completion of the threading, continues downward until the nut is released.   The spindle then returns to its upper position with the tapped nut on its shank.   When the shank has been filled with nuts, the tap is removed and the nuts emptied into a container.

Threading Machines

In Figure 17 is shown a double-spindle threading machine of $1\frac{1}{2}$-inch capacity.   Tangential threading tools provided with eight rotational speeds are used.   Parts to be threaded are held by vises mounted on the carriage, clearly illustrated in the figure.   Of the two hand wheels shown, one controls the double-acting vises while the other regulates the movement or positioning of the carriage.   The lever at the rear locks the carriage to the lead screw located beneath and feeds the work to the revolving die.   For absolute accuracy in pitch of threads, the use of a lead screw for feeding the carriage is recommended, especially for threads of large diameter.   Very often no mechanical positive feed is used for small threads, as the lead on the threading die feeds the work at the proper rate.   With proper adjustment of the rod stop at the side of the carriage, the die opens automatically when the required length of thread is cut.

Machines of this type may also be used for tapping nuts by using a tap chuck in the die head and gripping the nut in the vise. A long-shank tap is used for this work and, as the nuts are tapped, they feed back onto the shank of the tap.

Threading machines are made with as many as six dies for use on large production jobs. The size of the machine is generally controlled by the number of heads that can be attended by one operator. For small-lot or jobbing work, each head can be set up for a different-sized thread, and thus frequent die changes are eliminated. Other machines have a tandem construction, permitting the threading of both ends of a pipe or conduit simultaneously.

Courtesy The Hill Acme Company, Acme Machinery Division.

Fig. 17.   1½-Inch Double-Spindle Threading Machine.

Thread Milling

Accurate threads of large size, both external and internal, can be cut with standard or hob-type cutters. For long external threads a threading machine is used which is similar in appearance to a lathe. Work is mounted either in a chuck or between centers, and the milling attachment is at the rear of the machine. In cutting a long screw a single cutter is mounted in the plane of the thread angle and fed parallel to the axis of the threaded part.

Courtesy Murchey Machine & Tool Company.

Fig. 18. (Left) Enlarged View of Cutter Head for Outside Threads. (Below) Thread-Milling Machine Shown with Hob for Internal Threads.

A thread-milling machine intended for mass production of short internal or external threads, 1 inch in diameter up to 4 inches inclusive 3 inches long, and up to 1/2-inch pitch, is shown in Figure 18. Annular milling cutters covering the full length of cut are used on this machine. The figure shows cutter head setups for both internal and external threads. In operation, a work-holding fixture is mounted on the machine table which holds the work rigidly. The milling head carrying the hob is revolved eccentrically about the work and is simultaneously rotated on its own axis, advancing by means of a lead screw for a sufficient distance to produce the desired thread. The cutter spindle, after completing the milling operation, automatically returns to center position. A reversing switch is then contacted, and the sleeves are brought back to the original starting position. The depth of the thread is controlled by adjustment in the eccentricity of the spindle sleeve. After proper adjustment the entire cycle of operation is automatic.

Rolling Screw Threads

A large proportion of the standard bolts and screws manufactured have their threads formed by being rolled between suitable dies. In this process both cold- and hot-working methods are employed. Cold rolling of threads up to 1 inch in diameter has proved successful. In this process the metal on the cylinder is cold-forged under considerable pressure into the desired shape; and, as a result of this cold working of the metal, such threads have greater strength than cut threads. Slightly less material is required for bolts made in this fashion, as the outer parts of the threads are forced into the die. The diameter of the stock used should be approximately equal to the pitch diameter of the screw. The process of forming threads on heated blanks is similar to that just described.

There are two different methods employed in the rolling of threads. In one case the bolt is rolled between two flat dies, each being provided with parallel grooves cut the size and shape of the thread. One die is held stationary while the other reciprocates and rolls the blank between the dies. This operation is shown diagrammatically in Figure 19. The process may be illustrated by rolling a screw between two soft boards under pressure. On examination of the boards, we note that each has impressed into its surface a series of angular, parallel lines. By reversing this illustration and starting with similar grooves in hardened steel, threads will be rolled into a piece of soft

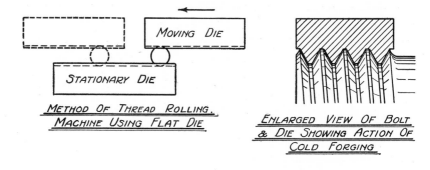

METHOD OF THREAD ROLLING.
MACHINE USING FLAT DIE

ENLARGED VIEW OF BOLT
& DIE SHOWING ACTION OF
COLD FORGING

FACE OF FLAT DIE FOR
ROLLING RIGHT HAND THREADS

Fig. 19. Sketch Illustrating the Principle of Rolling Threads with Flat Dies.

Courtesy The Hill Acme Company, Acme Machinery Division.

Fig. 20. Acme Roll Threading Machine, 1½-Inch Die Head.

steel rod placed between them. The other process employs three grooved rollers held in a radial position with reference to the stock. In appearance the die used resembles an ordinary die head with circular chasers. As the stock is fed between these rolling dies under pressure, the metal is forced into the grooves of the dies, thus forming the threads. A machine of this type is shown in Figure 20. Both these processes will produce threads at a faster rate but with less accuracy than ordinary cutting methods.

Thread Grinding

Grinding is used either as a finishing or a forming operation on many screw threads where accuracy and smooth finish are required. This process is particularly applicable for threads which have been heat-treated to eliminate possible errors resulting from the treatment. A production machine developed for the grinding of screw threads is shown in Figure 21. Provision is made on this machine for grinding both right-hand and left-hand threads in pitches ranging from 2 to 48, and both single- and multiple-threaded screws can be ground. Great accuracy must be built into grinding machines of this type, and, as a result, it is possible to hold the thread angle to plus or minus 5 minutes on the half angle, and to hold lead plus or minus 0.0002 inch per inch with accumulative error not to exceed 0.001 inch in 12 inches. The form of the grinding wheel is kept true by an automatic diamond truing device. The movement of the diamonds is controlled either from template formers or through a pantograph mechanism. Although the truing action is automatic, it must be predetermined by the operator.

The entire operating cycle of the machine is automatic, no matter how many cuts are used from the time the grinding wheel starts until the size is reached. The machine stops automatically at completion of the work cycle. As a result, on jobs where production is 40 pieces per hour or less on each machine, one operator can run two machines or perform other useful functions while tending a single machine. Threads may be ground from the solid by first rough grinding and then finish grinding. Whether on not this is desirable depends on size, length, and type of threads and should be specially considered for each job. Slender work of this type is supported by a roller-back rest to prevent springing of the work under the cut. Ground threads are preferred for many purposes, and, with the improved equipment available, they are being used more widely.

Fig. 21. Thread-Grinding Machine.

REVIEW QUESTIONS

1. Name and sketch five standard screw threads. Indicate angle on each thread.

2. To what various uses may screw threads be put?

3. Distinguish between lead and pitch on a screw thread.

4. What advantages does an Acme thread have over a square thread?

5. List the various methods by which external threads can be made.

6. Show by sketches how to set up a lathe for cutting V threads.

7. How do the American National Pipe threads differ from the National Coarse threads?

8. What methods can be used for cutting internal threads?

9. How are threads cut on a turret lathe?

10. Sketch the gears on the end of an engine lathe and, assuming the pitch of the lead screw to be 8, indicate the number of teeth in the driving and lead-screw gears necessary to cut 11 threads per inch.

11. What is a thread dial indicator, and how does it work?

12. What is a tap? Name six kinds.

13. What methods are employed in rolling threads?

14. How would you cut an internal square thread?

15. Sketch and label the gears on the end of an engine lathe.

16. How are threads cut on a turret lathe?

17. What type of threading equipment should be used for cutting internal threads on a drill press?

BIBLIOGRAPHY

FLANDERS, RALPH E., "American Thread Grinding Practice," *Machinery*, September 1939.

Machinery Handbook, Industrial Press, 1941.

PETERKA, A. E., *Bolts, Nuts, and Screws*, Lamson & Sessions Company, 1941.

SMITH, A. M., "Screw Threads — The Effect of Method of Manufacture on the Fatigue Strength," *Iron Age*, August 22, 1940.

CHAPTER 14

SHAPERS AND PLANERS

A *shaper* is a machine having a reciprocating cutting tool, of the lathe type, which takes a straight-line cut. By moving the work across the path of this tool, a plane surface is generated, regardless of the shape of the tool. This method of producing a flat surface has the advantage that its perfection is not dependent on the accuracy of the tool, as is the case in using a milling cutter for the same type of work. By the use of special tools, attachments, and devices for holding the work, a shaper can also be used for cutting external and internal keyways, spiral grooves, gear racks, dovetails, T slots, and other miscellaneous shapes.

Figure 1 shows a *plain* horizontal type of shaper commonly used for production and general-purpose work. This shaper, consisting of a base and frame which supports a horizontal ram, is quite simple in construction. The *ram*, which carries the tool, is given a reciprocating motion equal to the length of the stroke desired. The *quick-return mechanism* driving the ram is designed so that the return stroke of the shaper is faster than the cutting stroke. The purpose of such an arrangement is to reduce the idle time of the machine to a minimum. The tool head at the end of the ram can be swiveled through an angle and is provided with means for feeding the tool into the work. On it is fastened the *clapper-box tool holder* which is pivoted at the upper end to permit the tool to rise on the return stroke so as not to dig into the work.

The work table is supported on a crossrail in front of the shaper. By means of a lead screw in connection with the crossrail, the work can be moved crosswise or vertically by either hand or power drive. A *universal* shaper has these same features, and, in addition, is provided with swiveling and tilting arrangements to permit accurate machining at any desired angle. The swiveling adjustment takes place about an axis that is parallel to the motion of the ram. The tilting feature is in the table top and provides a means to set the table at an angle to the swiveling axis. Both adjustments are equipped with protractors to assist the operator in setting the table at any desired angle.

The work on a shaper is held by bolting it to the work table, or by fastening it in either a vise or some special fixture. The table is provided with T slots which permit bolts to be inserted to facilitate the holding of work or fixtures. Reference to the illustrations shows clearly how this is accomplished.

Courtesy The Hendey Machine Company.

FIG. 1. Plain Horizontal Shaper.

The tools used in shaper and planer work are the same type as those used on a lathe. The only difference is in the amount of end or front clearance, which is increased for these tools. Flat work requires more clearance than round work, since the natural curvature of the latter tends to increase the amount of clearance at the end of the tool. Otherwise, tools are ground with the same angles and shapes as for lathes. A heavy-duty shaper is shown at work in Figure 2. This cut clearly illustrates the methods of supporting the tool and work.

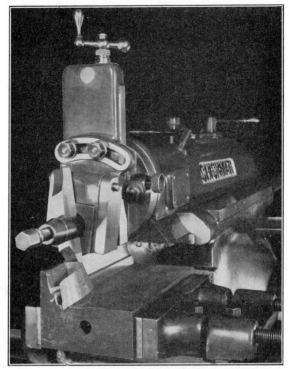

Courtesy The Cinncinnati Shaper Company.

FIG. 2. A Heavy-Duty Shaper Taking a Deep Cut.

Classification of Shapers

According to general design, shapers can be classified as follows:

1. Horizontal — push cut
   (*a*) Plain (production work).
   (*b*) Universal (toolroom work).
2. Horizontal — draw cut.
3. Vertical
   (*a*) Slotter.
   (*b*) Keyseater.
4. Special purpose, as for cutting gears.

Power can be applied to the machine by an individual motor either through gears or belt, or by step-cone pulley from the lineshaft. The reciprocating drive of the tool can be arranged in several ways. Some of the older shapers were driven by gears or by feed screw, but most shapers are now being driven by an oscillating arm and crank mechanism, as illustrated in Figure 3. A recent development is the hydraulic drive for controlling the reciprocating movement of the ram.

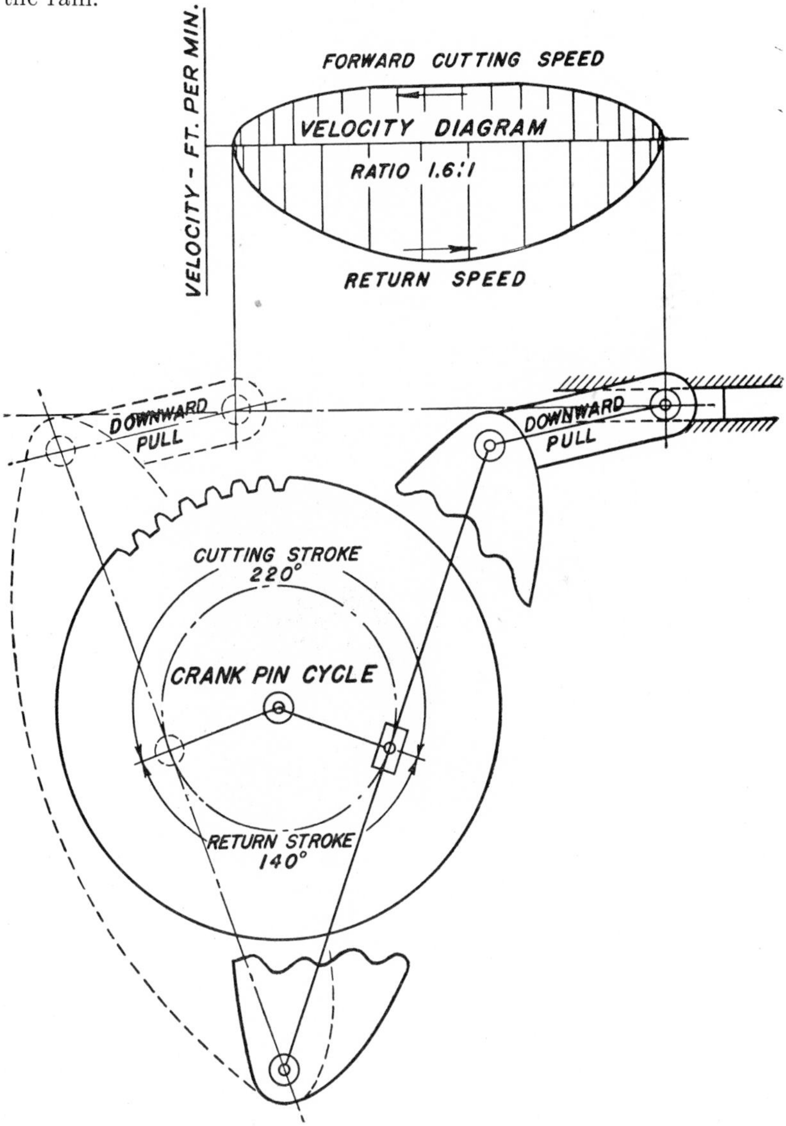

FIG. 3. Pillar-Type Quick-Return Mechanism for Shaper.

Quick-Return Mechanism

Several types of quick-return mechanisms have been developed for shapers, but the most common type is the *pillar* or *oscillating-arm* type shown diagrammatically in Figure 3. It consists of a rotating crank driven at a uniform speed which is connected to an oscillating arm by a sliding block. Referring to the figure, we note that the cutting stroke takes up 220 degrees of the crank revolution while the return is made through only 140 degrees movement of the crank. Hence the ratio,

$$\frac{\text{Cutting stroke}}{\text{Return stroke}} = \frac{220}{140} = \frac{1.57}{1}$$

The quick return is due to the crank end, with the sliding block, being close to the arm fulcrum during the lower half of rotation. The stroke length is varied by changing the length of the crank.

Courtesy The Hendey Machine Company.

FIG. 4.   Quick-Return Drive Mechanism Used on a Hendey Horizontal Push-Cut Shaper.

The quick return and drive mechanism of the Hendey Shaper is illustrated in Figure 4. In this construction the sliding block works in the center of the rather massive oscillating arm. The crank is contained in the large gear and may be varied by a screw mechanism. The crank gear is driven by the transmission gears, as shown in the figure. To change the position of the stroke, the clamp holding the connecting link to the screw is loosened, and the hand wheel at the end of the screw is turned. As the screw is fastened to the ram, it can be moved backward or forward by turning this wheel.

Hydraulic Shapers

Hydraulic shapers are similar in appearance to those driven by some form of mechanism, as may be noted in Figure 5. One of the

Courtesy Rockford Machine Tool Company.

FIG. 5. Hydraulic Shaper.

principal advantages claimed for this type of shaper is that the cutting speed and pressure in the ram drive are constant from beginning to end of the cut. Both the cutting-stroke length and its position relative to the work may be changed quickly without stopping the machine, by the use of two small handles at the side of the ram. Another feature is that the ram movement can be reversed instantly anywhere in either direction of travel. The hydraulic controlled feed is accomplished while the tool is clear of the work. The maximum ratio of cutting stroke to return stroke is about 2 to 1.

Hydraulic drives, because of their smooth action and other advantages, are rapidly being incorporated in many other machine tools.

Courtesy Morton Manufacturing Company.

FIG. 6. High-Duty Draw-Cut Shaper.

Horizontal Draw-Cut Shaper

This shaper is so named, because the tool is pulled across the work by the ram instead of being pushed. In Figure 6 is shown a machine of this type equipped with a small jib crane and hoist. Horizontal draw-cut shapers are especially recommended for heavy cuts, being widely used for cutting large die blocks and machining large parts in railroad shops. During the cut, the work is drawn against the adjustable back bearing or face of the column, thereby reducing the strains on the crossrails and saddle bearings. There is little tendency for vibration, as a tensile strength is exerted in the ram during the cut. This permits the use of large forming tools without resulting tool chatter marks on the work.

Courtesy Morey Machinery Company, Inc.

FIG. 7. Vertical Shaper for Production and Tool Work.

Vertical Shapers

Vertical shapers or *slotters* (see Figure 7) are used principally for internal cutting and planing at angles, and for operations that require vertical cuts because of the position in which the work must be held. Operations of this nature are frequently found on die work, metal molds, and metal patterns. The shaper ram operates vertically and has the usual quick-return feature, similar to the horizontal-type machines. Work to be machined is supported on a round table having a rotary feed in addition to the usual table movements. The circular table feed permits the machining of curved surfaces, which is particularly desirable for many irregular parts that cannot be turned on a lathe. Plane surfaces are cut by using either of the table cross-feeds. An interesting special machine, known as a universal vertical miller-shaper, has been developed for machining irregular punch and die shapes and other parts requiring both milling and shaping operations.

A special type of vertical shaper known as a *keyseater* is especially designed for cutting keyways in gears, pulleys, flywheels, and similar parts. The work is clamped to a horizontal table and the tool is reciprocated in a vertical position through its center. The work is fed to the cutter by table adjustments.

The Planer and Its Work

A *planer* is a machine tool designed to remove metal by moving the work in a straight line against a single-edged cutting tool. The type of work it does is very similar to that done on a shaper except that a planer is adapted to much larger work, illustrated in Figure 8, in the planing of the ways on a large machine table. The cuts are all plain surfaces, but they may be horizontal, vertical, or at an angle. In addition to machining large work, this planer is frequently used in production work to machine multiple small parts held in line on the platen. The *size* of a planer is designated by three figures, the first two indicating the width and height of work it will take and the last figure the length of the work.

Differences between Planer and Shaper

Although both the planer and the shaper are adapted to the machining of flat surfaces, there is not much overlapping in their fields of usefulness; they differ widely in construction and in method of operation. When the two machines are compared as to construction, operation, and use, the following differences may be seen:

1. The planer is specially adapted to large work; the shaper can do only small work.

2. On the planer the work is moved against a stationary tool; on the shaper the tool moves across the work, which is stationary.

3. On the planer the tool is fed into the work; on the shaper the work is usually fed across the tool.

4. The drive on the planer table is either by gears or by hydraulic means. The shaper ram can also be driven in this manner, but in most cases a quick-return link mechanism is used.

5. Most planers differ from shapers in that they approach constant-velocity cuts.

Courtesy G. A. Gray Company.

Fig. 8.   Planing a Machine Table on a Double-Housing Planer.

Classification of Planers

Planers may be classified in a number of ways, but according to general construction there are three types:

1. Double-housing planer (see Figure 8).
2. Open-side planer (see Figure 9.)
3. Plate planer (see Figure 12).

Each of these types may vary according to the method of drive. In such a classification there are gear drive (both spur gear and spiral), hydraulic drive, screw drive, and crank drive. The first two mentioned are the types most generally used. The screw drive is used principally on plate planers, whereas the crank drive does not have wide application and is found only on some small planers.

Another variation in design is the manner in which the power is brought to the planer. Most old planers are belt-driven from overhead countershafts. Two belts are used, one open and one crossed, for the forward and return stroke of the platen. At each end of the travel the belts are shifted automatically from tight to loose pulley. The belt controlling the return stroke is driven faster to give the platen a quick-return motion while the tool is not cutting. Variable-speed-reversing motors, controlled by stops at each end of the stroke, are now used in many installations. This arrangement is satisfactory and eliminates the noise and trouble so characteristic of belt drives. The hydraulic planer has a direct-connected motor which drives a displacement pump controlling the platen movement.

Planer Construction

The standard type of planer consists of a long heavy base upon which the table or platen is reciprocated. At the side of the base near the center is located the upright housing. This supports the crossrail upon which the tools are fed across the work. Figure 8, a double-housing planer, illustrates clearly how the tools are supported and the manner in which they can be adjusted for angle cuts. These tools may be fed manually or by power in either a vertical or crosswise direction. The motor drive is usually at one side or the planer near the center, and the drive mechanism is located under the platen.

The accuracy of a planer is determined largely by its rigidity and the manner in which the ways in the bed are machined. Most medium-sized planers have one flat and one double-V way which allow for unequal bed and platen expansions. Large planers having three ways will have a double-V way at the center and flat ways at each side. The controls for operation are all at the upright housing. Adjustable dogs at the side of the bed control the stroke length of the platen.

A variation in housing construction is shown in Figure 9. This type, having the housing on one side only, is known as an *open-side* planer and is adapted to handle wide work. A jib crane and electric hoist are frequently provided with planers for handling the work where they are not served by a traveling crane. The planer shown in the figure is hydraulically driven.

FIG. 9. Open-side Planer Equipped with Hoist and Jib Crane.

Large planers, as shown in Figure 10, are available for handling heavy and massive machine work. Planers of this type are similar

FIG. 10. Large Heavy-Duty Planer.

to other planers and differ principally in size. On the machine shown there are four tool supports, two located on the upper crossrails and two on the side housings.

Hydraulic Drive for Planer

Hydraulic drives have proved most satisfactory for planers for several reasons. First, uniform cutting speed is attained throughout the entire cutting stroke. The acceleration and deceleration of the table take place in such a short distance of travel that they need not be considered as a time element. A second advantage is that the inertia forces to be overcome are less in a hydraulic planer than in the conventional gear-driven planer. The gear-driven planer, with its fast-revolving parts, including the rotor of the drive motor, has several times more inertia force to overcome than has the simple

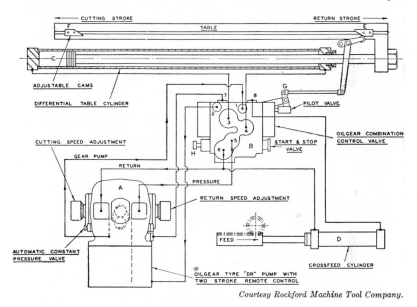

Courtesy Rockford Machine Tool Company.

Fig. 11.   Hydraulic Planer Circuit.

piston rod and piston of the hydraulic drive. Overcoming inertia consumes energy, and, with rapid short strokes, the difference in power consumption is noticeable. Further advantages of hydraulic drives are uniform cutting pressure, quick table reversal, rapid means of varying the stroke, and less noise in operation.

A hydraulic circuit, developed and patented by the Rockford Machine Tool Company, is illustrated in Figure 11.* In the diagrammatic sketch the motor-driven oil pump at *A* is shown delivering oil

* P. S. Jackson, *Planer Comparative Performance Data*, Rockford Machine Tool Company.

Fig. 12.   Plate or Edge Planer.

to the control valve at B. The oil enters at 5 and leaves at 4, causing the piston in cylinder C to move the table through its cutting stroke. Adjustable dogs E and F cause reversal of table after cutting and return strokes, respectively. The stroke may be changed while the planer is in operation.

The circuit used is known as a regenerative circuit. The oil entering the suction side of the pump, being under a back pressure, supercharges the pump and helps to drive it as a motor. Less electric energy is consumed by the pump because of this feature. The back-pressure oil from the cylinder enters the control valve at 3, and thence from 6 to the suction side of the pump. The tool feed is controlled by the feed cylinder D.

The cutting-speed adjustment on the gear pump makes it possible to control the cutting speed of the tool. On the return stroke the pressure from 5 connects with 3 and 4, causing a fast return by the differential principle. Reference to the figure will show that the area of the rod determines the return speed, while the area around it governs the cutting speed for a given pump displacement. During the cutting stroke the piston rod is in tension. The work pressure in the system rises only high enough to overcome the resistance of the cut, and, as the cut gets heavier, the pressure automatically increases. A relief valve is provided which functions when the hydraulic load is equal to the electric load rating of the motor.

Plate Planers

For the fabrication of heavy steel plates for pressure vessels and armor plate, a special type of planer, known as a *plate* or *edge planer*, has been devised. Such a planer is shown in Figure 12. The plate is stationary and is clamped to a large bed on one side of the housing. To insure further having the work securely held, a series of clamps come down from the cross housing and hold the plate edge in place. The cutting tool is attached to a carriage which is supported on the heavy ways of the planer. A large screw drive is used for moving the carriage carrying the operator and tools along the work.

Holding Work on the Planer

All planer tables or platens are constructed with T slots and reamed holes on their surfaces to provide means for holding and clamping down parts that are to be planed. It is very important that work be securely held because of the heavy cuts that are taken. A heavy-duty vise held by bolts engaging T slots in the platen is suitable for small objects. For light cuts on long thin objects a magnetic chuck can be used with stop pins at the end to take care of the tool thrust. Most work is held by clamping it directly to the platen, and a wide variety of clamps, stop pins, and holding devices

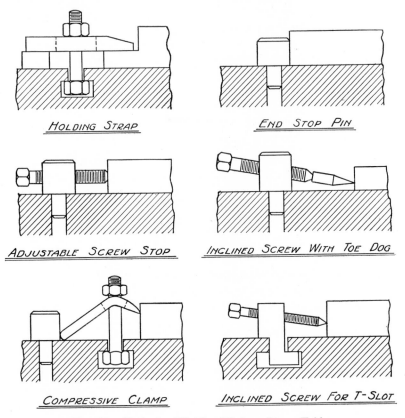

HOLDING STRAP

END STOP PIN

ADJUSTABLE SCREW STOP

INCLINED SCREW WITH TOE DOG

COMPRESSIVE CLAMP

INCLINED SCREW FOR T-SLOT

FIG. 13. Methods of Holding Work on Planer Table.

has been developed for this purpose. Figure 13 illustrates several common ways of holding work on planers and shapers. Note that several of these arrangements are adapted to holding down plates so that the entire surface may be machined.

REVIEW QUESTIONS

1. How is the position of the stroke adjusted on a shaper? On a planer?
2. Classify the different types of shapers and planers.
3. What are the principal differences between a shaper and a planer?
4. Show by sketch how the quick-return motion is accomplished on a shaper.

5. How is the feed accomplished on a shaper?

6. Describe a slotter, and state what kind of work it will do.

7. If a shaper makes 36 complete strokes a minute and the length of stroke is 9 inches, what is the cutting speed in feet per minute? The ratio of return stroke to cutting stroke is 2 to 3.

8. What are the advantages of a hydraulically driven shaper or planer?

9. What is a planer, and what type of work will it do?

10. List or illustrate by sketches five methods of holding work on a planer.

11. How is the feed accomplished on a planer?

12. What is the main feature of an open-side planer, and why is it so constructed?

13. Does a planer have a quick-return motion? If so, describe how it is accomplished.

14. What is a plate planer, and how does it operate?

15. Describe the method of supporting the tool on a planer.

16. What is the difference between a lathe tool and the tools used on planers and shapers?

17. How does a tool used on a keyseater differ from a broaching tool?

18. Discuss the various types of planer drives.

BIBLIOGRAPHY

Jackson, Paul S., *Planer, Comparative Performance Data*, Rockford Machine Tool Company.

"Planing versus Milling," *Mechanical Engineering*, April 1934.

CHAPTER 15

DRILLING AND BORING MACHINES

One of the simplest machine tools used in production and toolroom work is the ordinary *drill press*. It consists of a spindle which imparts rotary motion to the drilling tool, a mechanism for feeding the tool into the work, a table upon which the work rests, and a frame. It is essentially a single-purpose machine, although a number of similar machine operations can be performed with the addition of appropriate tools.

The operation of *drilling* consists of producing a hole in an object by forcing a rotating drill against it. The same results are accomplished in some machines by holding the drill stationary and rotating the work: an example of this is drilling on a lathe with the work held and rotated by the chuck.

Other methods of producing a hole are by punching, flame cutting, and coring. The punching process is very rapid and specially adapted to thin materials. It produces accurate holes, but the punches and dies are expensive. Oxyacetylene cutting or the oxygen lance will cut holes through any thickness of commercial material, but these holes are not accurate in either size or shape. Coring is used principally on large holes in castings to save metal and reduce machining costs.

Boring is the operation of enlarging a hole that has already been drilled or cored; it is principally an operation of truing a hole that has previously been drilled. A single-point lathe-type tool is used. To perform this operation on a drill press requires a special holder for the boring tool.

Counterboring refers to enlarging one end of a drilled hole. The enlarged hole, which is concentric with the original one, is flat on the bottom. The tool for this operation is similar to an end mill and is provided with a pilot pin which fits into the drilled hole to center the cutting edges. Counterboring is used principally to set bolt heads and nuts below the surface. When it is required to finish off a small surface around a drilled hole, the operation is known as *spot facing*. This is a customary practice on rough surfaces to provide smooth seats for bolt heads. If the top of a drilled hole is

420

beveled to accommodate the conical seat of a flat-head screw, the operation is called *countersinking*.

Reaming is the operation of enlarging a machined hole to accurate size with a smooth finish. A reamer is an accurate tool and is not designed to remove much metal; hence the allowance for reaming should not exceed 0.015 inch. Although this operation and those previously mentioned can be done on a drill press, other machine tools are equally well adapted to perform them.

Classification of Drilling Machines

Drilling machines are classified according to their general construction:

1. Portable drill.
2. Sensitive drilling machine
   (a) Bench mounting.
   (b) Floor mounting.
3. Upright drilling machine
   (a) Light duty.
   (b) Heavy duty.
4. Radial drilling machine
   (a) Plain.
   (b) Semiuniversal.
   (c) Universal.
5. Multiple-spindle drilling machine
   (a) Vertical.
   (b) Horizontal.
   (c) Radial or combination.
6. Gang drilling machine
7. Automatic-production drilling machine.
8. Deep-hole drilling machine
   (a) Vertical.
   (b) Horizontal.

These drilling machines vary considerably in size, method of feeding the drills, and application of power.

Portable and Sensitive Drills

Portable drills are small compact drilling machines used principally for such drilling operations as cannot be conveniently done on a regular drill press. The simplest of these is the hand-operated drill. Most portable drills, however, are equipped with a small electric motor. These drills operate at fairly high speeds and accommodate drills up to 1/2 inch in diameter. Similar drills, using compressed air as a means of power, are used in cases where sparks from the motor may constitute a fire hazard.

The *sensitive* drilling machine is a small high-speed machine of simple construction similar to the ordinary upright drill press. It consists of an upright standard, a horizontal table, and a vertical spindle for holding and rotating the drill. Machines of this type are hand-fed, usually by means of a rack and pinion drive on the sleeve holding the rotating spindle. These drills may be driven directly by a motor, by a belt, or by means of a friction disk. The friction-disk drive has considerable speed regulation, although it is not suitable for slow speeds and heavy cuts. Sensitive drill presses are suitable only for light work and are seldom capable of rotating drills over 5/8 inch in diameter. A bench-type sensitive drill is shown in Figure 1.

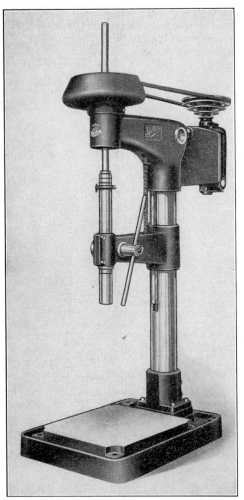

Courtesy Buffalo Forge Company.

FIG. 1. Bench-Type Sensitive-Drill Press.

Upright Drills

Upright drills are similar to sensitive drills except that they have a power-feeding mechanism for the rotating drill and are designed for heavier work. Figure 2 shows a 21-inch machine with a box-type upright. Box-column machines are more rigid than the round-column machines and consequently adapted to heavier work. This machine is provided with nine spindle speeds offered in several speed ranges from 75 to 3500 rpm. Feed rates of 0.004, 0.008, 0.014, and 0.020 inch per revolution are controlled by a single feed lever. The feed clutch is automatically controlled so that the spindle will be disengaged when it reaches its upper or lower limit of travel. It also can be set to disengage at any predetermined depth if the feed trip dial on the left of the sliding head is set. A separate motor is used for reversing the rotation. This machine can be used for tapping as well as for drilling.

Radial Drilling Machine

The *radial* drilling machine is designed to be used for large work where it is not feasible for the work to be moved around if several holes are to be drilled. Such a machine is shown in Figure 3. It consists of a vertical column supporting an arm which carries the drilling head. The arm may be swung around to any desired position over the work bed, and the drilling head has a radial adjustment along this arm. These adjustments permit the operator to locate the drill quickly over any desired point on the work. *Plain* machines of this type will drill only in the vertical plane. On *semi-universal* machines the head may be swiveled on the arm to drill holes at various angles in a vertical plane. *Universal* machines have an additional swiveling adjustment in either the head or the arm and can drill holes at any angle.

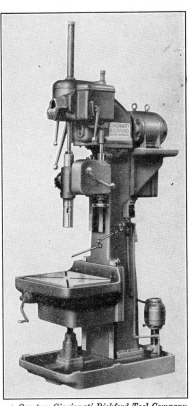

Courtesy Cincinnati Bickford Tool Company.

FIG. 2. 21-Inch Upright Drill.

The machine illustrated has 32 spindle speeds in geometrical progression, ranging from 20 to 1600 rpm. Sixteen selective power feeds from 0.003 to 0.087 inch per revolution are available, and the machine is also equipped with a tapping attachment. There is an elevating mechanism for raising the arm as well as a rapid traversing means for moving the drilling head.

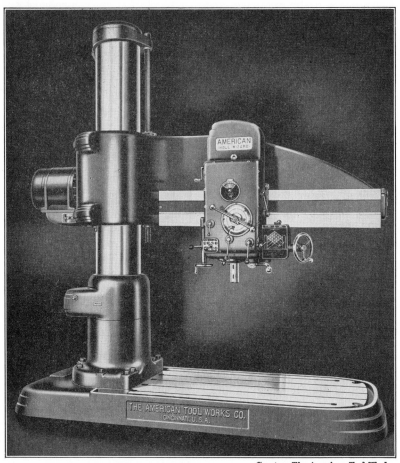

Fig. 3.   Radial Drilling Machine.

Gang Drilling Machine

When several drilling spindles are mounted on a single table it is known as a *gang drill*.   There are two types: those with spindle units permanently spaced along the table, and those with an adjusting feature permitting the spindles to be spaced at various distances.   The first and most common type is adapted to production work where several operations must be performed.   The work is usually held in a jig which can be easily slid on the table from one spindle to the next. If several operations must be performed, such as drilling two different-sized holes and reaming them, four spindles are set up for this purpose. With automatic feed control, two or more of these operations may be going on simultaneously, attended by only one operator. The arrangement is similar to operating several independent drill presses, but much more convenient because of its compactness. A four-spindle machine of this type is shown in Figure 4. This drilling machine has separate motors for each spindle with

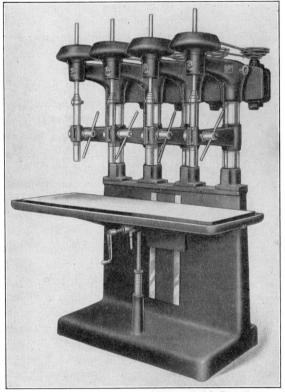

Courtesy Buffalo Forge Company.

FIG. 4. Four-Spindle Gang Drilling Machine.

a V-belt drive. The speed range is 400 to 1600 rpm, suitable for 1/2-inch or smaller drills. This drill is of the sensitive type, although it is also available with power feed.

When the job demands that several holes be drilled in line on a long piece, it is necessary to have spindle units that can be adjusted to give the desired hole spacing. Machines of this type are used for any straight-line multiple-hole drilling applications, as in pipes, channels, castings, angles, and plates.

Multiple-Spindle Drilling Machine

Multiple-spindle drilling machines have been developed for the purpose of drilling several holes simultaneously. These machines are essentially production machines and, when once set up, will drill many parts with such accuracy that all parts are interchangeable. In many applications drilling jigs are unnecessary, but in some cases a plate provided with hardened bushings is essential to guide the drills accurately into the work. Multi-spindle drilling machines differ principally in the way the drills are held and in the way the feed is accomplished.

Most machines are vertical machines, as shown in Figure 5. This machine is provided with a maximum of 16 drill spindles covering an area 30 inches in diameter. Each lower drill spindle is driven from the upper drill spindle through two universal joints and a tubular driving shaft, the tube being splined at the upper end to permit maximum adjustment of the drill. The head assembly carrying the spindles is mounted on a carriage which travels on vertical double-V ways. The feed is hydraulically operated, and this is accomplished by feeding the head assembly with the drills to the work. In operation the drilling cycle

Courtesy Moline Tool Company.

Fig. 5. Hydraulic Feed Driller with 16 Universal Joint-Type Spindles and a 30-Inch-Diameter Drilling Area.

consists of rapid advance of drills to the work, proper feed, and rapid return of drills to the starting position. The approximate capacity of this machine is sixteen 7/8-inch holes in soft steel. Tapping can

Courtesy The Foote-Burt Company.

FIG. 6. Three-Way Semiautomatic-Production Drilling Machine.

be done by adding feed in reverse equal to feed in forward direction so that the taps can be backed out of the threaded holes.

Multiple-drilling machines frequently use a table feed in place of the one just described, thus eliminating the movement of the heavy geared-head mechanism which rotates the drills. This may be done in several ways: by rack and pinion drive, by lead screw, or by a rotating plate cam. The latter method is well adapted to provide varying motions which give rapid approach, uniform feed, and quick return to the starting position.

Way-type semiautomatic drilling machines are used extensively in production work. These are usually two-, three-, or four-way drilling

FIG. 7.   Special Three-Way Horizontal and One-Way Vertical Drilling Machine. Drilling 25 Holes in Truck Axle Mounting.

machines designed principally for single-purpose jobs. Engine block castings and similar parts, requiring the drilling of many holes, are typical examples of the work done on these machines. A large three-way drilling machine of this type is illustrated in Figure 6. The part to be drilled is clamped in the drilling fixture, and the drills enter simultaneously through hardened guide bushings. One feature of these machines is that the drill-spindle assembly with the motor mounting is made up as a separate unit. These units can be assembled to make up special machines or to accommodate changes in design. Movement of the drills to position and feeding is hydraulic.

Another way-type machine having drills in both horizontal and vertical position is shown in Figure 7. In this illustration a total of 25 holes are drilled in a truck axle mounting from four directions. Approximately 54 pieces per hour can be drilled with this setup. Other machines, principally used in aircraft-engine work, have drills radially spaced to take care of jobs where holes must be drilled at different angles around the circumference of the part. Most way-type production machines can be adapted to tapping, boring, and similar operations if desired.

Deep-Hole Drilling Machine

Several problems not encountered in ordinary drilling operations arise in the drilling of long holes in rifle barrels, long spindles, connecting rods, certain oil-well drilling equipment, and many other similar applications. As the hole length increases, it becomes more and more difficult to support the work and the drill properly. The rapid removal of chips from the drilling operation becomes necessary to insure the proper operation and accuracy of the drill. Rotational speeds and feeds must be carefully determined, since there is greater possibility of deflection than when using a drill of ordinary size.

To overcome these problems, deep-hole drilling machines have been developed which are especially adapted to this type of machine work. In design these machines may be of either the horizontal or vertical type; they may be of single-spindle or multispindle construction; and they may vary as to whether the work or drill is caused to revolve. In Figure 8 is illustrated a two-spindle machine of the horizontal type. The work is supported at one end in the headstock and on the other end by the work carriage at the center of the machine. Rotation is given to the work from the headstock spindles. The work carriage supports the drills by means of hardened bushings at a point just adjacent to where they enter the work. The other end of the drill is supported by the drill carriage at the right, and, if necessary, center supports are also used. The feeding of the drill is obtained from the lead screw which forces it slowly into the rotating work. The drill feed must be very light to avoid deflecting the drill.

A typical drill used in this type of machine is shown in Figure 9. This drill has only a single cutting edge. In the lip of the drill is a hole to carry the oil to the drill point. The chips are carried

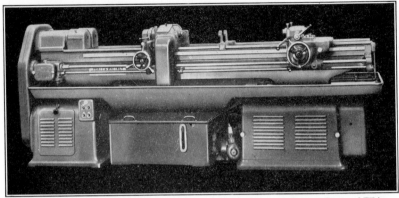

Courtesy Pratt and Whitney.

Fig. 8.   Two-Spindle Horizontal Deep-Hole Drilling Machine.

out of the hole along the flute of the drill as rapidly as they are formed. Twisted drills with oil holes and two flutes are also available for this type of drilling. In most cases the drill is held stationary

Courtesy Pratt and Whitney.

Fig. 9.   Single-Fluted Drills for Deep-Hole Drilling.

to facilitate pumping the oil through the drill. Holes drilled in these machines are accurate and concentric with the diameter. Additional finish may be given by special reamers or broaching tools.

Kinds of Drills

The most common type of drill used is the *twist drill*, having two flutes and cutting edges. In Figure 10 the nomenclature of this drill, as well as the usual point, clearance, and end angles, is shown. The drill is held and properly centered in the socket of the drilling-machine spindle by means of the *tapered shank*. This has a Morse

Taper of approximately 5/8 inch per foot, which is standard for drills, reamers, and other similar tools. The *tang* at the end of the taper fits into a slot in the socket to prevent slipping of the tapered surface. Straight-shank drills are held and properly centered in a

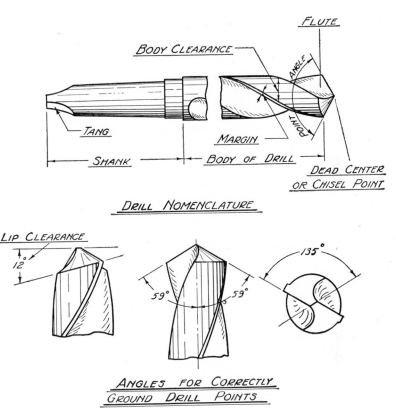

FIG. 10. Twist-Drill Nomenclature and Angles.

drill chuck. These drills are cheaper than those having a tapered shank; and such construction is common for small drills.

A number of different kinds of drills is shown in Figure 11, the principal variation among them being the number and angle of the flutes. Straight-fluted drills have several special applications, one of them being the drilling of soft metals where there is a tendency for a regular drill to " dig in." These are successful, because there is no rake angle to the cutting edge. Similar drills of this type are used in drilling brick or tile and also in deep-hole drilling. Two-fluted drills with either interior or exterior oil channels are frequently used on turret-lathe-production drilling. Three-fluted drills are used principally for enlarging holes previously punched or drilled. Various drills with different flute angles have been developed to give

Taper-Square-Shank Drill

Taper-Shank Twist Drill

Three-Groove Drill

Straight-Shank Twist Drill

Drill For Molded Plastics

Courtesy National Twist Drill and Tool Company.

Fig. 11.   Types of Drills.

improved drilling to special materials and alloys.   In addition, there are some drills made in combination with other tools, as, for example, the combination drill and tap or the drill and countersink.

For drilling large-size holes in pipe or sheet metal twist drills are not suitable.   Either the cut is too large, or the drill tends to dig into the work, or the hole is too large to be cut by a standard-size drill.   An adjustable hole cutter, made up of several lathe-type tool bits, is illustrated in Figure 12.   This tool can be readily adjusted to the size of the hole desired and is very successful in cutting through thin sections of metal.   For very large holes in thin metal a cutter known as a *fly cutter* is used.   Such a cutter, shown in Figure 13, consists of tool bits held in a horizontal holder and capable of being adjusted to accommodate a wide range of diameters.   Both cutters cut in the same path, but one is set slightly below the other.

A drill developed by the Black Drill Company is designed for hardened steel.   The drill operates at high speed and develops sufficient friction to anneal the steel and permit cutting without softening the

FIG. 12.   Adjustable Hole Cutter Cutting Hole in Side of Heavy Pipe.

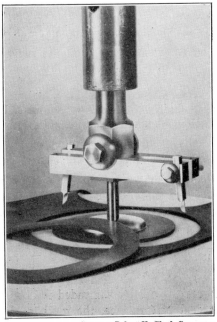

FIG. 13.   Cutting with a Fly Cutter.

drill point. This drill has a triangular section with three flutes, although for counterboring work a flat end can be used. The point angle for best results is 130 degrees. Drills of this type are used for carburized stock, spring steel, die sections, knives, and similar hard materials.

Drill Performance

To obtain good service from a drill, it must be properly ground. The point angle should be correct for the material that is to be drilled: for steel, aluminum, brass, and most materials, 118 degrees has proved very satisfactory; for plastics, this angle should be reduced to 90 degrees or under; whereas some of the harder steel alloys use angles greater than 118 degrees. In grinding this angle, care must be exercised to get the lips the same length as well as to have the angle the same on each side of the drill center line. The clearance angle (see Figure 10) should be 12 degrees.

The cutting speed, expressed in feet per minute, is a measure of the peripheral speed of the drill. For high-speed drills on ordinary steels this value is about 80 feet per minute. Cutting speeds vary from 20 to 300 feet per minute, depending on the material hardness. Carbon-steel drills should be operated at about one-half the speeds recommended for high-speed steel drills.

To obtain best performance and long life for the cutting edges, some coolant should be used. A few of the suggested coolants are listed below:

> Cast iron, magnesium, and brass — dry.
> Steel — soda water or sulphurized mineral oil.
> Aluminum — kerosene or soda water.
> Plastics — dry or soda water.

Drill performance is also affected by the helix angle of the flutes. Although this angle may vary from 0 to 45 degrees, the usual standard for steel and most materials is 30 degrees. The smaller this angle is made, the greater is the torque necessary to operate a given feed. As the angle is increased appreciably, the life of the cutting edge is reduced. Some materials are drilled more efficiently by drills with special helix angles. For example, an angle of 45 degrees works very satisfactorily for zinc alloys and aluminum, whereas a 20-degree angle is recommended for Bakelite.

In evaluating drill performance, the material of which the drill is made must not be overlooked. It has been previously stated that high-speed steel tools (18–4–1 type) will stand about twice the cutting speed of carbon-tool steel. For hard and extremely abrasive materials, drills tipped with tungsten carbide give excellent service. Stellite and other nonferrous hard-surfacing alloys are also being used for similar difficult materials. Many drills are now chrome-plated to provide a hard wearing surface.

Thrust and torque figures for standard drills, 1/8 to 1½ inches in diameter, in drilling steel and cast iron are given in Table 9. Values for horsepower are given, both at average commercial speeds and at 100 rpm. The power values allow for average machinability and for a 20% friction loss in the machine.

Procedure for Producing Accurate Holes

Drilling accuracy depends to a large extent on the proper grinding of the drill point. Slight variations in the angles and lip lengths cause the hole to be oversize or off center. Because of the bluntness

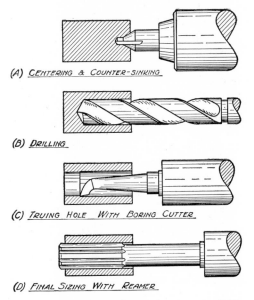

(A) *Centering & Counter-sinking*

(B) *Drilling*

(C) *Truing Hole With Boring Cutter*

(D) *Final Sizing With Reamer*

Fig. 14. Procedure for Producing Accurate Holes.

of the drill point, it is difficult to start a drill properly or accurately without a centering hole. Furthermore, when finish and accuracy to size are paramount, a drilled hole is not suitable. To eliminate these difficulties, it is frequently necessary to employ several operations in producing an accurate hole. Figure 14 illustrates the pro-

TABLE 9*

THRUST AND POWER FOR DRILLING

Thrust and torque figures are given for drills from 1/8-inch to $1\frac{1}{2}$-inch diameters when drilling steel, and from 1/2-inch to $1\frac{1}{2}$-inch diameters when drilling cast iron. Values for horsepower are given both at average commercial speeds and at 100 rpm The latter figure is convenient for conversion purposes.

The complete set of values follows:

Diameter of Drill	Feed per Revolution	Thrust Lb Cast Iron	Steel
$\frac{1}{8}$	0.004		130
$\frac{5}{32}$	0.0045		185
$\frac{3}{16}$	0.005		245
$\frac{1}{4}$	0.006		380
$\frac{3}{8}$	0.008		550
$\frac{1}{2}$	0.010	200	750
$\frac{5}{8}$	0.011	275	1000
$\frac{3}{4}$	0.012	325	1275
$\frac{7}{8}$	0.013	400	1590
1	0.014	500	1910
$1\frac{1}{4}$	0.016	700	2650
$1\frac{1}{2}$	0.016	850	3200

Diameter of Drill	Feed per Revolution	HorsePower Cast Iron 100 Rpm	100 Surface Ft per min	Steel 100 Rpm	60 Surface Ft per min
$\frac{1}{8}$	0.004			0.011	0.200
$\frac{5}{32}$	0.0045			0.018	0.260
$\frac{3}{16}$	0.005			0.028	0.340
$\frac{1}{4}$	0.006			0.053	0.490
$\frac{3}{8}$	0.008			0.210	1.308
$\frac{1}{2}$	0.010	0.26	1.60	0.375	1.717
$\frac{5}{8}$	0.011	0.40	2.30	0.580	2.133
$\frac{3}{4}$	0.012	0.59	2.90	0.855	2.620
$\frac{7}{8}$	0.013	0.80	3.44	1.180	2.078
1	0.014	1.06	4.02	1.550	3.550
$1\frac{1}{4}$	0.016	1.83	5.58	2.680	4.918
$1\frac{1}{2}$	0.016	2.44	6.22	3.740	5.718

These values allow for average machinability and for a 20% friction loss in the machine.

* Courtesy of *Engineering Bulletin* 91, July 15, 1941. National Twist Drill & Tool Company.

cedure recommended. Four steps are involved: Locating and centering the hole, the actual drilling, truing the hole with a boring tool, and reaming the hole to accurate size. This is common practice for producing accurate holes on lathes and milling machines as well as on drill presses. If drilling jigs are used, the centering and boring operation may be omitted. A jig to hold and guide the drill during the drilling operation should be used in the manufacture of interchangeable parts.

Reamers and Miscellaneous Tools

A *reamer* is a tool used to finish a hole previously drilled or bored. The material removed by this process should be around 0.015 inch and for very accurate work should not exceed 0.005 inch. Because of the small stock removed by this process, reamed holes are perfectly round and have a smooth surface. Any tolerance is above the nominal size. In some cases it is desirable to have the hole a fraction of a thousandth oversize in order to produce certain fits.

A number of types of reamers are available for different materials and applications. They are:

1. Hand reamer.	5. Expansion reamer.
2. Chucking reamer.	6. Adjustable reamer.
3. Shell reamer.	7. Special-purpose reamer.
4. Taper reamer.	

Illustrations of several types of reamers are shown in Figure 15. The *hand reamer* is a finishing tool for very accurate holes. Only a few

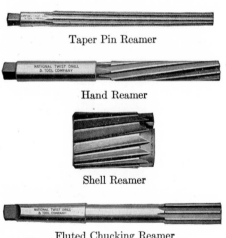

Taper Pin Reamer

Hand Reamer

Shell Reamer

Fluted Chucking Reamer

Courtesy National Twist Drill and Tool Company.

FIG. 15. Types of Reamers.

thousandths of an inch of metal should be removed. It is slightly tapered at the end to facilitate starting and has very little clearance on the flutes. This type, as well as most of the others, is made with both straight and spiral flutes. *Chucking reamers* are designed to be power-driven at slow speeds and are made in two general types, rose and fluted reamers. *Rose reamers* do all their cutting on the

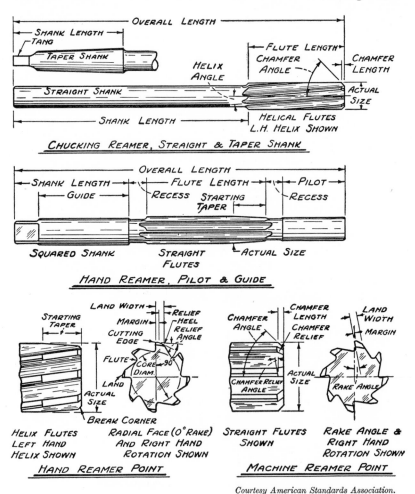

Courtesy American Standards Association.

FIG. 16. Sketch Illustrating Terms Applying to Reamers, Proposed American Standard 1940.

beveled end. There is no relief on the lands of the flutes, and they have a very slight taper towards the shank to prevent binding. *Chucking fluted reamers* do their cutting on the straight flutes, which are backed off or relieved the entire length. Both these reamers are made with straight or taper shanks. A reamer, when set up on a machine, should be " floating " so as properly to center itself with the hole. *Shell reamers* are made in both types. They consist of a shell end mounted on an arbor. This construction results in an economy where high-priced alloys are used, since the arbor can be salvaged when the reamer is worn out. *Expansion reamers* are those reamers which can be adjusted either to compensate for wear or purposely to ream oversized holes. *Adjustable reamers* differ from them in that they can be manipulated to take care of a considerable range in sizes. *Taper* and other special-purpose reamers are similar to those described, except that they are shaped for some special job. The sketches shown in Figure 16 illustrate terms applying to reamers as proposed by the American Standards Association.

Boring Machines

Several machines have been developed that are especially adapted to boring work. One of them, known as a *jig borer*, is constructed for precision work on jigs and fixtures. This machine, which is similar in appearance to a drill press, will do both drilling and end-milling work in addition to boring. Two other machines, the *vertical boring mill* and the *horizontal boring machine*, are adapted to large work. Although the operations which these machines perform can be done on lathes and other machines, their construction is justified by the ease and economy obtained in holding and machining the work.

Jig-Boring Machine

In Figure 17 is shown a machine designed for boring holes in jigs, fixtures, dies, gages, and accurate machine parts. Holes can be located within one one-thousandth of an inch, using the precision lead screws on the work table. These lead screws are hardened, seasoned, ground, and finally lapped to attain this high degree of accuracy. The machine is equipped with ball bearings throughout and has ample rigidity to maintain its accuracy. Like a universal milling machine, a jig borer to give best service must be well equipped with proper small tools and accessories. The time in boring holes is so short that setting up is often the longest part of the job. Much time can be saved by using spotters or locating tools, reamer drills, and end reamers with the same-size straight shank. These tools can be held in a collet and quickly changed. This method represents the fastest system of locating, drilling, and reaming holes. A variety of boring bits and eccentric boring chucks should also be available. Eccentric chucks have a dial graduated in thousandths, permitting the tool to be quickly adjusted to the

Courtesy The Moore Special Tool Company.

FIG. 17. Precision Jig Borer.

proper diameter. Other accessories include a rotary table, microsine table, precision drill chuck, assorted collets for tools, and the necessary bolts and straps for holding the work.

Vertical Boring Mill

The vertical boring mill is so named because the work rotates on a horizontal table in a fashion similar to the old potter's mill. The cutting tools are stationary, except for feed movements, and are mounted on the adjustable-height crossrail. These tools are of the lathe and planer type and are adapted to horizontal facing work, vertical turning, and boring. This machine is sometimes called a

Courtesy Steel Founders Society.

FIG. 18. Vertical Boring Mill, 20-Ft. Capacity, Machining Cast-Steel Tank Turret.

rotary planer, and its cutting action on flat disks is identical with that of a planer. It may also be compared to a lathe placed in a vertical position with the rotating chuck or face plate horizontal. The cutting action would be the same in turning the outside diameter of a large cylinder. These machines, rated according to their table diameter, vary in size from 3 to 40 feet. The large machine shown in Figure 18 has a table diameter of 20 feet.

The vertical boring mill is able to hold large heavy parts, since the work can easily be placed on the table with a crane and does not require much bolting down to hold it in place. It also takes up very little floor space as compared with other machines that might do the same work. Examples of the type of work machined on a vertical boring mill are large pulleys, grinding disks for glass plants, large flange fittings, vertical housings for pumps and motors, flywheels, and numerous other circular-shaped parts. Very accurate work can be done on these machines because of their extreme rigidity and simplicity of design.

Horizontal Boring Machine

The horizontal boring machine differs from the vertical boring mill in that the work is stationary, and the tool is revolved. Furthermore, it is adapted to the boring of horizontal holes, as can be seen by reference to Figure 19. The horizontal spindle for holding the tool is supported in an assembly at one end which can be adjusted vertically within the limits of the machine. This movement and the rotary motion given the tool are the only movements the tool usually has. A work table having longitudinal and crosswise movements is supported on ways on the bed of the machine. In some cases the table is capable of being swiveled to permit indexing the work and boring holes at desired angles. At the other end of the machine is an upright to support the outer end of a boring bar when boring through holes in large castings. On some machines designed for work on extremely large parts, the parts are bolted to a large face plate permanently mounted on the floor. The upright carrying the boring spindle is then mounted on ways to provide means of crosswise adjustment with the work. The longitudinal feed of the rotating spindle is accomplished by having two spindles, one inside the other, the inside one having an independent traverse feed.

For facing work and in boring short holes, the tool is held only by the rotating spindle at the head end of the machine. The operation is similar to boring or end-mill work on the ordinary knee-type milling machine. For long holes the outer bearing is used to give support to the rotating tool. Such work is similar to mounting a casting on the carriage of a lathe and using a boring bar supported between the headstock and the tailstock. These machines have wide application in boring large engine blocks and turbine and motor housings.

Boring Tools

The process of boring is the enlarging of holes previously drilled or cored. Drilled holes are frequently bored to eliminate any possible eccentricity and to enlarge the hole to a reaming size. Boring

FIG. 19. Horizontal Boring Machine.

LIGHT BORING TOOL WITH BENT SHANK

A.

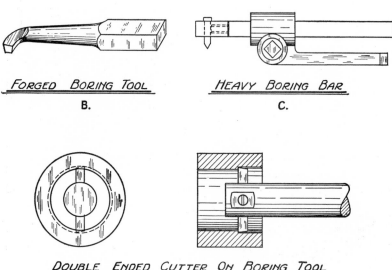

FORGED BORING TOOL

B.

HEAVY BORING BAR

C.

DOUBLE ENDED CUTTER ON BORING TOOL

D.

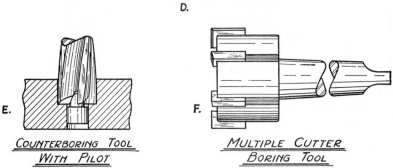

E.

COUNTERBORING TOOL
WITH PILOT

F.

MULTIPLE CUTTER
BORING TOOL

FIG. 20.   Types of Boring Tools.

tools may also be used to finish holes to correct size. This is frequently done on large holes or on odd-sized holes for which there is no reamer available. Also, where the expense of core drills or reamers is great, a boring cutter may prove to be the economical tool to use.

The boring tool most commonly used is a single-pointed tool supported in a manner that permits its entry into a hole. A boring tool of this type is shown in Figure 20A. This tool is forged at the end and then ground to shape. It is supported in a separate holder which fits into a lathe tool post. For turret lathes slightly different holders and forged tools, similar to the one shown at B, are used. A modification of this tool is the boring bar shown at C, which is designed to hold a small high-speed steel tool bit at the end. The bar supporting the tool is rigid and may be adjusted according to the hole length. While the clearance, rake, and cutting angles of these tools should be similar to those recommended for lathe work, these angles cannot be used if the holes are small. Greater end clearance is necessary, owing to the curvature of the hole surface, and back rake is almost impossible to attain because of the position of the tool. This may be seen by reference to the illustration showing the tool in working position. The side-rake and side-clearance angles have no restrictions placed upon them and may be ground correctly. As the internal diameter is increased and large boring machines of the vertical and horizontal type are used, turning tools and holders of the lathe type are possible. Because of the decrease in surface curvature, properly shaped tools with correct angles can be used.

In production work boring cutters with multiple-cutting edges are widely used. These cutters, shown at F, somewhat resemble shell reamers in appearance, but are usually provided with inserted tooth cutters. The cutters may be adjusted radially to compensate for wear and variations of diameter. Boring tools of this type have longer life than single-pointed tools and hence are more economical for production jobs. The counterboring tool at E is designed to recess or enlarge one end of a hole. These tools are provided with pilots to insure concentric diameters.

An adjustable counterboring tool of somewhat different design is shown in Figure 21. This tool can be set up to cut a variety of diameters by changing the position of the blades in the holder. Spot-facing operations can also be done if desired.

For precision boring work on milling machines, jig borers, or drill presses, it is necessary to use a tool having micrometer adjustment. This tool differs from those previously described in that its position is fixed in the machine and it rotates. Hence any increase in hole size must be obtained by adjusting the tool radially from its center. A tool of this type, manufactured by the Moline Tool Company, has

Courtesy Robert H. Clark Company.

Fig. 21. Adjustable Counterboring and Spot-Facing Tool.

a single-pointed tool with micrometer adjustment as close as 0.0001 inch. An important feature of the tool is that it recedes several thousands of an inch upon completion of the cut to prevent scoring of the hole surface as it is withdrawn.

REVIEW QUESTIONS

1. Distinguish among drilling, boring, counterboring, and reaming.
2. Prepare a classification of drilling machines.
3. What is a sensitive drill press, and for what type of work is it used?
4. How does an upright drill differ from a vertical milling machine?
5. Describe the type of work done on a gang drill.
6. What is the difference between a plain and a universal radial drilling machine?
7. Describe the process of deep-hole drilling.
8. Sketch a taper-shank twist drill, and label all principal parts.
9. How are straight-shank drills held in a drill press?
10. What are the angles for correctly ground drill points?

11. How is the cutting speed of a drill determined? What should it be for high-speed drills?

12. State the procedure for producing an accurate hole in a gear blank. Assume the gear blank to be held in a lathe chuck.

13. How much material should be removed by a reamer?

14. Name five types of reamers.

15. How should a reamer be held on a turret lathe?

16. Describe a jig-boring machine.

17. What type of cutting tools are used on a vertical boring mill?

18. What advantage does a vertical boring mill have over a lathe in boring work?

19. How is the feed accomplished on a horizontal boring machine?

20. Describe a horizontal boring machine, and state how it operates.

21. Name and describe three types of boring tools.

22. How does a counterboring tool differ from an end mill?

23. How may a tapered hole be bored?

24. What type of boring tool would you recommend for use with a jig-boring machine?

BIBLIOGRAPHY

Burghardt, H. D., *Machine Tool Operation*, Part II, 1st edition, McGraw-Hill Book Company, 1922.

Hinman, C. W., *Practical Design for Milling and Drilling Tools*, McGraw-Hill Book Company, 1938.

Hirvonen, Eric, "Automatic Step Drilling of Deep Holes," *Mechanical Engineering*, May 1940.

Hoagland, F. O., "Drill Points for Deep-Hole Drilling," *Machinery*, October 1940.

Schneider, A. W., "Finishing Internal Surfaces," *Mechanical Engineering*, April 1936.

Stewart, G. D., "Precision Boring — Plus," *American Machinist*, June 11, 1941.

CHAPTER 16

MILLING MACHINES AND CUTTERS

A milling machine is a machine tool which removes metal as the work is fed against a rotating cutter. Except for rotation, the circular-shaped cutter has no other motion. It is called a *milling cutter* and has a series of cutting edges on its circumference, each of which acts as an individual cutter in the cycle of rotation. The work is held on a table which controls the feed against the cutter. In most machines there are three possible table movements — longitudinal, crosswise, and vertical; but in some the table may also possess a swivel or rotational movement.

The milling machine is the most versatile of all machine tools. Flat or formed surfaces may be machined with excellent finish and great accuracy. Angles, slots, gear teeth, and recess cuts can be made by using various cutters. Drills, reamers, and boring tools can be held in the arbor socket by removing the cutter and arbor. As all table movements have micrometer adjustments, holes and other cuts can be accurately spaced. Most operations performed on shapers, drill presses, gear-cutting machines, and broaching machines can likewise be done on the milling machine. It produces a better finish and holds to accurate limits with greater ease than a shaper. Heavy cuts can be taken with no appreciable sacrifice in finish or accuracy. Cutters are efficient in their action and can be used a long time before resharpening. In most cases the work is completed in one pass of the table. These advantages plus the availability of a wide variety of cutters make the milling machine indispensable in the shop and toolroom.

Classification of Milling Machines

Milling machines are made in a great variety of types and sizes. The drive may be either a cone-pulley-belt drive or an individual motor. The feed of the work may be by hand, by mechanical means, or by a hydraulic system. There is also a variety of possible table movements. The usual classification is in accordance with the general design, but even in this classification there is some overlapping. According to design, the distinctive types are:

1. Column and knee type
   (a) Hand miller.
   (b) Plain milling machine.
   (c) Universal milling machine.
2. Vertical milling machine
   (a) Reciprocating table.
   (b) Rotating table.
3. Planer milling machine.
4. Fixed-bed type
   (a) Simplex milling machine.
   (b) Duplex milling machine.
   (c) Triplex milling machine.
5. Special types
   (a) Drum milling machine.
   (b) Thread milling machine.
   (c) Cam milling machine.
   (d) Offset milling machine.
   (e) Duplicator or profiling machine.
   (f) Pantograph milling machine.

Hand and Plain Milling Machines

The simplest type of milling machine is *hand-operated*. It may have either the column and knee construction or the table mounted on a fixed bed. Machines operated by hand are used principally in production work for light and simple milling operations, such as cutting grooves, short keyways, and slotting. These machines have a horizontal arbor for holding the cutter and a work table which is usually provided with three movements. The work is fed to the rotating cutter either by the hand movement of a long lever, or by a hand screw feed.

The *plain milling machine* is similar to the hand machine except that it is of sturdier construction and is provided with power-feeding mechanism to control the table movements. Plain milling machines of the column and knee type have three motions, longitudinal, transverse, and vertical. Those of the fixed-bed type, as shown in Figure 1, have only longitudinal table travel, but have provision for transverse and vertical adjustments on the spindle which holds the milling cutter arbor. The machine is a fast-cutting manufacturing-type unit having complete electric control for all table and spindle movements. Twenty-eight feeds are provided, ranging from 1/2 to 35 inches per minute, which may be operated in either direction. Fast table travel, up to 300 inches per minute, is provided for quick positioning of the table. Plain milling machines are especially adapted to form milling and other types of production work.

A hydraulically operated plain milling machine of the fixed-bed type is shown in Figure 2. This machine, which is designed for

Courtesy Brown & Sharpe Manufacturing Company.

Fig. 1.   Plain Milling Machine.

simple production jobs, is equipped with a fully automatic table cycle. The table can be rapidly traversed in either direction, automatically shifted from rapid traverse to feed, reversed, or stopped at any point. The table has only a longitudinal movement while vertical and transverse movements of the cutter are accomplished in the head assembly. This machine can be used for many production jobs involving slotting, facing, plain and form cutting.

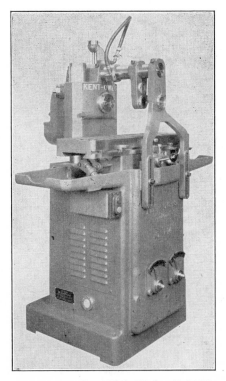

FIG. 2. Kent-Owens No. 1-14 Plain Hydraulic Milling Machine.

Universal Milling Machine

The *universal* milling machine is the most versatile of all the various types. It is essentially a toolroom machine, constructed for accurate work. In appearance it is quite similar to the plain type of milling machine, but differs in that (1) the work table is provided with a fourth movement which permits the table to swivel horizontally, and (2) it is equipped with an index or dividing head. The machine was originally designed with the swiveling feature to permit the cutting of spirals, such as are found on drills, milling cutters, and cams. In Figure 3 is shown a universal machine of modern design. In addition to being equipped with the dividing-head equipment, universal millers may also be provided with vertical milling attachment, rotary-table attachment, vise, and other similar accessories, all of which add to its utility as a toolroom machine. Aside from doing all types of milling operations, these machines will also do practically any type of operation that can be done on a shaper or a drill press.

Courtesy Cincinnati Milling Machine and Cincinnati Grinders, Inc.

FIG. 3: Universal Dial-Type Milling Machine.

Vertical Milling Machine

A typical *vertical* milling machine is shown in Figure 4. It is so called because of the vertical position of the cutter spindle. The table movements in this type of machine are the same as in plain machines. Ordinarily no movement is given to the cutter other than usual rotational motion. However, the spindle head may be swiveled, which permits setting the spindle in a vertical plane at any angle from vertical to horizontal. This machine is also provided with a short axial spindle travel to facilitate step milling. Some vertical milling machines are provided with rotary attachments or rotating · work tables to permit the milling of circular grooves or continuous

Courtesy Brown & Sharpe Manufacturing Company.

FIG. 4. Vertical Milling Machine.

milling of small production parts. Cutters used on vertical milling machines are all of the end-mill type.

Drilling, boring, and reaming can also be done on these machines. Accurate spacing of holes is possible because of the micrometer adjustment of the table. The machine is especially adapted to taking facing cuts and finishing in recesses. Profiling and die-sinking machines are very similar to vertical milling machines in their operation.

A special hydraulically operated vertical milling machine having four spindles is shown in Figure 5. This machine is equipped with a 360-degree automatic profiling unit (shown at the right of the spindles) and has two banks of holding fixtures. The operator can load and unload one bank while the parts in the other bank are

Courtesy Cincinnati Milling & Grinding Machines, Inc.

FIG. 5. Vertical Hydro-Tel Milling Machine Equipped with Four-Spindle Head and Automatic Profiling Unit.

being profile milled. This operation proceeds automatically with no attention from the operator. The bed of the machine has only a longitudinal movement and is built extra long to provide ample work space. This machine can perform a wide variety of production milling jobs and is equipped to handle automatically die-sinking and profiling operations.

Planer-Type Milling Machines

This type of milling machine receives its name from its resemblance to a planer. The work is carried on a long table, having only a longitudinal movement, and is fed against the rotating cutter at the proper speed. The variable table-feeding movement and the rotating cutter are the principal features that distinguish this machine from a planer. Transverse and vertical movements are provided on the cutter spindle carried. These machines are designed for milling large work requiring heavy stock removal and for accurate duplication of contours and profiles. A hydraulically operated unit of this type is shown in Figure 6.

Courtesy Cincinnati Milling & Grinding Machines, Inc.

Fɪɢ. 6.   Planer-Type Milling Machine.

Fixed-Bed Types of Milling Machines

Machines of this type are essentially production machines and are of rugged construction. The names *simplex, duplex,* and *triplex* indicate that the machine is provided, respectively, with single, double, and triple spindle heads. Figure 7 illustrates a duplex machine set up for rough-milling the master rod of an airplane engine. This machine, known as a Duplex Hydromatic, is equipped with an automatic dog-controlled hydraulic table feed. An infinite variety of feeds is available within the range of the machine, and it is also possible automatically to change the feed during the cut.   Any feed cycle needed can be obtained automatically.

FIG. 7.   Hydromatic Duplex Milling Machine.

Special Milling Machines

 A large variety of machines has been developed to take care of special types of milling operations.   In Figure 8 is shown a machine for milling plate cams to the proper shape.   A master cam held in contact with a fixed roller is mounted on the work table and controls the table movements.   The work, which is mounted on a face plate, is slowly fed against the rotating cutter, its outline being controlled by the master cam.

Drum-type millers are special machines designed for production work.   Most machines of this type have a large drum fixture, similar to the turret on a turret lathe, upon which the work is mounted.   As illustrated in Figure 9, the drum fixture rotates slowly, carrying the work against the rotating cutters.   Usually there are four cutter spindles.   The operation is continuous, since the parts are removed and new ones are added after the work has completed its cycle.

Figure 9 shows a large machine of this type designed for machining the top faces of cylinder blocks.   Much of the work formerly done on drum millers is now done on broaching machines.

Thread milling machines, discussed in Chapter 13, are used to cut threads on taps, lead screws, worms, and other parts, where accuracy and good finish are required.

The *offset miller* is another production-type machine.   In most of these millers the cutter is mounted on a vertical spindle which extends into a fixture.   This fixture is offset, or mounted eccentrically with

Courtesy Automatic Machinery Manufacturing Company.

Fig. 8.   Cam Milling Machine.

the cutter, so that as it revolves the work is milled as it passes by. The operation is similar to internal grinding with the work being rotated.   Only single milling operations are performed in this manner such as slotting or straddle cutting.

A large variety of machines has been developed for die and mold cutting, engraving, and profiling.   Machines for these purposes are known as *duplicators, die sinkers, profiling machines, pantograph machines,* and so on.   Most of them are a special adaptation of a vertical milling machine, although some few operate with the spindle in a horizontal position.   The hand-profiling machine is perhaps the simplest, having a rotating cutter whose motion is controlled by hand movements of the work table.   These movements are guided by moving the table so that the guide pin is in contact with some form or template.   This, in general, is the principle involved in all machines of this type, except that in many of them the movement is automatically controlled from the template although most of these machines operate only in one plane, some machines are three dimen-

Courtesy Ingersoll Milling Machine Company.

FIG. 9:   Drum-Type Milling Machine.

sional in their operation.   Those known pantograph machines will reproduce from a template at either an enlarged or a reduced scale.

A two-dimensional pantograph used for engraving work is shown in Figure 10.   The template is mounted at the upper rear of the machine and is engaged by a hand-guided pointer.   This pointer controls the

Courtesy George Gorton Machine Company.

FIG. 10. Two-Dimensional Pantograph Machines Used for Engraving Work.

movement of the rotating cutter through the pantograph mechanism shown at the top of the machine. By changing the link arrangement of the pantograph, any figure or design can be enlarged or reduced in size according to desired proportions. The machine shown is for light work in one plane and is used for engraving and light-metal die work. For larger work and for jobs requiring three-dimensional machining, a machine such as is shown in Figure 11 can be used. This machine is capable of reproducing from a model of any shape or

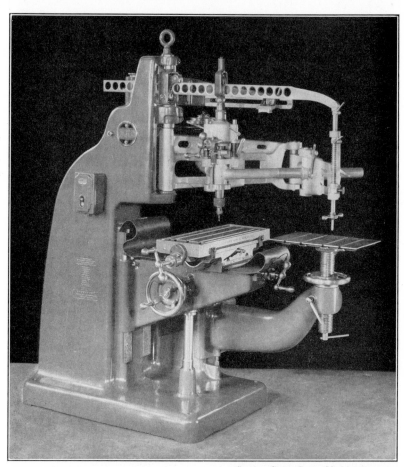

Courtesy George Gorton Machine Company.

Fig. 11. Front View of Three-Dimensional Pantograph Machines for Modeling Direct from Model of Any Shape or Contour.

contour. Because it is equipped with a pantograph mechanism, an increase or reduction in size can be obtained.

The production of large forming dies for automobile fenders, tops, and panels is also an important use of duplicator machines. Such a machine, of large capacity, is illustrated in Figure 12. The machine itself rests on a track and is entirely automatic in its operation. Both the template and the die block are supported on the large angle bracket in front of the machine. A template of the part to be produced is first prepared and is mounted above as shown in the figure.

Courtesy Ingersoll Milling Machine Company.

FIG. 12. Automatic Duplicator Machine for Die Work.

These templates are made of hard wood, plaster of Paris, or other easily worked materials, as the only purpose they serve is to guide the pointer which controls the tool position. The method of cutting and general operation of the machine can be seen by reference to the figure.

The Index Head and Its Operation

An *index* or *dividing head* is used to rotate the work through a certain number of degrees or through a fraction of a revolution. It is a regular piece of equipment on a universal milling machine, but may be used on some of the other types as well. In Figure 13 is shown an index head and its footstock mounted on the work table of a machine. Since much of the work to be machined has to be supported between centers, both units are necessary.

Courtesy Cincinnati Milling & Grinding Machines, Inc.

Fig. 13.   Index Head Assembly.

The index head is nothing more than a worm-gear reducer having a ratio of 40 to 1 — that is, 40 turns of the crank will rotate the work one complete revolution. The work spindle which is attached to the worm gear has a tapered hole to receive the live center and is also threaded to hold a chuck. A U-shaped piece is ordinarily on the spindle to give positive motion to the work through a dog. Just back of this piece is a direct index plate having 24 equally spaced holes. The entire spindle assembly may be swiveled from horizontal to vertical position. The shaft which operates the worm gear extends to the side of the mechanism and through the index plate and sector. On its end is a crank, having an index pin, which fits into the holes on the index plate. Since the plate does not turn, the crank is locked when the index pin is engaged in one of the holes. The index plate is provided with several concentric circles of equally spaced holes to assist in determining the proper number of revolutions of the handle where a fraction of a revolution is involved. The sector arms on the plate are to eliminate the counting of spaces when the handle is being turned between cuts.

The three types of indexing used are *simple, direct,* and *differential.* Most indexing is of the first type and is accomplished by turning the crank a number of turns to rotate the work the desired amount, the index plate being held in a fixed position. With a ratio of 40 to 1, one revolution of the crank will rotate the work 1/40 of a revolution. Hence, in cutting a gear with 40 teeth the crank would be locked to the plate by the index pin and a cut would be made. After the cut, the handle would be turned one revolution and another cut taken, and so on. To cut a gear with 20 teeth would require 2 turns of the handle, or to cut 8 flutes on a reamer 5 turns would be required. As long as the number of cuts to be taken is a factor of 40, it is a simple matter to calculate the number of handle turns. By following these simple calculations it is quite obvious that an expression can be set up to compute the number of handle turns for a given condition. The rule to use is:

$$\text{Number of turns of index handle} = \frac{\text{turns of handle to produce one turn of work (usually 40)}}{\text{number of cuts to be made in one revolution of work}}$$

When the number of cuts to be made is not evenly divisible into 40, it follows that fractional turns must be made, and for this it is necessary to use the sector device. Assume that it is desired to cut 32 teeth on a spur gear. If the aforementioned rule is used, the number of turns of the index handle would be 40/32 or $1\frac{1}{4}$ turns. First select a circle on the index plate that is divisible by 4. If a 24-space circle is available, set the sector arms so that there are 6 spaces between them. Hence 1 revolution of the crank plus 6 spaces would give the work the required movement. The same results could be obtained by using a circle with 48 spaces and turning one turn plus 12 spaces. To cut a gear with 72 teeth, the number of turns would be in the ratio of 40/72 or 5/9. If a 54-space circle was available, the sector arms should be set for 30 spaces. In setting the arms, care must be exercised to count spaces and not holes.

Direct indexing is accomplished by using the index plate attached to the work spindle. This plate has 24 divisions and is engaged by a plunger pin on the head. The worm is disengaged from the worm gear, and the plate is turned the required amount by hand. This system of indexing is limited only to those divisions which are factors of 24. It is a quick method of indexing and used when only a few cuts are required in a revolution.

When indexing is done by degrees with a 40-to-1 index head, each turn of the handle represents 360/40 or 9 degrees. If a 27-space circle were selected a movement of three spaces of the handle would move the work 1 degree. Four spaces on a 36-hole circle would also represent 1 degree, while one space would move the work only 1/4 degree. By similar calculations the work may be moved any number of degrees or through most common fractions of a degree.

Differential indexing is used when the work has to be turned an amount which cannot be obtained by simple indexing, owing to the lack of a circle on the index plate with the correct number of spaces. For such conditions the index plate is unlocked and geared to a train of gears which receive their motion from the worm-gear spindle. As the handle is turned, the index plate also turns, but at a different rate and perhaps in the opposite direction. Its movement depends on the gears used to drive it. Information for calculating the correct gears to use for a given condition may be found in machine instruction books. After the gears are set up, the operation is similar to simple indexing. Differential indexing makes it possible to rotate the work any fraction of a revolution with the usual index plates furnished with the equipment.

Spiral milling is accomplished by rotating the work as it moves against the rotating cutter. This is done with the use of connecting gears from the lead screw of the work table to the handle spindle of the index head. The lead of the spiral, the distance it advanced in one revolution, is controlled by these gears. With a 40-to-1 reduction in the index head and a 4-pitch lead screw, the lead would be 10 inches, if no increase or reduction in the outside gears is assumed. The ratio for computing these gears is:

$$\frac{\text{Product of the driving gears}}{\text{Product of the driven gears}} = \frac{10}{\text{desired lead}}$$

Spiral milling is used in cutting spiral gears, flutes on various tools, screws, worm gears, and some types of cams.

Types of Milling Cutters

The milling machine is most versatile because of the large variety of milling cutters available. These cutters are usually classified according to their general shape, although in some cases they are classified by the way they are mounted, the material used in the teeth, or the method used in grinding the teeth.

There are three methods of mounting cutters on a milling machine, giving rise to three general types or designs of cutters, namely:

1. Arbor cutters. These cutters have a hole in the center for mounting on an arbor.

2. Shank cutters. Cutters of this type have either a straight or tapered shank integral with the body of the cutter. When in use, these cutters are mounted in the spindle nose or in a spindle adapter.

3. Face cutters. These cutters are bolted or held on the end of short arbors, and are generally used for milling plane surfaces.

Classification according to materials follows similar classifications of other types of cutting tools. Milling cutters are made of high-carbon steels or various high-speed steels, with sintered-carbide tips, or of certain cast nonferrous alloys. High-carbon steel cutters have a limited use, as they dull quickly if high cutting speeds and feeds are used. These cutters will rapidly lose their hardness at temperatures above 400 F. Most general-purpose cutters are made of high-speed steels, as such steels maintain a keen cutting edge at temperatures around 1000 to 1100 F. Consequently they may be used at cutting speeds 2 to $2\frac{1}{2}$ times those recommended for carbon-steel cutters. Cast nonferrous metals such as Stellite, Crobalt, or Rex-alloy, and carbide-tipped cutters, have even greater resistance to heat and are especially adapted to heavy cuts and high cutting speeds. These materials are either used as inserts held in the body of the cutter or brazed directly on the tips of the teeth. Sintered-carbide inserts or tips cannot be softened by heat treatment and have a red hardness surpassing any other tool material. At 1300 F they maintain a hardness equal to that attained in high-speed steels at room temperature. Cutting speeds for these cutters range from two to five times those recommended for high-speed steel.

Teeth in milling cutters are made in two general styles according to the method used in sharpening them. *Profile cutters* are sharpened by grinding a small land back of the cutting edge of the tooth. This also provides the necessary relief at the back of the cutting edge. *Formed cutters* are made with the relief back of the cutting edge the same contour as the cutting edge. To sharpen these cutters the face of the cutter is ground so as not to destroy the tooth contour.

The types of cutters most generally used are shown in Figure 14. These cutters are classified principally according to their general shape or the type of work they will do.

1. Plain milling cutter. A plain cutter is a disk-shaped cutter having teeth only on the circumference. The teeth may be either straight or spiral, but they are usually spiral if the width exceeds 5/8 inch. Wide spiral cut-

Courtesy Brown and Sharpe Manufacturing Company.

FIG. 14.   Types of Milling Cutters.

(a) End-mill cutter.   (f) Plain milling cutter.
(b) T-slot cutter.   (g) Inserted tooth cutter.
(c) Spiral plain milling cutter.   (h) Metal slitting saw cutter.
(d) Angle milling cutter.   (i) Gear cutter.
(e) Woodruff keyseat cutter.   (j) Side milling cutter.

ters of this type used for heavy slabbing work may have notches in the teeth to break up the chips and facilitate their removal.

2. Side milling cutter. This cutter is similar to a plain cutter except that it has teeth on the sides also. In some cases, where two cutters operate together, the cutter is plain on one side with teeth on the other. Cutters of this type may have straight, spiral, or staggered teeth.

3. Metal slitting saw cutter. This cutter resembles a plain or side cutter except that it is made very thin, usually 3/16 inch or less. Plain cutters of this type are relieved by grinding the sides to afford clearance for the cutter.

4. Angle milling cutter. Any cutter, angle-shaped, comes under this classification. Cutters of this type are made into both single- and double-angle cutters. The single-angle cutters have one conical surface whereas the double-angle cutters have teeth on two conical surfaces. Angle cutters are used for such purposes as cutting ratchet wheels, dovetails, flutes on milling cutters, and reamers.

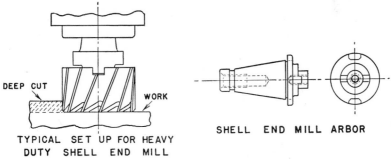

DEEP CUT

WORK

SHELL END MILL ARBOR

TYPICAL SET UP FOR HEAVY
DUTY SHELL END MILL

Fig. 15. Shell End Mill and Arbor.

5. Form milling cutters. This name is applied to any cutter on which the teeth are given a special shape. This group includes convex and concave cutters, gear cutters, fluting cutters, corner-rounding cutters, and many others.

6. End-mill cutters. These cutters have an integral shaft for driving and have teeth on both periphery and ends. The flutes on these cutters may be either straight or helical. Large cutters of this type, called shell end mills, have the cutter part separate and are held to a stub arbor, as shown in Figure 15. Owing to the cost of high-speed steel, this construction results in a considerable saving in material cost. End mills are used for surfacing projections, squaring ends, cutting slots, and recess work such as in die making.

7. T-slot cutters. Cutters of this type resemble small plain or side milling cutters which have an integral straight or tapered shaft for driving. They are used for milling T slots. A special form of this cutter is the Woodruff keyseat cutter. They are made in standard sizes for cutting the round seats for Woodruff keys.

8. Inserted tooth cutter. As cutters increase in size, it is economical to insert the teeth made of expensive material into less expensive ordinary steel. Teeth in such cutters may be replaced when worn out or broken.

Milling-Cutter Teeth

A typical milling cutter, with various angles and cutter nomenclature, is shown in Figure 16. For most high-speed cutters positive *radial rake angles* of from 10 to 15 degrees are used. These values are satisfactory for most materials and represent a compromise between good shearing or cutting ability and strength. Milling cutters made for softer materials, such as aluminum, can be given much greater rake with improved cutting ability.

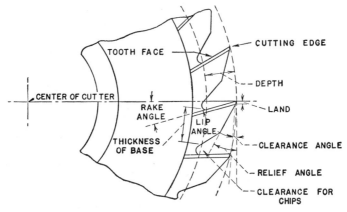

FIG. 16. Milling Cutter with Various Angles Indicated.

Usually only saw-type and narrow plain milling cutters have straight teeth with zero axial rake. As cutters increase in width, a positive-axial-rake angle is used to increase cutting efficiency.

For high-speed milling with carbide-tipped cutters, negative-rake angles (both radial and axial) are generally used. Improved tool life is obtained by the resultant increase in the *lip angle;* also, the tooth is better able to resist shock loads. Plain milling-type cutters with teeth on the periphery usually are given a negative rake of from 5 to 10 degrees when steel is being cut. Alloys and medium-carbon steels require greater negative rake than soft steels. Exceptions to the use of negative-rake angles for carbide cutters are made when soft nonferrous metals are milled.

The *clearance angle* is the included angle between the land and a tangent to the cutter from the tip of the tooth. It is always positive and should be small so as not to weaken the cutting edge of the tooth. For most commercial cutters over 3 inches in diameter the clearance angle is around 4 to 5 degrees. Smaller-diameter cutters have increased clearance angles to eliminate tendencies for the teeth to rub on the work. Clearance values also depend on the various work materials. Cast iron requires values of from 4 to 7 degrees, whereas soft materials such as magnesium, aluminum, and brass are cut efficiently with clearance angles of from 10 to 12 degrees. The width of the land should be kept small; its usual values are from 1/32 to 1/16 inch. A *secondary clearance* is ground back of the land to keep the width of the land within proper limits.

Much research on cutter form and size has proved that coarse teeth are more efficient for removing metal than fine teeth. A coarse-tooth cutter takes thicker chips, has freer cutting action, and has more clearance space for the chips. As a consequence, these cutters result in increased production and decreased power consumption for a given amount of metal removed. Also, fine-tooth cutters have a greater tendency to chatter than those with coarse teeth. However, such cutters are recommended for saw cutters for the milling of thin materials.

Cutting Speed and Feed

The cutting speed of a milling cutter is determined by the peripheral or surface speed of the cutter. The movement of the work past the cutter is not considered in this calculation. The cutting speed may be expressed by the following equation:

$$CS = \frac{\pi DN}{12} = \text{feet per minute}$$

D is diameter of cutter in inches and N is rpm.

Since the cutting speed is seldom the unknown, the equation is generally expressed in terms of spindle revolution.

$$N = \frac{CS \times 12}{\pi D} = \text{rpm of spindle}$$

If the cutter diameter and cutting speed of the given material are known, this expression gives the proper rotational speed of the spindle. In the selection of the proper cutting speed the following factors should be considered:

1. **Cutter material.** Cutting speeds are usually given in values for high-speed steel cutters. These values are twice those for carbon-steel cutters and one-half to one-fifth those recommended for carbon-tipped cutters.

2. Kind of material being cut. Approximately, roughing cuts on cold rolled steel are 80–100 feet per minute, cast iron 50–60 feet per minute, brass 150–200 feet per minute, bronze 80–120 feet per minute, aluminum 400–500 feet per minute, and magnesium 600–800 feet per minute. Speeds about 20% higher can be used for finish cuts. Additional cutting speeds are given in Table 7.

3. Rigidity and size of machine. This factor is of particular importance when old or lightly constructed equipment is used.

4. Type of finish desired. Best finishes are obtained with light feeds and high cutting speeds.

5. Size of cut. Heavy cuts which accumulate heat rapidly must be taken more slowly than shallow cuts.

6. Kind of coolant used. The heat generated in milling must be dissipated to protect the cutter and the work. To accomplish this, the tool and work are usually flooded with an appropriate coolant such as soluble, sulfurized, or mineral lard oil. An exception to this is cast iron, which is milled dry because of the abrasive action of the chips. Kerosene is frequently used as a coolant for aluminum. Since water mixtures present a fire hazard with magnesium, only straight cutting oils should be used in machining this metal.

There are two possible methods of feeding the work to the cutter, as shown in Figure 17. Feeding the work against the cutter, as in *A*, is usually recommended, as each tooth starts its cut in clean metal

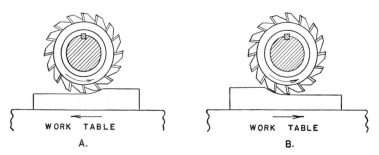

Fig. 17. Methods of Feeding Work on Milling Machine.

and does not have to break through possible surface scale. However, tests have proved that, when the work is fed in the same direction as cutter rotation (as shown at *B*), the cutting is more efficient. Larger chips are removed, and there is less tendency for chatter. This method is frequently used in production work, where large cuts are to be taken and the surface of the work is free from scale.

Feed on milling machines is expressed in either of two ways. On some machines it is expressed in thousandths of an inch per revolution of cutter. Such machines have feed changes ranging from 0.006

to as high as 0.300 inch. The other method is to express the feed of table in inches per minute, the usual range being from 1/2 to 20 inches per minute.

REVIEW QUESTIONS

1. Show by sketch how work should be fed against a rotating cutter in ordinary milling. State the reasons for your answer.

2. What is meant by a fixed-bed milling machine?

3. How does a universal milling machine differ from a plain milling machine?

4. How does a vertical milling machine differ from an upright drill press?

5. Sketch and name eight types of milling cutters.

6. What is the approximate spindle speed for a 5-inch cutter to mill cast iron?

7. Describe an index head, and state its purpose.

8. Distinguish among simple, direct, and differential indexing.

9. If it is desired to cut 32 teeth on a spur gear, how many turns of the index crank should be made between each two cuts? Same for 66 teeth?

10. Explain how a spiral gear would be cut on a universal milling machine.

11. How does a planer-type milling machine differ from a planer?

12. What factors determine the feed on a milling machine?

13. What is an offset miller, and how does it operate?

14. Explain the operation of a duplicator machine.

15. What type of machines are used for machining metal dies and molds?

16. What advantage does a coarse-tooth milling cutter have over a fine-tooth cutter?

17. What is the purpose of notching the teeth on some milling cutters?

18. Discuss the relative merits of a shaper and a milling machine for machining plane surfaces.

BIBLIOGRAPHY

ARMITAGE, J. B., and SCHMIDT, A. O., " Rake Angles in Face Milling," *Mechanical Engineering*, June, July, August, 1945.

HINMAN, C. W., *Practical Design for Drilling and Milling Tools*, McGraw-Hill Book Company, 1938.

LUCHT, FRED W., " Face-Milling with Carbide," *Mechanical Engineering*, March 1945.

MARTELLOTTI, M. (*a*) " Milling Cutters and How to Use Them," *American Machinist*, August 17, 1944. (*b*) " Special Rake Angles Used for Carbide Tools," *American Machinist*, February 15, 1945. (*c*) " Clearance and Relief Angles for Milling Cutters," *American Machinist*, March 15, 1945. (*d*) " Selection and Application of Milling Cutters," *American Machinist*, March 16, 1944.

MEYER, A. W., and ARCHIBALD, F. R., " Carbide Milling of Steel," *Mechanical Engineering*, October 1945.

CHAPTER 17

GEARS AND GEAR-CUTTING MACHINES

Gears are a common method of transmitting power or rotary motion from one shaft to another. They have the advantage over friction and belt drives in that they are positive in their action. Most machinery requires positive action, since exact speed ratios are essential; however, friction and surface-contact drives have some use in industry where high speeds and light loads are required and in a few cases where loads subjected to impact are transmitted. A gear differs from a friction disk in that it has projections or teeth built up on its circumference so that it may transmit motion through the meshing teeth without slippage.

If teeth were to be built up on the circumferences of two rolling disks in contact with each other, recesses would have to be provided between the teeth so as to eliminate interference. The original diameter of each disk would still figure in the gear calculations, however. This dimension is known as the *pitch diameter*. It is only an imaginary circle and cannot be seen by inspecting a gear. A portion of a gear is shown in Figure 1 with the pitch circle indicated.

Kinds of Gears

The gears most commonly used are those which transmit power between two parallel shafts. Such gears having their tooth elements parallel to the rotating shafts are known as *spur gears*. If the elements of the teeth are twisted or helical, as shown in Figure 2, they are known as *helical gears*. These gears may also be made for connecting shafts that are at an angle in the same or different planes. The advantage of helical gears is that they are smooth-acting, because there is always more than one tooth in contact. Some power is lost because of end thrust, and provision must be made to compensate for this thrust in the bearings. The *herringbone gear* is equivalent to two helical gears, one having a right-hand and the other a left-hand helix.

Usually when two shafts are in the same plane, but at an angle with one another, a *bevel gear* is used. Such a gear is similar in appear-

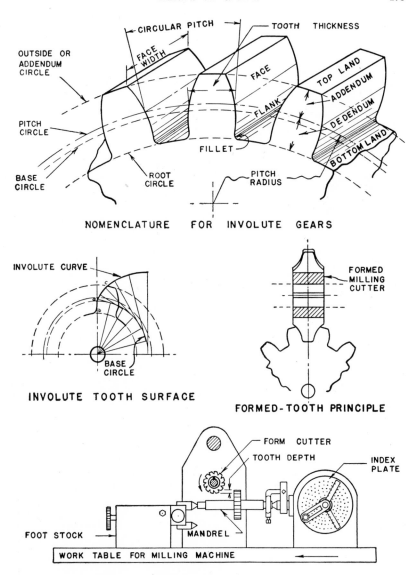

FIG. 1. Nomenclature for Involute Gears.

ance to the frustum of a cone having all the elements of the teeth intersecting at a point, as shown in Figure 3. Bevel gears are made with either straight or spiral teeth. When the shafts are at right angles and the two bevel gears are the same size, they are known as *miter gears.* *Hypoid gears,* an interesting modification of bevel gears (see Figure 4), have their shafts at right angles but they do not intersect as do the shafts for bevel gears. Correct teeth for these gears are difficult to construct, although a generating process has been developed which produces teeth with satisfactory action. One of the principal uses of hypoid gears is that between the driving shaft and rear axle of automobiles.

FIG. 2. Helical Gears.

Worm gearing is used where a large speed reduction is desired. The small driving gear is called a *worm* and the driven gear a *wheel.* The worm in appearance resembles a large screw and is set in close to the wheel circumference, the teeth of the wheel being curved to conform to the diameter of the worm. The shafts for such gears are at right angles but not in the same plane. These gears are similar to helical gears in their application, but differ considerably in appearance and method of manufacture. Several worm-gear sets are shown in Figure 5.

In addition to the various kinds of gears just discussed, there are several special types that deserve mention. *Rack gears* are straight, having no curvature, and represent a gear of infinite radius. Such gears are used in feeding mechanisms and for reciprocat-

FIG. 3. Miter Bevel Gears.

ing drives. They may have either straight or helical teeth. If the
rack is bent in the form of a circle, it becomes a bevel gear having a

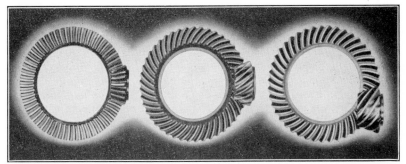

Courtesy Gleason Works.

Fig. 4. Helical Bevel and Hypoid Gears.

Courtesy Foote Brothers Gear and Machine Corporation.

Fig. 5. Worm-Gear Sets.

cone apex angle of 180 degrees, which is known as a *crown gear*. The
teeth all converge at the center of the disk and mesh properly with a
bevel gear of the same pitch. A gear with internal teeth is known as

an *annular gear*. It can be cut to mesh with either a spur or bevel gear, depending on whether the shafts are parallel or intersecting.

Gear Nomenclature

The system of gearing used in the United States is known as the *involute system*, since the profile of a gear tooth is principally an involute curve. An *involute* is a curve generated on a circle, the normals of which are all tangent to this circle. The method of generating an involute is shown in Figure 1. Assume that a string, having a pencil on its end, is wrapped around a cylinder. The curve described by the pencil as the string is unwound is an involute, and the cylinder upon which it is wound is known as the *base circle*. The portion of the gear tooth, from the base circle at A to the outside diameter at C, is an involute curve and is the portion of the tooth that contacts other teeth. From B to A the profile of the tooth is a radial line down to the small fillet at the root diameter. The location of the base circle upon which the involute is described is inside the pitch circle and dependent upon the angle of thrust of the gear teeth. The two systems most commonly used have their lines of action at $14\frac{1}{2}$ and 20 degrees, respectively. Other angles may be used, but the larger the angle is made the greater will be the radial force component tending to force the gears apart. If a common tangent is drawn to the pitch circles of two meshing gears, the line of action or angle of thrust is drawn at the proper angle ($14\frac{1}{2}$ degrees) to this line. The base circles, upon which the involutes are drawn, are tangent to the line of action. Involute gears fulfill all the laws of gearing and have the advantage over some other curves in that the contact action is not affected by slight variations of gear center distances.

Referring again to Figure 1, we see the nomenclature of a gear tooth clearly illustrated. The principal definitions and tooth parts for standard $14\frac{1}{2}$-degree involute gears are:

The *addendum* of a tooth is the radial distance from the pitch circle to the outside diameter or addendum circle. Numerically, it is equal to one divided by the diametral pitch.

The *dedendum* is the radial distance from the pitch circle to the root or dedendum circle. It is equal to the addendum plus the tooth clearance.

Tooth thickness is the thickness of the tooth measured on the pitch circle. For cut gears the tooth thickness and tooth space are equal. Cast gears are provided with some *backlash* — the difference between the tooth thickness and tooth space, measured on the pitch circle.

The *face* of a gear tooth is that surface lying between the pitch circle and the addendum circle.

The *flank* of a gear tooth is that surface lying between the pitch circle and the dedendum circle.

Clearance is a small distance provided so that the top of a meshing tooth will not touch the bottom land of the other gear as it passes the lines of centers.

Table 10 gives the proportions of standard $14\frac{1}{2}$-degree and 20-degree involute gears expressed in terms of diametral pitch and number of teeth.

TABLE 10

AESC STANDARD FOR INVOLUTE GEARING

	$14\frac{1}{2}$ Pressure Angle	20 Pressure Angle
Addendum	$\dfrac{1}{P_d}$	$\dfrac{0.8}{P_d}$
Clearance	$\dfrac{0.157}{P_d}$	$\dfrac{0.2}{P_d}$
Dedendum	$\dfrac{1.157}{P_d}$	$\dfrac{1}{P_d}$
Outside Diameter	$\dfrac{T+2}{P_d}$	$\dfrac{T+1.6}{P_d}$
Pitch diameter	$\dfrac{T}{P_d}$	$\dfrac{T}{P_d}$

Pitch of Gears

The *circular pitch* P_c is the distance from a point on one tooth to the corresponding point on an adjacent tooth, and it is measured on the pitch circle. Expressed as an equation,

$$P_c = \frac{\pi D}{T}$$

where D is diameter of pitch circle and T is number of teeth.

The *diametral pitch* P_d is the ratio of the number of teeth to the pitch diameter. It may be expressed by the following equation:

$$P_d = \frac{T}{D}$$

Multiplying these two equations, we obtain the following relationship between circular and diametral pitch:

$$P_c \times P_d = \frac{\pi D}{T} \times \frac{T}{D} = \pi$$

Hence, knowing the value of either pitch, we may obtain the other by dividing it into π.

All gears and gear cutters are standardized according to diametral pitch, as this pitch can be expressed in even figures or fractions. Circular pitch, being an actual distance, is expressed in inches and thousandths of an inch. A 6-pitch gear indicates one that has 6 teeth per inch of pitch diameter. If the pitch diameter is 3 inches, the number of teeth is 3×6 or 18. The outside diameter of the gear is equal to the pitch diameter plus twice the addendum distance, or 3 inches $+ 2 \times 1/6$, which is 3.333 inches.

Any involute gear of a given pitch will mesh properly with a gear of any other size of the same pitch. However, in cutting gears of various diameters, a slight difference in the cutter is necessary, to allow for the change in curvature of the involute as the diameter increases. The extreme case would be a rack tooth, which would have a straight line as the theoretical tooth profile. For practical reasons, the number of teeth in an involute gear should be not less than 12.

Methods of Making Gears

Most gears are produced by some machining process. Accurate machine work is essential for high-speed long-wearing quiet-operating gears. Gears operating at slow speeds and under exposed conditions may be sand-cast, but such gears are not efficient in their power transmission. Die casting of small gears carrying light loads has proved very satisfactory. The materials for such gears are limited to low-temperature melting metals and alloys; consequently, these gears do not have the wearing qualities of heat-treated steel gears. Stamping, while reasonably accurate, can be used only in making thin gears from sheet metal.

The various commercial methods employed in producing gears may be summarized as follows:

1. Casting
   - (a) Sand and plaster casting.
   - (b) Die casting.
2. Stamping.
3. Machining
   - (a) Formed-tooth process
     - (1) Form cutter in milling machine.
     - (2) Form cutter in broaching machine.
     - (3) Form cutter in shaper.
   - (b) Templet process.

(c) Cutter generating process
    (1) Cutter gear in shaper.
    (2) Hob cutter.
    (3) Rotary cutter.
    (4) Reciprocating cutters simulating a rack.
4. Powder metallurgy.

Formed-Tooth Process

A *formed* milling cutter, as shown in Figure 1, is commonly used for cutting a spur gear. Such a cutter is used on a milling machine, and the setup is shown in the lower part of the figure. The cutter is formed according to the shape of the tooth space to be removed. Theoretically, there should be a different-shaped cutter for each-sized gear of a given pitch, as there is a slight change in the curvature of the involute. However, one cutter can be used for several gears having different numbers of teeth without much sacrifice in their operating action. Each pitch cutter is made in eight slightly varying shapes to compensate for this change. They vary from no. 1, which is used to cut gears from 135 teeth to a rack, to no. 8, which cuts gears having 12 or 13 teeth. Special cutters are available where greater accuracy is necessary.

The method of setting up a milling machine to cut gears is discussed in Chapter 16 on milling machines and milling cutters. The formed milling process may be used with accurate results for cutting spur, helical, worm, and worm gears. Although sometimes used for bevel gears, the process is not accurate because of the gradual change in the tooth thickness. When it is used for bevel gears, at least two cuts are necessary for each tooth space. The usual practice is to take one center cut of proper depth and about equal to the space at the small end of the tooth. Two shaving cuts are then taken on each side of the tooth space to give the tooth its proper shape.

The formed-tooth principle may also be utilized in a broaching machine by making the broaching tool conform to the tooth space. Small internal gears can be completely cut in one operation by having a round broaching tool made with the same number of cutters as the gear has teeth. Broaching is limited to large production because of the high cost of the cutters.

A recent development in the roughing and semifinishing of spur and spiral gears is a new machine known as a " shear speed " gear shaper. This machine is designed so that all teeth on a gear are cut simultaneously by a ring of form-cutting tools or blades surrounding the gear blank. In Figure 6 is shown a view looking up inside the cutter head of this machine after a spur shoulder gear has been cut.

The cutting action of the machine is as follows: The gear is clamped on an arbor and the cutting head lowered and locked into position. A ram holding the gear blank is reciprocated, and at each up stroke the radial blades are fed into the work an equal amount. At the upper end of the stroke the blades are retracted slightly to

Courtesy Michigan Tool Company.

FIG. 6. Looking up Inside the Cutter Head of a "Shear-Speed" Gear Shaper. The Cutter Head Has Moved to Loading Position after Completing a Spur Shoulder Gear.

provide clearance as the work returns to starting position. As the blades approach the proper tooth depth, the feed is reduced by the controlling cam.

This method of gear cutting is very rapid; in many cases the actual cutting cycle is less than one minute. Accurate finishing of gears produced by this method is done on a gear-shaving machine.

Templet Gear-Cutting Process

In the *templet* process for cutting gear teeth, the form of the tooth is controlled by a templet instead of a formed tool. The tool itself is similar to a side-cutting shaper tool and is given a reciprocating motion in the process of cutting. The process is especially adapted to cutting large teeth which would be difficult to cut with a formed cutter, and also to cutting bevel-gear teeth. The principle involved in a bevel gear planer is shown in Figure 7. The frame carrying the reciprocating tool is guided at one end by a roller acting against a templet, while the other end is pivoted at a fixed point corresponding to the gear being cut. Three sets of templets are necessary, one for the center cut and one for each side of the

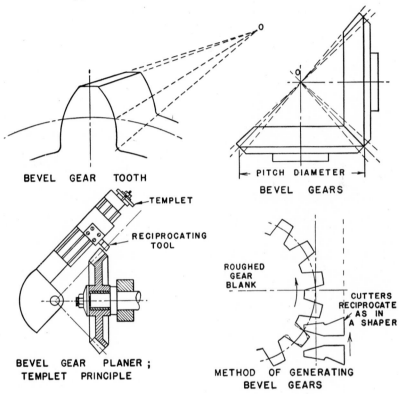

BEVEL GEAR TOOTH

PITCH DIAMETER

BEVEL GEARS

TEMPLET

RECIPROCATING TOOL

ROUGHED GEAR BLANK

CUTTERS RECIPROCATE AS IN A SHAPER

BEVEL GEAR PLANER; TEMPLET PRINCIPLE

METHOD OF GENERATING BEVEL GEARS

Fig. 7. Bevel-Gear Cutting.

tooth space. The gear blank is held stationary during the process and is moved only when indexed. This method of cutting produces an accurately formed tooth having the proper taper. Machines of this type are fully automatic in their operation.

For cutting gears with straight teeth, such as spur gears, the action is similar to that just described. In this case the cutter head is adjusted so that the tool travel is parallel to the elements of the teeth.

Cutter-Gear Generating Process

The *cutter-gear generating* process for cutting involute gears is based upon the fact that any two involute gears of the same pitch will mesh together. Hence, if one gear is made to act as a cutter and is given a reciprocating motion, as in a shaper, it will be capable of cutting into a gear blank and generating conjugate tooth forms. A gear-shaper cutter of this description is shown in Figure 8, and the adjacent Figure 9 shows how it is mounted in a Fellows

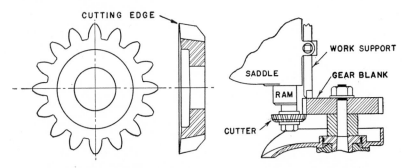

FIG. 8.   Gear-Shaper Cutter.   FIG. 9.   Gear Shaper.

gear shaper. At the start of the operation the gear is held stationary, and the reciprocating cutter is fed into the blank to the correct tooth depth. The feed is then stopped, and both cutter and gear blank are given intermittent rotary motion of the same pitch-line velocity. The rotation, or cutter feed, takes place at the end of the stroke.

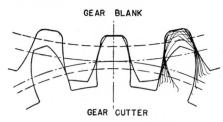

FIG. 10.   Generating Action of Fellows Gear-Shaper Cutter.

The generating action of the cutter and gear blank is shown in Figure 10, the fine lines indicating the amounts of metal removed by each cut in a given tooth space. The usual practice is to have the cut take place on the up stroke. However, in many cases, the cutting action is of necessity on the down stroke. A complete view of a Fellows gear shaper is shown in Figure 11.

The cutter-gear method of generating gears is not limited to involute spur gears, but has many other applications. By using a spiral cutter and giving it a twisting motion on the cutting stroke, spiral or helical gears may be generated. Worm threads may be cut in a similar fashion. Figure 12 illustrates the application of this process in the cutting of internal gears. In addition, this process may be used in cutting sprocket wheels, splines, gear-type clutches, cams, ratchet wheels, and many other straight and curved forms.

Another interesting application of the cutter-gear generating principle is the Sykes gear-generating machine, well known for its ability

Courtesy The Fellows Gear Shaper Company.

FIG. 11.   Gear Shaper Setup for Cutting Spiral Gear.

to cut continuous herringbone teeth. One of these machines is illustrated in Figure 13. This machine employs two cutter gears mounted in a horizontal position, as shown in Figure 14. In the cutting of herringbone gears the cutters are given a reciprocating motion, one cutting in one direction up to the center of the gear blank and the other cutting to the same point when the motion is reversed. The cutters not only reciprocate but also are given a twisting motion according to the helix angle. Both the gear blank and cutters slowly revolve, generating the teeth in the same fashion as is done in the Fellows shaper. Machines of this type are built in various sizes, up to those capable of cutting gears 22 feet in diameter.

Courtesy The Fellows Gear Shaper Company.

FIG. 12.   Shaping an Internal Gear.

An important feature of the cutter gear process is its ability to cut double and single helical gears, internal gears, spur gears, worms, racks, pump rotors, and a large variety of special forms. Two members of a cluster gear may be cut simultaneously, even though they are not the same pitch or same type of gear. The machine generates tooth contours of true involute shape with teeth uniformly spaced and smoothly finished.

Bevel-Gear Generators

Correctly formed bevel gears can also be cut with reciprocating tools, utilizing the generating principle. The principle involved is based on the fact that any bevel gear will mesh with a crown gear of the same pitch having its center coinciding with the cone apex of the gear. In the lower right-hand corner of Figure 7 the two cutting tools used represent the sides of adjacent teeth of a crown gear.

Courtesy Farrel-Birmingham Company, Inc.

FIG. 13. Sykes Gear-Generating Machine for Continuous-Tooth Herringbone Gears.

Courtesy Farrel-Birmingham Company, Inc.

FIG. 14. View of the Cutter Gears on Sykes Gear-Generating Machine.

These cutting tools rotate about the point which would be the axis of the crown gear, and at the same time are given a reciprocating motion. The gear blank is also rotated about its axis at the rate it would have were it meshing with the crown gear. As the tools are simulating the respective positions taken by the crown gear, the correct form of tooth is cut. Both sides of a single tooth are cut on

Fig. 15.   Straight-Bevel-Gear Generator.

a single generating roll of the tools and blank, and, at the end of the cut, the blank is withdrawn and indexed, while the tools return to correct position for the next cut.

Prior to the generating process, the tooth spaces are roughed without generating roll, as shown in the figure, so that only a small amount of metal is removed by the reciprocating tools when finishing. Both operations can be done on the straight-bevel-gear generator, although in some cases separate roughing machines are used. Frequently special form tools are used to produce a closer shape in roughing. A front view of a 12B straight-bevel-gear generator is shown in Figure 15. This machine is fully automatic and, when started, will complete a gear without further attention being given it. An important feature of this machine is its ability to obtain localized tooth bearing in straight-bevel gears. The slight crowning on the tooth surface localizes the tooth bearing on the center three quarters and eliminates load concentrations on the ends of the teeth.

Courtesy Gleason Works.

Fig. 16. Hypoid Gear Generator.

The method of cutting spiral-bevel gears also uses the generating principle, but the cutter in this case is circular and rotates as a face milling cutter. The cutter is similar to the one in Figure 17, which is shown cutting a hypoid pinion. The spiral teeth on gears cut by this process are curved on the arc of a circle, the radius being equal to the radius of the cutter. The blades of the cutter have straight cutting profiles to correspond with the tooth profile of a crown gear. The revolving cutters move through the same space as would be occupied by a crown gear tooth. As in the previous method, the teeth are first roughed out before the true shape is generated. The rotating cutters may be designed to cut only one or both sides of the tooth space, the latter-type cutter having the advantage of more rapid production. Spiral-bevel gears have an advantage over straight-bevel gears in that the teeth engage with one another gradually, eliminating any shock or noise in their operation.

Courtesy Gleason Works.

FIG. 17.   Close-up of Work Head Shown in Position for Cutting a Hypoid Pinion.

A special type of gear, known as a hypoid gear, can be cut in the machine just described. The hypoid gear, widely used in drives for automobiles, streetcars, motorcycles, and similar applications, has the axis of the pinion offset and does not intersect the axis of the gear. Such a gear is shown in Figure 4 with two types of bevel gears. As the cutter rotates (see Figure 17) it is fed into the gear blank and then withdrawn. There is an accompanying rolling generating movement of the cutter cradle and gear blank to produce the correct tooth profile. The rolling motion corresponds to the meshing action between the gears and a crown gear of which the cutter represents a tooth. This operation is repeated until all teeth are cut. Figures 16 and 17 show, respectively, a hypoid-gear generating machine and a close-up of the work head with the rotary cutter. Heat-treated spiral-bevel and hypoid gears may be ground by using a cup wheel and employing the same principles as just described.

Generating Gears with a Hob Cutter

Any involute gear of a given pitch will mesh with a rack of the same pitch. One form of cutting gears utilizes a rack as a cutter.

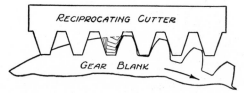

FIG. 18.   Rack-Type Cutter Generating Teeth of Spur Gear.

If it is given a reciprocating motion, similar to cutting on a Fellows shaper, involute teeth will be generated on the gear as it rotates intermittently in mesh with the rack cutter. This method is shown

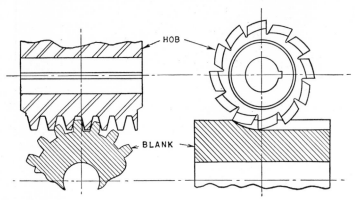

FIG. 19.   Cutting Gear with Hob.

diagrammatically in Figure 18. Such machines require a long rack cutter in order to cut all the teeth on the circumference of a large gear, and for this reason are little used in the United States.

The hobbing system of generating gears is somewhat similar to the principle just described. A rack is developed into a cylinder, the teeth forming threads and having a lead as in a large screw. Flutes are cut across the threads, forming rack-shaped cutting teeth. These cutting teeth are given relief, and, if the job is viewed from one end, it looks the same as the ordinary form gear cutter. This cutting tool, known as a *hob*, may be briefly described as a fluted steel worm. In Figure 19 is shown a hob in section and end view, as it appears when cutting a gear blank.

Courtesy The Cleveland Hobbing Machine Company.

FIG. 20. Hobbing a 1½ Diametral-Pitch Spur Gear to Full Depth in One Cut.

Hobbing, then, may be defined as a generating process consisting of rotating and advancing a fluted steel worm cutter past a revolving blank. This action is clearly illustrated in Figure 20, where the teeth on a spur gear are being cut to full depth by a rotating hob. In this process all motions are rotary, there being no reciprocating or indexing movements. In the actual process of cutting, the gear and hob rotate together as in mesh. The speed ratio of the two depends upon the number of teeth on the gear and on whether the job is single-threaded or multi-threaded.

At the start of operations the gear blank is moved in toward the rotating hob until the proper depth is reached, the pitch-line velocity of the gear being the same as the lead velocity of the hob. The action is the same as if the gear were meshing with a rack. As soon

FIG. 21.   Cutting a Worm Gear in a Cleveland Single-Spindle Hobbing Machine

as the depth is reached, the hob cutter is fed across the face of the gear until the teeth are complete, both gear and cutter rotating during the entire process.

Inasmuch as the hob teeth have a certain amount of lead, the axis of the hob cannot be at right angles to the axis of the gear when cutting spur gears but must be moved an amount equal to the lead angle. For spiral gears, the hob must be moved around an additional angle equal to the helix angle of the gears. Worm gears may be cut

Courtesy Gould and Eberhardt.

FIG. 22. Hobbing Machine Setup for Cutting Three Helical Gears.

with the axis of the hob at right angles to the gear and the hob fed tangentially as the gear rotates. This operation is clearly shown in Figure 21.

In Figure 22 is shown a hobbing machine set up for cutting three spiral gears. This is a universal machine and can be used for the production of spur gears, single- or double-helical gears, worm gears, worms, sprockets, or splines. A close-up view of this same machine set up for cutting splines on a stub shaft, is shown in Figure 23.

Courtesy Gould and Eberhardt.

FIG. 23.   Cutting Splines on a Stub Shaft with a Hob Cutter.

It is interesting to note that gear hobs based on the rack principle will cut gears of any diameter. This eliminates the need for a variety of hobs for gears having the same pitch but varying in diameter. Special hobs, such as those used for cutting splines, will cut only the one part.

Finishing Operations Used on Gears

The object of any finishing operation on a gear is to eliminate slight inaccuracies in the tooth form, spacing, and concentricity so that the gears will have conjugate tooth forms and give quiet operation at high speeds. These inaccuracies are very small dimensionally, frequently not exceeding 0.0005 inch, but even this amount is sufficient to increase wear and set up undesirable noises at high speeds.   In spite of the accuracy of various gear-forming and gear-generating processes, slight errors enter into gears as a result of wear in machine bearings, lead screws, or gear trains; faulty mounting of cutter on work; use of improper material; heat treatment; and the like.

To remedy these errors in gears that are not heat-treated, such operations as *shaving* or *burnishing* are used.   Burnishing is a cold-working operation accomplished by rolling the gear in contact and under pressure with three hardened burnishing gears.   Although the gears may be made accurate in tooth form, the disadvantage of this process is that the surface of the tooth is covered with amorphous or " smear " metal rather than metal having true crystalline structure, which is desirable from a long-life standpoint.   More accurate results may be obtained by a shaving process which removes only a few thousandths of an inch of metal.   This process is strictly a cutting and not a cold-working process.

Two methods of shaving are in use: one rolling the gear in contact with a rack cutter and the other using a rotary cutter.   Either method will produce accurately formed teeth.   Both external and internal spur and helical gears can be finished by this process.

In Figure 24 is shown a rack-type gear finisher, the insert above showing a close-up view of the rack shaving cutter finishing a spur gear.   The generating rack consists of straight replaceable blades, each tooth having a number of small vertical grooves separated by lands which form the parallel cutting edges.   It is mounted on the table of the machine, which reciprocates similarly to a planer table. The gear to be finished is mounted above the rack on live centers and is driven by contact with the cutting rack.   The gear shaving or finishing is accomplished by the gear rolling on the rack cutter and at the same time being reciprocated back and forth across the face of the cutter.   At each stroke the gear is fed down into the rack cutter until the correct tooth depth is obtained.   Spur gears are shaved with racks having blades at a slight angle, while straight-bladed racks are used for gears having helix angles up to 30 degrees. Only one rack is required for gears of the same pitch.

A close-up view of the rotary crossed-axis gear finisher in operation is shown in Figure 25.   The rotary cutter is a gearlike tool having a plurality of cutting edges on the teeth, which in turn are conjugate to

Finishing a Spur Gear with a Rack Shaving Cutter.

Courtesy Michigan Tool Company.

Rack-Type Gear Finisher.

FIG. 24.

the teeth to be produced on the gear in the machine. On the surfaces of the rotary cutters are small cutting edges similar to those used on rack-type cutters. Improved cutting action is obtained by having the axis of the gears and cutter at some angle ranging from 3 to 15 degrees. The cutter is the driver, and at the same time there is an axial movement of the cutter so as to finish the full width of the gear. Both types of gear-finishing machines may be arranged for curve shaving, a process which produces teeth slightly thicker at the center to eliminate load concentrations on the ends of the teeth. This operation is commonly known as *crowning*.

Courtesy Michigan Tool Company.

Fig. 25. Rotary Shaving Cutter Finishing Transmission Gears.

The time required for gear shaving is very short, and many thousand gears may be cut before the cutter must be resharpened. Rack cutters are more expensive than rotary cutters, but the tool cost per gear is much lower by reason of the increased life of the cutter. Rotary finishing machines, being less expensive, are economical for finishing varieties of gears in relatively smaller quantities. Also, if the gears are extremely large, or if there is close interference (as in cluster gears), rotary shaving is economical. The process of gear shaving is widely used in accurately finishing gears for transmissions, reduction units, machine tools, pumps, and numerous other high-speed applications.

Heat-treated gears can be finished either by *grinding* or by *lapping*. Grinding may be done either by the forming or by the generating process. In the forming process the grinding wheel conforms to the tooth space to be ground. Three diamonds are mounted on the machine and controlled by templates through a pantograph mechanism to give the correct contour to the wheel. The generating process, as shown in Figure 26, uses a flat-faced grinding wheel which corresponds to the face of an imaginary rack meshing with the gear. One side of a tooth is ground at a time, the gear rolling on its pitch circle past the revolving wheel as if meshing with a rack. Another machine, operating on the same principle, will grind both surfaces of a tooth simultaneously. The disadvantage of gear grinding is that considerable time is consumed in the process. Also the surfaces

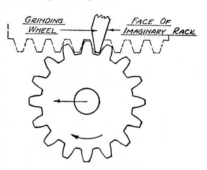

FIG. 26.   Grinding Involute Gears.

of the teeth have small scratches or ridges which increase both wear and noise. To eliminate the latter defect, ground gears are frequently lapped a short time.

Gear lapping is accomplished by having the gear in contact with one or more cast-iron lap gears of true shape. A two-lap gear finishing machine, operating on the crossed-axis principle similar to the rotary gear shaver, is shown in Figure 27. The work is mounted between centers and is slowly driven by the rear lap. It, in turn, drives the front lap, and at the same time both laps are rapidly reciprocated across the gear face. Each lap has individual adjustment and pressure control. A fine abrasive is used with kerosene or a light oil to assist in the cutting action.

The machines may be adjusted to provide an automatic lapping cycle of from 4 seconds to 20 minutes, operating first in one direction, then reversing for the same length of time in the opposite direction. By adjusting the axis of the laps it is possible to crown the tooth slightly so that the major bearing is at the center.

The entire operation of lapping is essentially a corrective and finishing one, and very little material is removed in the process. The time consumed for average-sized gears is 1/2 to 2 minutes per side of gear teeth. The results of lapping are demonstrated by longer-wearing and quieter-operating gears.

Fig. 27.   Gear-Lapping Machine Using Two Laps.

Gear-Testing Equipment

Although modern gear-cutting and gear-generating machines are capable of producing correctly formed gears, it is advisable to subject gears required for high-speed and accurate machinery to some tests prior to their assembly.   These tests vary, but include such things as concentricity, size, noise, tooth bearing, and spacing.   All tests should be made with accurate and rigid equipment and in the shortest possible amount of time.   The best method of testing and inspection is that which most nearly simulates the actual working conditions of the gears.

A universal gear-testing machine with spiral-bevel gears mounted for testing is shown in Figure 28.   This machine will test straight bevels, spiral bevels, hypoids, helicals, herringbones, worms, worm wheels, and spur gears either at no load or under light load and at speeds simulating operating conditions. Accurate adjustments are provided, and any size variation, run out, or noise is quickly noted.

A simple gear-testing fixture used for making a quick check on large quantities of gears is shown in Figure 29. The fixture is hinged

Courtesy Gleason Works.

Fig. 28. Universal Gear Testing Machine.

and is used to check concentricity and size of gears. The dial indicator, located just above the handles, will return to the same point each time a gear of the same size is rolled with a master gear. The gears are rolled by hand.

Numerous other inspecting and testing devices have been developed for checking such factors as the pitch diameter, tooth form, eccentricity, tooth spacing, and helix angle. Minor defects, such as those caused by distortions during heat treatment, can be detected readily by these measuring devices. Similar equipment is available for checking gear hobs and cutters of other types. Although individual tests are necessary to locate some specific error, the final test as to whether a gear is satisfactory or not is to run it under conditions as near the actual service operation as possible.

Fig. 29.   Quick-Acting Fixture for Testing Eccentricity and Size of Gears.

REVIEW QUESTIONS

1. What advantage does a helical gear have over a spur gear?

2. How does a hypoid gear differ from a bevel gear?

3. What are rack gears used for?

4. What type of gears are used for large speed reductions?

5. Illustrate by sketch how an involute curve is drawn.

6. Sketch an involute gear tooth, and indicate addendum, pitch circle, base circle, circular pitch, and tooth face.

7. What is the base circle of an involute gear, and how is its diameter determined?

8. A 14½-degree standard involute gear has a pitch diameter of 4 inches. If a 6-pitch cutter is used, how many teeth will the gear have? What should be the outside diameter of the gear?

9. A spur gear is cut with a 12-pitch cutter. Determine the circular pitch, addendum, and outside diameter, if the gear has 36 teeth.

10. List the various processes used in making gears.

11. Describe three methods of making a spur gear by the formed-tooth process.

12. For what purposes are sand-cast gears used?   Die-cast gears?

13. How are bevel gears cut?   Briefly describe the process.

14. Explain how to set up a milling machine to cut spur gears.

15. What is a Fellows gear shaper, and how does it operate?

16. How are continuous herringbone gears cut?

17. What kinds of gears can be cut on a gear shaper?

18. What is a gear hob?

19. Why are finishing operations necessary on some gears?

20. Describe the operation of a rack-type gear-shaving machine.

21. Show by sketch the setup used for grinding gears.

22. How are gears lapped?

23. What are the important things to check in gear inspection?

BIBLIOGRAPHY

BUCKINGHAM, EARLE, *Spur Gears*, McGraw-Hill Book Company, 1928.

COLVIN, F. H., and STANLEY, F. A., *Gear Cutting Practice*, McGraw-Hill Book Company, 1937.

Gears — Cutting, Finishing, Checking, Michigan Tool Company, 1945.

Hobbing, Bulletin 149, Michigan Tool Company, 1939.

STAUB, C. R., and ANDERSON, M. R., "Shaving and Lapping Gears," *Tool Engineer*, December 1940.

STRAUCHEN, D., "Producing Precision Gears in a Machine Tool Plant," *Machinery*, September 1939.

The Practical Art of Generating, Fellows Gear Shaper Company.

TRAUTSCHOLD, R., *Standard Gear Book*, McGraw-Hill Book Company, 1935.

WILDHABER, ERNEST, (1) "Precision Gears Cut Quickly," *American Machinist*, June 7, 1945. (2) "Basic Relationships of Bevel Gears," *American Machinist*, September 27, 1945. (3) "Special Analysis of Gear Mesh Clarifies Curvature Conditions," *American Machinist*, October 25, 1945.

YOUNG, GARDNER, "Modern Methods in Gear Manufacture," *Machinery*, October 1940.

CHAPTER 18

METAL SAWING

An important operation in any shop is the sawing of materials and bar stock for subsequent machining operations. Although most machine tools can do cutting-off operations to a limited extent, special machines are necessary for mass-production work and for miscellaneous work which requires a wide variety of shapes and sizes. Metal sawing is similar to wood sawing, except that the type of saw used has teeth specially designed for metal work, with the proper spacing and angles for efficient cutting.

Hand sawing, used on many simple jobs and in situations where the work cannot be brought to a power saw, is done with a thin flexible blade, usually 8 to 12 inches in length, held in a hacksaw the kind of material being cut and its thickness. An average pitch of such saws will vary from 14 to 32 teeth per inch. While coarse-tooth saws allow more chip space, the spacing will vary according to the kind of material being cut and its thickness. An average pitch for handsaws is around 18 teeth per inch, but for thin materials and tubing a finer pitch is advisable.

Metal saws for power machines are made in *circular, straight,* or *continuous* shapes, depending upon the type of machine with which they are to be used. The various types of power sawing machines are listed in the following classification:

METAL-SAWING MACHINES

1. Reciprocating saw
   (a) Horizontal hacksaw machine.
   (b) Vertical sawing and filing machine.
2. Circular saw
   (a) Metal slitting saw.
   (b) Steel disk.
   (c) Abrasive disk.
3. Band saw
   (a) Work held on table.
   (b) Work held in vise.

Reciprocating Sawing Machines

The reciprocating hacksaw, which may vary in design from light-duty crank-driven saws to large heavy-duty machines hydraulically driven, has long been a favorite because of its simplicity in design and low operating cost. It consists of a saw frame, a means for reciprocating the saw and frame, a work table and vise, a supporting base, and a source of power. Machines of this type vary as to the manner in which the saw is fed into the work and the type of drive used.

The simplest type of feed is the *gravity feed*, in which the saw blade is forced into the work by the weight of the saw and frame. Uniform pressure is exerted in the work during the stroke, but some provision is usually made to control the depth of feed for a given stroke. Some machines of this type have weights clamped on the frame to give additional control to the cutting pressure. This may also be accomplished by means of springs with suitable adjustment. Positive-acting screw feeds, with some provision for overloads, provide a means of obtaining a definite depth of cut for each cutting stroke. Hydraulic feeds are now widely used, since they afford excellent control of the cutting pressures. Several machines with this type of feed are discussed in subsequent paragraphs of this chapter.

In general, methods of feeding can be classified as either *positive* or *definite pressure* feeds. A positive feed has an exact depth of cut for each stroke, and the pressure on the blade will vary directly with the number of teeth in contact with the work. Therefore, in cutting a round bar the pressure is light at the start and maximum at the center. A disadvantage of this method is that the saw is prevented from cutting fast at the start and finish where the contact stroke is short. With definite pressure feeds, the pressure is uniform at all times, regardless of the number of teeth in contact. This condition prevails in gravity or friction feeds. Here, the depth of cut varies inversely with the number of teeth in contact, so that the maximum pressure which can be used depends upon the maximum load that a single tooth can stand. Many machines of recent design have incorporated both these systems into their design with automatic control. In all cases the pressure is released on the return stroke to eliminate wear on the saw blade.

The simplest drive for the saw frame is that with a crank rotating at a uniform speed. With this arrangement the cutting action is taking place only 50% of the time, since the time of the return stroke equals that of the cutting stroke. An improvement of this design provides a link mechanism which gives a quick-return action on the cutting stroke. Several such link mechanisms are used, including the Whitworth mechanism found on some shapers. These designs reduce the idle time to about one third of the total and result in faster cutting than the crank saws without increasing the cutting speed.

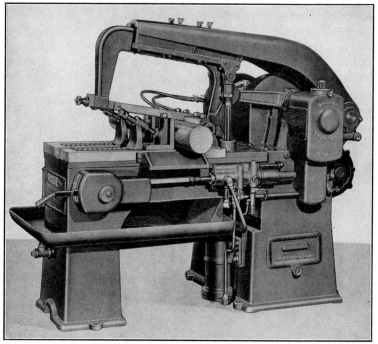

Courtesy Racine Tool and Machine Company.

FIG. 1.   Racine Hydraulic Shear-Cut Production Saw.

In Figure 1 is shown a Racine hydraulic shear-cut production saw. The term " shear-cut " describes a cutting action used in this machine, whereby the cutting edge of a metal saw is fed progressively and uniformly in a manner to give the most effective cutting. The feed is hydraulically operated and is equipped with valve control to give either a positive progressive feed or a flexible constant-pressure feed. The positive feed is recommended for production runs in the sawing of tough steels, such as stainless steel, die blocks, and numerous other alloys. The flexible constant-pressure feed is used for automatically increasing or decreasing the feed in accordance with the area, shape, or density of the material being cut. In both cases the hydraulic feed pressure is applied progressively during the cut, and each tooth produces a long curling chip. The wear on the teeth is evenly distributed over the entire cutting surface of the blade. This machine can be provided with an automatic stock feed. With this device a single bar, or a bundle, can be cut to the desired lengths, leaving the operator free to do other work. The entire cycle is hydraulically controlled and positive in its action.

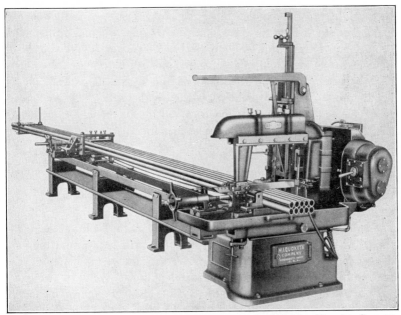

Courtesy Maquoketa Company.

FIG. 2. Sawmore Metal-Cutting Machine with Automatic Bar Feed Carrier.

Another reciprocating saw, the Sawmore metal-cutting machine with automatic bar feed carrier, is shown in Figure 2. One feature of this machine is that it will swivel on its base for angle cutting. This means that the bars being cut remain in the same straight line whether they are cut at angles or straight across. On a fully automatic machine, the operator needs only to set his scale on the spacer bar, put the stock in the machine, turn the counting dial to the number of pieces required, and then start the machine. The machine will cut off exact duplicate pieces up to the number dialed or until the stock is depleted. A positive compensating feed applies pressure on the blade during its complete cutting stroke and assures a maximum amount of steel removal for each tooth of the blade.

Hacksaw Blades

Power hacksaw blades are similar to those used for hand sawing. High-speed steel blades vary from 12 to 24 inches in length and are made in various thicknesses up to 0.065 inch. The pitch is coarser than for hand sawing, ranging from 6 to 18 teeth per inch, since the material being cut is usually much larger. For efficient cutting of ordinary steel and cast iron, as coarse a pitch as possible should be used to provide ample chip space between teeth. High-carbon and alloy steels require a medium-pitch blade, whereas thin metal, tubing, and brass require a fine pitch. To provide ample clearance for the blade while cutting, the teeth are set to cut a slot or *kerf* slightly wider than the thickness of the blade. This is done by bending certain teeth slightly to the right or left. A coolant is recommended for all power hacksaw cutting to prevent overheating the blade and to wash away the small chips accumulating between the teeth.

Circular Sawing Machines

Machines using circular saws are commonly known as *cold sawing machines*. The saws are fairly large in diameter and operate at low rotational speeds. The cutting action is the same as that obtained with a milling cutter. The machine consists of a rotating saw, means for feeding the saw into the work, a vise for clamping the part to be cut, and a supporting frame.

In Figure 3 is shown a hydraulic-feed cold sawing machine capable of sawing round stock up to $9\frac{5}{8}$ inches in diameter. Nine changes of speed in geometric progression are provided for the saw blade. This affords a wide selection of cutting speeds, ranging from a speed of 18 feet per minute for hard materials to 134 feet per minute for softer materials. The feed for the saw carriage is hydraulically operated and provides a " stepless " variable feed as well as a quick return. A low-pressure system is used to avoid leakage and excessive maintenance. Incorporated in the hydraulic unit are two adjustable valves, one of which may be set to limit the total feed pressure while the other insures clamping the work. The feed pressure may be set at a point which will protect the saw, regardless of the rate for which the feed is set. Hence, if the saw encounters a change in section or hardness of material which overloads the blade, the rate of feed is automatically decreased until the overload is eliminated. All controls for starting and stopping, speed and feed changes, and clamping are mounted on the front of the machine. Included on the panel is an ammeter to measure power input to the saw blade, which indicates to the operator when the saw needs resharpening. Additional features

FIG. 3. Hydraulic-Feed Cold Sawing Machine.

are the hydraulic vise and a pressure gage to measure the feed and clamping pressure. A 28-inch-diameter saw, with inserted teeth, is furnished as standard equipment. Another circular-sawing machine made by the Cochrane-Bly Company is illustrated in Figure 4. The inserted tooth saw and saw carriage are clearly shown in working position. A large bundle of rods is being cut to length on a typical production job.

Circular Saws

Saws for rotating cutter machines are the same as the *metal slitting saws* used with milling machines. Metal slitting saws are made in

Courtesy Henry Disston and Sons, Inc.

FIG. 4. Inserted-Tooth Disk-Saw in Use on Production Job.

diameters up to 8 inches and are provided with side clearance either by dish-grinding the sides or by providing the teeth with side-cutting edges. Coarse teeth are recommended, because they are stronger and provide more space for chips. Zero to four degrees rake is used for most steels and cast iron, while softer materials such as aluminum, copper, and plastics make use of rake angles up to 20 degrees. In the larger cutters, commonly called *cold metal saws*, it is common practice to use inserted teeth of either high-speed steel or carbide. This design, in addition to being economical from the standpoint of cutter-material cost, permits the replacement of broken or worn-out teeth. In these cutters the tooth sides are frequently beveled on alternate teeth to assist in breaking up the chips and to reduce the load per tooth.

Friction and Abrasive Disks

Steel *disks* operating at high peripheral speeds provide a rapid means of cutting through structural-steel members and other steel sections. When the disk is rotating at rim speeds from 18,000 to 25,000 feet per minute, the heat of friction quickly melts a path through the part being cut. Figure 5 shows a Ryerson friction disk at the start of a cut through an I beam. About one-half minute is required to cut through a 24-inch I beam. Large-diameter disks are used in this work, ranging in diameter from 24 to 60 inches. They are usually furnished without teeth, although in some cases nicked edges or V-shaped teeth are specified. The disks are slightly hollow ground to provide side clearance in cutting through a large member. Water cooling is recommended in this type of cutting. Large machines have the disk wheel mounted on a carriage which is controlled by either hand or power feed. Light machines ordinarily use a hand- or foot-pedal feed.

A small dry-cutting machine equipped with a thin abrasive wheel is illustrated in Figure 6. This machine will cut solid stock up to 3/4 inch in diameter as well as tool bits, drills, reamers, and other hardened materials. Also, small tubing used in aircraft manufacture and plastic rods and tubes can be cut rapidly and with excellent finish.

Abrasive-disk machines operate either wet or dry, but where heavy cutting is involved a coolant should be used. The disks are either rubber- or resinoid-bonded, since, being thin, they must have some flexibility. The wheels recommended for this machine are 10 by 1/16 rubber bond or 10 by 3/32 resinoid bond. With a 10-inch

Courtesy Henry Disston and Sons, Inc.

FIG. 5. Friction Disk Cutting a Structural Member.

Courtesy A. P. De Sanno and Son, Inc.

FIG. 6. Radiac Cutoff Machine Equipped with Abrasive-Disk Wheels.

wheel the surface speed is 13,000 feet per minute. Abrasive disks can be operated safely at surface speeds up to 16,000 feet per minute.

Band-Sawing and Filing Machine

The sawing machines described thus far are designed for taking straight cuts and are used primarily for cutting-off purposes. Cutting saws of the *band type* can also be used for this work, but in addition can cut irregular curves in metal. This greatly widens the field of usefulness for the band saw, since it enables the machine to do a great variety of work that formerly had to be done with other machine tools. Contour sawing of dies, jigs, cams, templates, and numerous other parts that formerly had to be made entirely on other machine tools or by hand at much greater expense, is now done with band saws. Recent developments of these machines include also suitable and accurate arrangements for continuous filing and polishing, both necessary operations in contour finishing.

Band-sawing machines for metals are very similar in appearance to those used for wood, but differ in the saw-cutting speed and the type of saw. Most machines are designed with the saw running in a vertical position, the work being supported on a horizontal table having a tilting adjustment for cutting angles. Another type is quite similar in design to the ordinary hacksaw machine. The work is held in a vise while a small band saw operates above the work in nearly a horizontal position. Both types are widely used for cutting off stock in the same way that such work is done on circular- and reciprocating-sawing machines.

A contour machine for *band sawing, filing,* and *polishing* is shown in Figures 7 and 8. This machine employs saw blades 1/16 to 1/2 inch in width, file bands in three different sizes, and emery bands for polishing work. It has an infinitely variable-cutting-speed range of 50 to 375 feet per minute. An indicating dial showing the exact operating speed of the saw in feet per minute is mounted on the face of the column. Just below this dial, in a convenient position, is a resistance-type butt welder which automatically provides the proper welding current for any saw 1/16 to 5/8 inch wide. This procedure is followed when doing internal cutting: the saw blade is first cut, then inserted through a hole drilled in the work, and finally butt-welded together. A small built-in grinder is provided for dressing the ends of saws and grinding the welds.

A new development in band sawing is the design of a high-speed machine, for friction or standard cutting, having a surface speed range of from 1500 to 10,000 feet per minute. Saws otherwise too dull for regular cutting can be used in friction cutting. This machine has special application in cutting lightweight aircraft metals, sheet metal, plastics, and other nonmetallic materials.

Courtesy Continental Machines, Inc.

Fig. 7. DoAll Contour Machine for Band Sawing, Filing, and Polishing.

Types of Band-Saw Blades

An important step in precision sawing is the selection of the proper saw for each job. The width is determined by the feed that is to be used and by the curvature to be cut.   It is a good rule always to use the widest blade possible.

The temper or hardness of the saw varies with the type of material to be cut.   A hard temper should be used for cutting alloy steels; a softer temper is recommended for nonferrous materials.

The number of teeth per inch is also a function of the material being worked upon.   This factor does not vary directly with the hardness of the material, but is determined from actual cutting

⌊ Courtesy Continental Machines, Inc.

Fig. 8.   Band-Sawing Breech Rings for Small Guns.   Three 8-Inch Cuts Are Taken in Addition to the Contour at Top.

experience; manufacturers' recommendations should be followed in this respect.

The *set* refers to the type of tooth construction on a saw.   A *straight-tooth saw* has one tooth set to the right and the next tooth to the left.   This type of saw is used for brass, copper, and plastics. On the *raker-tooth saw* one straight tooth alternates with two teeth set in opposite directions.   It is this tooth construction that is used for most steel and iron cutting.   A *wave set* consists of an alternate arrangement of several teeth set to the right, followed by several teeth set to the left. This design is used in cutting tubes and light sheets of metal.

Band Filing

When the machine is to be used for filing work, the saw band is removed and a file band put in its place. The file band is made up of 26 files mounted on a flexible Swedish steel band. A snap joint, shown in Figure 9, is provided for quick fastening and unfastening for internal filing. A light-to-medium pressure is used on contour

Courtesy Continental Machines, Inc.

FIG. 9.   Setting Up File Band for Internal Filing.

filing, and the filing speeds range from 50 to 200 feet per minute. An advantage of this type of filing is that it is accomplished with a continuous downward stroke. The absence of a back stroke greatly lengthens the life of the file and helps in holding the work on to the table.

A number of the principal file sections used in band filing are shown in Figure 10. These sections have the same shapes and styles

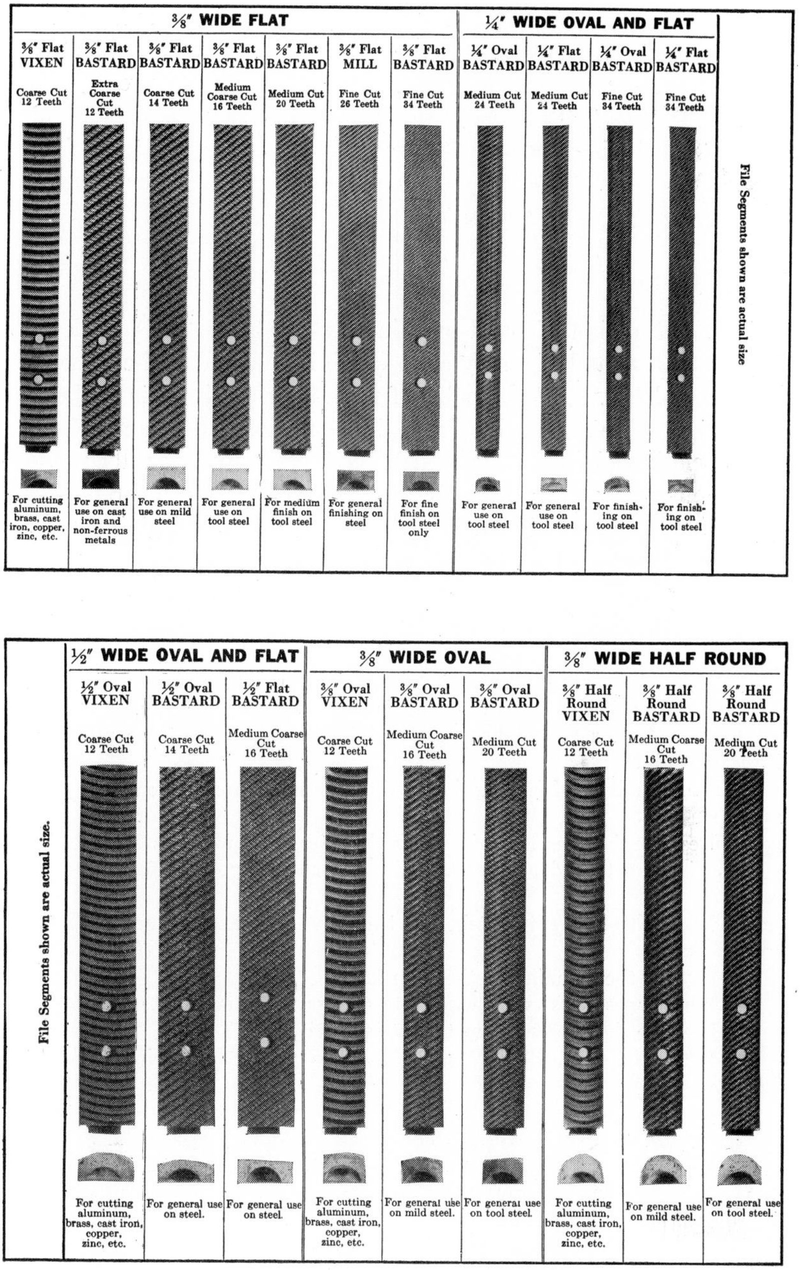

3/8" WIDE FLAT							1/4" WIDE OVAL AND FLAT				
3/8" Flat VIXEN	3/8" Flat BASTARD	3/8" Flat BASTARD	3/8" Flat BASTARD	3/8" Flat BASTARD	3/8" Flat MILL	3/8" Flat BASTARD	1/4" Oval BASTARD	1/4" Flat BASTARD	1/4" Oval BASTARD	1/4" Flat BASTARD	
Coarse Cut 12 Teeth	Extra Coarse Cut 12 Teeth	Coarse Cut 14 Teeth	Medium Coarse Cut 16 Teeth	Medium Cut 20 Teeth	Fine Cut 26 Teeth	Fine Cut 34 Teeth	Medium Cut 24 Teeth	Medium Cut 24 Teeth	Fine Cut 34 Teeth	Fine Cut 34 Teeth	File Segments shown are actual size
For cutting aluminum, brass, cast iron, copper, zinc, etc.	For general use on cast iron and non-ferrous metals	For general use on mild steel	For general use on tool steel	For medium finish on tool steel	For general finishing on steel	For fine finish on tool steel only	For general use on tool steel	For general use on tool steel	For finishing on tool steel	For finishing on tool steel	

1/2" WIDE OVAL AND FLAT			3/8" WIDE OVAL			3/8" WIDE HALF ROUND		
1/2" Oval VIXEN	1/2" Oval BASTARD	1/2" Flat BASTARD	3/8" Oval VIXEN	3/8" Oval BASTARD	3/8" Oval BASTARD	3/8" Half Round VIXEN	3/8" Half Round BASTARD	3/8" Half Round BASTARD
Coarse Cut 12 Teeth	Coarse Cut 14 Teeth	Medium Coarse Cut 16 Teeth	Coarse Cut 12 Teeth	Medium Coarse Cut 16 Teeth	Medium Cut 20 Teeth	Coarse Cut 12 Teeth	Medium Coarse Cut 16 Teeth	Medium Cut 20 Teeth
For cutting aluminum, brass, cast iron, copper, zinc, etc.	For general use on steel.	For general use on steel.	For cutting aluminum, brass, cast iron, copper, zinc, etc.	For general use on mild steel.	For general use on tool steel.	For cutting aluminum, brass, cast iron, copper, zinc, etc.	For general use on mild steel.	For general use on tool steel.

Courtesy Continental Machines, Inc.

Fig. 10. File Sections for Band Filing.

found on standard commercial files. *Single cut, double cut,* and *rasp cut* are the terms used in describing the cut of the file. Rasp cut differs from the other two in that the teeth are disconnected from each other, each tooth being made by a single punch. The coarseness of the teeth is described by the terms *rough, coarse, bastard, double cut,* and *smooth.* File cross sections are indicated by such terms as *flat, oval, half round,* and *mill.*

Band Polishing

A third function of band machines is polishing. This work calls for an endless band of emery cloth which is mounted in the same way as the band saws. At the point of work the cloth band is backed up by a rigid plate. The bands used are made of the same kind of emery cloth and in the same grits that are conventionally used in hand-polishing work. Polishing is used either as a preparatory operation before making a layout or as a means of giving a high polish to a surface that has been filed.

REVIEW QUESTIONS

1. Prepare a list of the various kinds of power saw machines.

2. What factors should be considered in the selection of the pitch for a hacksaw blade?

3. What methods of feeding the saw into the metal are used on reciprocating saws?

4. Describe the operation of a bar-feed carrier on a power hacksaw.

5. What are the advantages and disadvantages of positive feed on a hacksaw?

6. Make line-diagram sketches of two drive mechanisms used on reciprocating saws.

7. Explain the meaning of the term kerf.

8. What types of saws are used on a cold sawing machine?

9. Describe the process of metal cutting with a friction disk.

10. What kinds of disks are used in abrasive cut-off machines? At what surface speed are they operated?

11. How does a metal band saw differ from a wood band saw?

12. List the various operations that can be performed on a metal band saw.

13. What is the procedure for doing internal cutting on a band saw?

14. In metal-band-saw work what determines the width of the saw, the number of teeth per inch, and the set?

15. How are files used on a band-sawing machine?

16. What is the difference between a single-cut and a rasp-cut file?

17. What are the terms used to describe tooth coarseness of a file?

18. Why is a cutting fluid used in machine hacksawing?

19. Distinguish between straight set, raker set, and wave set as applied to metal hacksaws and band saws.

BIBLIOGRAPHY

Handbook on Contour Sawing, Continental Machine Specialties Company.

HOLLOWAY, R. C., " Friction Sawing," *American Machinist,* June 7, 1945.

WIESE, R. R., " Cutting with Abrasives," *American Machinist,* February 19, 1941.

CHAPTER 19

BROACHING MACHINES AND TOOLS

Broaching is the operation of removing metal by means of an elongated tool having a number of successive teeth of increasing size which cut in a fixed path. A part is completed in one stroke of the machine, the last teeth on the cutting tool conforming to the desired shape of the finished surface. In most machines the broach is moved past the work, but equally effective results are obtained if the tool is stationary and the work is moved. Although the process of broaching has been known for many years, it has not been used extensively in production work until recently. The first developments of broaching were confined principally to internal operations, such as the broaching of holes and keyways. Many cuts, both external and internal, can be made on recently designed machines at a high rate of production and with satisfactory accuracy and finish.

Types of Broaching Machines

A broaching machine consists of a work-holding fixture, a broaching tool, a drive mechanism, and a suitable supporting frame. Although the component parts are few, several variations in design are possible. A brief classification of broaching according to method of operation is as follows:

1. **Pull of draw broaching.** The broaching tool moves and the work is stationary.
2. **Push broaching.** The broaching tool moves and the work is stationary.
3. **Surface broaching.** Either the work or the broaching tool moves across the other.
4. **Continuous broaching.** The work is moved continuously against stationary broaches. The path of movement may be either straight or circular.

Some broaching machines are made in both vertical and horizontal designs. Vertical machines are used mostly for the broaching of external surfaces, with the tool supported on a suitable slide, although

518

both pull and push types are made in this design. Horizontal machines usually pull the broach, especially for small and medium-sized work. Surface broaching can be done on either the horizontal- or vertical-type design, as illustrated in Figure 1. This figure illustrates

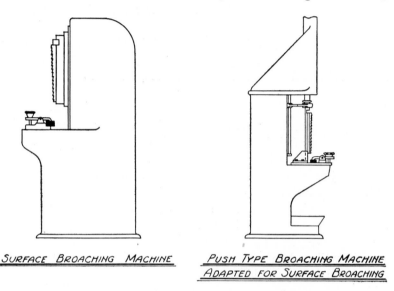

SURFACE BROACHING MACHINE

PUSH TYPE BROACHING MACHINE
ADAPTED FOR SURFACE BROACHING

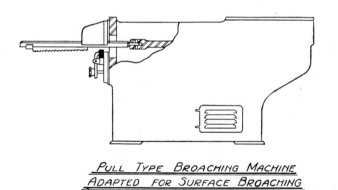

PULL TYPE BROACHING MACHINE
ADAPTED FOR SURFACE BROACHING

Courtesy Illinois Tool Works.

FIG. 1. Typical Machines for Surface and Internal Broaching.

the first three methods of operation as previously listed. Although the pull-type and push-type machines are shown set up for surface broaching, both these machines are equally well adapted to internal broaching.

·Another variation in broaching machines is the method of drive. Because a large force is required, most modern machines are hydraulically driven. Such a drive is smooth-acting, economical, and readily adjustable for both speed and length of stroke. Other drive systems are operated by means of gears, power screw, chain, or a link mechanism. The chain type of drive is especially adapted to the continuous broaching machine. On all intermittent-type machines, a quick-return feature is incorporated in the drive. The return stroke is two to five times the speed of the cutting stroke.

Advantages and Limitations of Broaching

Broaching machines have been rapidly adopted for mass-production work because of their exceptionally high rate of production. The actual cutting time is only a matter of seconds, since the operation is completed in one stroke of the machine. Rapid loading and unloading of fixtures keep the total production time to a minimum.

These machines can also be used for either internal or external surface finishing. Operations formerly performed on milling machines, planers, and the like, can frequently be done with broaches equally well and in a much shorter time. Any form that can be reproduced on a broaching tool can be machined. This includes a wide variety of irregular shapes as well as plane surfaces. Tolerances can be maintained which are suitable to interchangeable manufacture. Finishes comparable to milling work are obtained, and in some cases burnishing teeth are incorporated on the broach to improve the surface finish.

A limiting factor in broaching is the cost of the broaching tools, many of which are extremely expensive because of their size and irregular shape. High tooling cost eliminates broaching for short-run jobs, but it is not an important factor in mass production. Parts being broached must be capable of being rigidly supported and must be able to withstand the broaching forces set up. This is particularly true of surface-broaching work, where the force of the cut is not central. The surface of the work to be broached cannot have any obstruction. Likewise, on internal work, the broaching tool must have free passage through the opening.

Vertical Broaching Machines

A vertical pull-down broaching machine having a normal capacity of around 25 tons and broaching strokes from 30 to 66 inches is shown in Figure 2. The tool-handling carriage, which is mounted on the main slide, has an adjustable bracket to handle pull-down broaches of different lengths. In addition, the main slide is designed so that surface-broaching tools can be mounted if desired. Both types of broaches are shown in the figure. Cutting speeds up to 30 feet per minute are provided with return speeds as high as 80 feet per minute.

FIG. 2. Oilgear Type-XP Vertical Pull-Down Broaching Machine.

In actual operation, the operator has only to load and unload the work. When the operation is started, the upper carriage and tool move down to insert the tool shank through the work and table into the puller. The main slide then pulls the tool through the work where it stops while the work is removed. The main slide and tools then return upward, where the tool is received by the upper carriage and returned to starting position.

Illustrated in Figure 3 is an Oilgear type-XM 20-by-54-inch stroke vertical cyclomatic broaching machine with tools and fixtures for broaching 17 splines in universal joint yokes at the rate of 440 pieces per hour (when using three cutting tools and a cutting speed of 30

Courtesy The Oilgear Company.

Fig. 3.   Vertical Cyclematic Broaching Machine with Tools and Fixtures for Broaching 17 Splines in Universal Joint Yokes.

feet per minute). Two push buttons above the selector switch on the right provide manual control for the tool-handling and work-table slides for setup and test purposes. Setting the selector switch for semiautomatic operation and depressing the dual push-button control provide the following cycle. The tools are lowered by the upper ram and threaded through the work into the pullers below the slide. As the work table begins to move upward, the work is flooded with a cutting lubricant. The broaching tool is stationary and during a major portion of the stroke is secured on both ends. Near the top of the stroke the tool-handling slide rises with the work to disengage the tools, and as the work clears the ends of the tools a mechanical ejector removes it from the fixture. The work slide then returns quickly to the loading position while the tool-handling slide engages the broaches, lifting them upward to loading position. The entire operation is semiautomatic, the operator having only to load the work and depress the control button to start the cycle of operations.

Fig. 4. Lapointe 3-Ton 30-Inch-Stroke Single-Ram Surface-Broaching Machine with Special Setup for Broaching Airplane-Propeller Spiders.

Figure 4 shows a Lapointe 3-ton 30-inch-stroke vertical single-ram surface-broaching machine with special setup for broaching airplane-propeller spiders. The broaching tools are mounted on the slide, which is hydraulically driven and accurately guided by hardened and ground steel ways. Work is held stationary in a fixture on the platen

of the machine, and the cutting tools are forced downward past the
work at cutting speeds ranging from 10 to 30 feet per minute.   Return
speeds of the slide to the starting position are approximately twice
the cutting speeds.

Fɪɢ. 5.   Upper Part of Double-Slide Vertical Surface-Broaching Machine Equipped
with Tools and Fixtures for Broaching Rack Teeth on Small Parts.

The upper part of a double-slide vertical broaching machine is
shown in Figure 5.   In the operation shown, rack teeth are being
broached in a small part which is held in a six-station indexing
fixture.   The teeth are cut in two operations, first roughed out and
then finish cut.   Two passes are necessary to finish each part, so

that one part is finished with each down stroke of the tool slide. The production rate on this job is approximately 235 pieces per hour. With alternating tool-slide machines there is a minimum loss of time, since the operator loads and unloads the work as one side broaches and the other returns.

Surface-broaching machines combine economic production with a high degree of accuracy and finish. External slots, forms, contours, and flat surfaces are finished rapidly with one stroke of the machine. Connecting rods, bearings, universal joint yokes, and numerous small rifle parts are examples of parts finished to size by surface broaching.

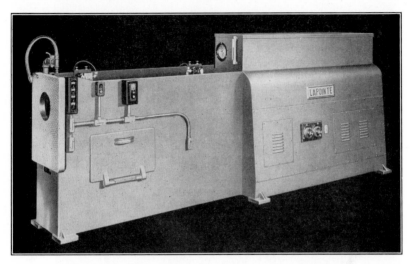

FIG. 6. Lapointe Horizontal Hydraulic Broaching Machine.

Horizontal Broaching Machines

Horizontal broaching machines of the pull type are adapted to both internal and surface broaching. A single-slide machine of 15 tons capacity and 52 inches maximum stroke is illustrated in Figure 6. The work is held in a fixture mounted on a face plate at the end of the machine. In operation the broach is threaded through the work and fastened to the slide just back of the face plate. The hydraulic cylinder which pulls the slide and the broach is housed in the end of the machine at the right. The normal broaching force is 1000 pounds per square inch of effective ram area; however, there is an ample factor of safety to compensate for overload because of dull tools, variations in material, or amount of stock removed. Variable cutting

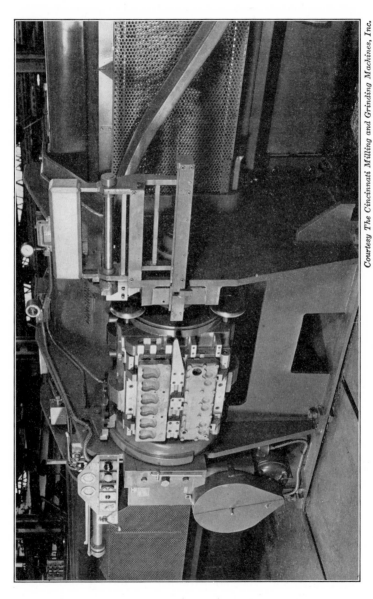

FIG. 7: Cincinnati Horizontal Broaching Machine Equipped with Drum-Type Fixture for Finishing Cylinder Heads.

speeds of 10 to 35 feet per minute are provided with return speeds up to 100 feet per minute.

Figure 7 illustrates a special horizontal broaching machine equipped for broaching top and bottom surfaces of cast-iron cylinder heads. This machine, equipped with a drum-type fixture for supporting the heads, removes 3/16 inch of stock at a production rate of 77 cylinder heads per hour. An inserted tooth broach, operating at a speed of 36 feet per minute, is used in this work. Chips are removed by an exhaust system located at one end of the machine.

The drum fixture has two work-holding units located opposite each other, each holding two cylinder heads. One head is completely finished with each stroke of the ram. After both heads have been broached on one side, the fixture indexes 90 degrees toward the operator and a pusher shoves the two heads out of the fixture. One cylinder head is moved into a chute which catches the finished heads while the other is moved into a hinged frame. This frame is then swung through 90 degrees while the drum fixture indexes another 90 degrees. The head in the frame can then be moved into the fixture again for broaching the unfinished side, which was turned over by the combined opposite rotation of frame and fixture. While this head is moved into position, a fresh head is placed in the lower position of the fixture. During the time in which these two pieces are being put into place, the two pieces on the other side are being broached. Broaching tools are mounted on a slide which carries them past the stationary cylinder heads held in the drum fixture.

Similar horizontal broaching machines are used in the surface broaching of cylinder blocks and crankcases. These special machines can perform all the broaching cuts on a cylinder block at a production rate of 60 blocks per hour. In most machines of this type the work is held stationary in a fixture and the broaching tools are power-operated.

A recent development in broaching* is the cutting of helical grooves or splines by pulling a broach through the part and at the same time either rotating the part or broaching tool according to the helix desired. This procedure has been adopted by many gun manufacturers in rifling small-caliber and light-cannon-gun barrels. Horizontal machines are used, equipped with either single or six-station-type fixtures. Both arrangements pull the broaches through the barrels, and a positive lead is usually provided which will turn

* Swindle, T. A., "Rifling Gun Barrels by Broaching," *American Machinist*, October 14, 1943.

either the barrel or the broach. Two passes are used in the single-fixture machines, one for roughing and one for finishing. In the multiple-type fixture having six stations, one is used for loading and five for reaming, broaching, and finishing the bore. Rifling produced by broaching is more economical than by the old hook-cutter method, and the rejections are much fewer.

Rotary-Broaching Machines

Rotary broaching consists in mounting the work in fixtures supported on a revolving table which moves past stationary broaches. These broaches are made in short sections so that they can be easily adjusted and sharpened. A rotary-broaching machine, known as a

Courtesy The Cincinnati Milling Machine and Grinding Machines, Inc.

FIG. 8. Rotary-Broach Setup for Brake Spreader Cams.

" mill broach," is illustrated in Figure 8. In this particular case the operation was so short that one piece could be completed in one half of a revolution. By duplicating the tooling and using automatic clamping and ejection fixtures, two operators stationed opposite each other attained a production rate of 4800 pieces per hour. The loading and unloading position is shown at the right of the machine. After the work is ejected, it falls into a chute and is carried away by a conveyor. An automatic control is provided to stop the machine instantly if a part is not properly located in the fixture. Rotary-broaching machines are used for squaring distributor shafts, slotting, straddle milling, form milling, and the facing of small parts.

Continuous Broaching Machine

Continuous broaching machines are adapted only for surface broaching. This type of broaching machine consists of a frame and driving unit with several work-holding fixtures, mounted on an endless chain which carries the work in a straight line past the stationary broaches. A view of a continuous machine equipped with a motor-driven conveyor for removing work from the machine is shown in

Courtesy The Foote-Burt Company.

Fig. 9. Continuous-Surface-Broaching Machine Equipped with Motor-Drive Conveyor Used in Removing Work from Machine.

Figure 9. Loading is done by an operator who drops the parts in the fixtures as they pass the loading station. The work is automatically clamped before it passes into the fixture tunnel in which the broaches are held. After the fixtures pass through the broach cut, they are automatically released by a cam, and at the unloading position the work falls out of the fixtures into the work chute. The broach holder raised up to show the tooling and fixtures, used in broaching two right-angle surfaces on a cast-iron part, is shown in Figure 10. Pans are provided around the fixtures for carrying cast-iron chips to a compartment at one end of the machine. Production is high, as the operator handles only the work in the loading position; the output of the machine varies with the ability of the operator to keep the machine loaded.

The design of this machine permits the use of long broaches, and each tooth has only a small amount of metal to remove. The first teeth are roughers and remove the major portion of the metal; the last teeth do the final finishing and sizing. Broaches are made up in short sections to facilitate replacement. A variable cutting speed of 20 to 40 feet per minute is provided by a gear drive.

FIG. 10. Broach Holder on Continuous Broaching Machine Raised Up to Show the Tooling Used in Broaching Two Right-Angle Surfaces on Cast-Iron Part.

Examples of surface-broaching operations are shown in Figure 11. Where there are several surfaces to be broached, such as on a hexagonal section, an indexing fixture can be used, with three pairs of broaches mounted in the broach holder and spaced so that the fixture can be indexed between each pair of broaches.

Broaching Tools

Broaching tools differ from most other production tools in that they are usually adapted to a single operation. The feed of the tool must be predetermined, and once a broach is made the feed remains a constant. These facts necessitate having complete information as to the job, the material, and the machine to be used before a broach can be made. A few types of flat or regular section broaches can be made up in advance, but these which finish unusual surfaces and shapes are

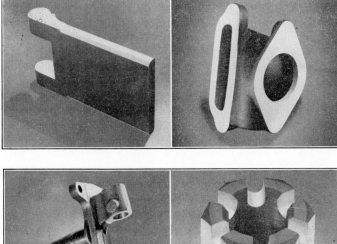

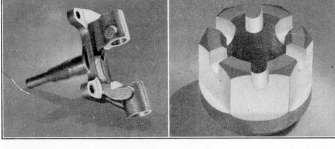

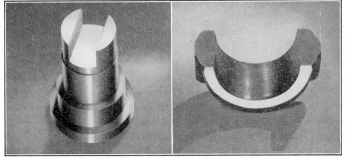

Courtesy The Foote-Burt Company.

Fig. 11. Examples of Surface Broaching.

specially designed. In designing and constructing a broach, the following information must be known:

1. Kind of material to be broached.
2. Size and shape of cut.
3. Quality of finish required.
4. Hardness of material.
5. Tolerance to be maintained.
6. Number of parts to be made.

7. Type of machine to be used.
8. Method of holding broach.
9. Pressure that the part will stand without breakage.

Fixtures on special jobs are frequently designed simultaneously with the broaching tools in order to work out the best arrangement. For successful operation of a broaching tool, the work must be supported rigidly with ample clearance and cooling facilities.

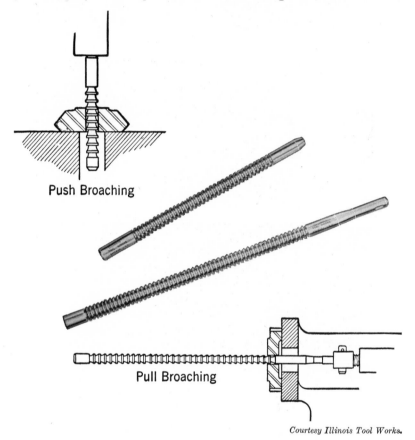

Push Broaching

Pull Broaching

Courtesy Illinois Tool Works.

FIG. 12.   Round Broaches for Push-and-Pull-Type Machines.

A common example of broaching is the finishing of round holes. This method of finishing is more rapid than reaming or boring and at the same time can be held to accurate limits.   The life of the tool is long, since the broach has a great number of cutting teeth with each taking only a very small cut.   Two round broaches, one *push-*

type and one *pull-type,* are shown in Figure 12. Finished holes can be made from holes previously drilled, reamed, bored, punched, or cored. At least 1/64 inch of stock should be allowed for the finish when one is broaching holes that have been previously machined. Either an ordinary *burnishing broach* or a broach which has several burnishing teeth on its end can be used if an extremely fine finish is desired. This is a cold-working action which produces a hard smooth surface.

A special broach designed for removing a large amount of metal is shown in Figure 13. At the start the teeth are somewhat hexagonal in shape instead of round. Points on each successive tooth cut at

Courtesy The Lapointe Machine Tool Company.

Fig. 13. Rotor-Cut Broach Designed for Removing a Large Amount of Material from Rough-Cored Holes.

different places; with such an arrangement relatively deep cuts can be made. Parallel portions from a segment of an internal surface are successively and separately removed with this type of broach. The teeth gradually approach a circular shape as they near the finishing end, and the final ten teeth, used for sizing and finishing, are round. This type of broach is especially recommended for holes that have a considerable amount of stock to be removed, such as cored or punched holes.

Square, hexagonal, and other uniformly shaped holes usually start from a round hole. The first teeth of such broaches conform to the original hole, but gradually the broach changes in section according to the final shape of hole desired. If the starting-hole diameter for a square hole can be slightly larger than the finished side, a more economical broach can be made.

The broaching of internal keyways is one of the oldest uses of this process. A keyway broach and its adapter are shown in Figure 14. The adapter guides the broach and also assists in holding and locating the work. Broaching tools for this purpose are extremely simple and can be obtained for general-purpose use. If multiple keyways of splines are to be cut, a single broach can be used with the work, the proper amount being indexed after each cut. This procedure is used only for large splines or in jobs where the production is small, since spline broaches can be obtained for any desired number of keys. Figure 15 shows a pull-type spline broach and an enlarged view of the broach entering a round piece. The form of this broach is ground parallel with the axis, and it will not be changed by repeated sharpenings. Most broaches for this type of work are long and are not adapted to the push method.

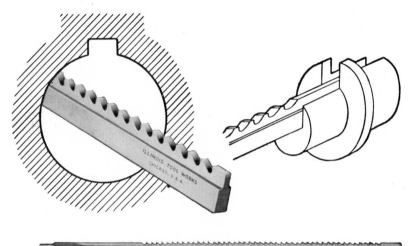

Courtesy Illinois Tool Works.

FIG. 14. Keyway Broaching.

Internal-gear broaches are similar to spline broaches except for the involute contours on the sides of the teeth. Broaches for internal gears can be made to cut any number of teeth and are used for broaching as small as 48-diametral pitch. This method of gear cutting is known as the form-tooth process, and the accuracy of the teeth is entirely dependent on the accuracy of the form cutter. External-gear teeth may also be broached, but external-gear cutting is usually limited to cutting teeth on sector gears where only a few teeth are involved.

Surface broaching, a relatively new development in the field, has grown rapidly to be a very important means of surface finishing. The simplest broaches are designed for flat surfaces. These can be made up with either straight or angular teeth, but the latter produce a smoother cutting action. Since the entire length of such broaches is supported on a slide, it is possible to make them up in short sections. Heavy-duty broaches frequently have inserted teeth to reduce the initial cost and facilitate replacements. Many irregular or intricate

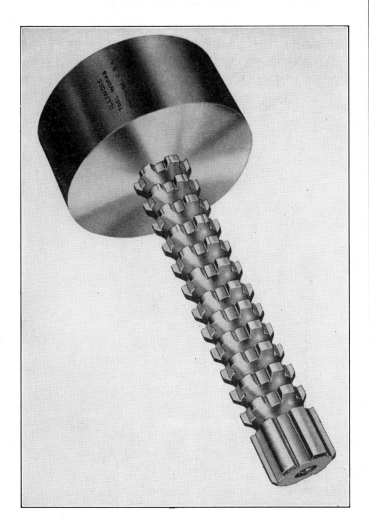

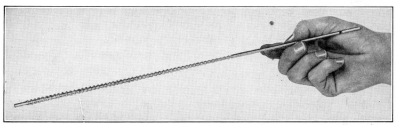

Fig. 15.   Pull-Type Broaches for Spline Broaching and Enlarged View of Broach
in Work.

535

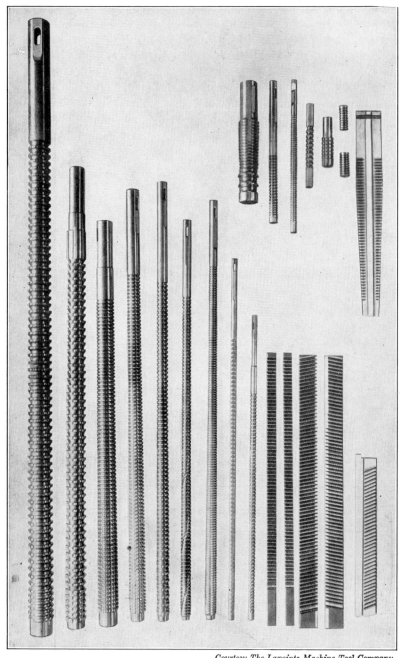

FIG. 16.   Pull, Push, and Surface Broaches of Various Shapes.

shapes can be broached, but the tools must be specially designed for each job.

A group of miscellaneous pull, push, and surface broaches of various shapes and sizes is shown in Figure 16. The method of holding these broaches in the machine depends on the type of broach as well as on its size and shape. Several methods are used in pull-type broaches, as shown in Figure 17. One requirement of all types of puller ends is that they must permit rapid insertion and removal of the broach. The attachment of surface-type broaches presents no difficult problem, as their entire length is supported on the slides of the machines.

Broach Terms and Angles

Reference to previous figures of pull-type broaches will illustrate some of the terms usually applied to broaches. Starting at the puller end, that portion of the tool up to the first teeth is known as the *shank*. It is made up of the *keyed* or *pull end* and the *front pilot*, which is a short section next to the teeth. The first teeth of the broach are the *cutting teeth;* the last few are called the *finishing teeth*. The distance from a point on one tooth to the corresponding point on an adjacent tooth is the *pitch*. This depends on the length of broach, chip thickness, and the kind of material being broached. The short end next to the finish teeth is the *rear pilot*.

The shapes or angles used on broaching tools are not necessarily the same throughout the length of the broaching tools. The top portion of a tooth is called the *land* and in most cases is ground to give a slight clearance. This angle, called *backoff* or *clearance angle*, is usually $1\frac{1}{2}$ to 4 degrees on the cutting teeth. Finish teeth have a smaller angle, ranging from 0 to $1\frac{1}{2}$ degrees. There should be no regrinding on the lands of most broaching tools, because this changes the size of the broach. Sharpening is done by grinding the face or front edge of the teeth. The angle to which this surface is ground corresponds to the rake angle on a lathe tool and is called the *face angle, hook angle, undercut angle,* or *rake angle*. The last term is probably the best, as it is the term used for this angle on other cutting tools. The rake angle varies according to the material being cut and, in general, increases as the ductility increases. Values of this angle range from 0 to 20 degrees, but for most steels a value of 12 to 15 degrees is recommended. This angle has considerable effect on the force required to make the cut and the finish. A large angle might give excellent results, but from the standpoint of lengthening

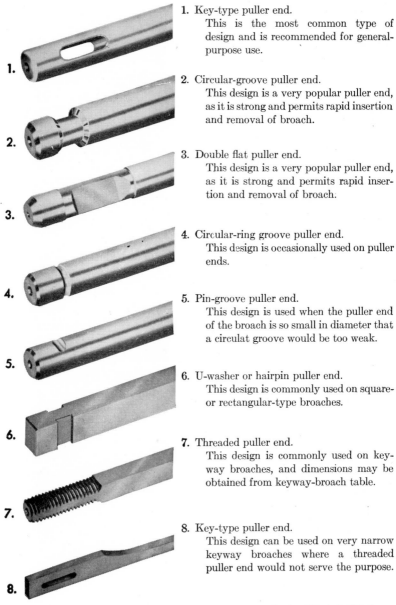

1. Key-type puller end.
   This is the most common type of design and is recommended for general-purpose use.

2. Circular-groove puller end.
   This design is a very popular puller end, as it is strong and permits rapid insertion and removal of broach.

3. Double flat puller end.
   This design is a very popular puller end, as it is strong and permits rapid insertion and removal of broach.

4. Circular-ring groove puller end.
   This design is occasionally used on puller ends.

5. Pin-groove puller end.
   This design is used when the puller end of the broach is so small in diameter that a circulat groove would be too weak.

6. U-washer or hairpin puller end.
   This design is commonly used on square- or rectangular-type broaches.

7. Threaded puller end.
   This design is commonly used on key-way broaches, and dimensions may be obtained from keyway-broach table.

8. Key-type puller end.
   This design can be used on very narrow keyway broaches where a threaded puller end would not serve the purpose.

Courtesy Illinois Tool Works.

FIG. 17.   Various Types of Broach Puller Ends.

tool life a smaller angle would be used. Frequently the first cutting teeth are rugged in shape and have a small rake angle, while the finish teeth are given a larger rake angle to improve the finish. Side-rake angles of 10 to 30 degrees are widely used on surface broaching to improve the finish of the cut.

REVIEW QUESTIONS

1. Define broaching, and state how it is done.
2. What are the advantages and limitations of the broaching process?
3. Classify broaching machines according to method of operation.
4. What is meant by surface broaching?
5. Distinguish between pull- and push-type broaching machines and describe the type of broach used by each.
6. Describe the process of machining cylinder blocks and heads by broaching.
7. What cutting speed is recommended for broaching mild steel?
8. How much metal should be removed per tooth on a broaching tool?
9. For what type of work is a continuous broaching machine used?
10. Briefly describe a rotary-broaching machine, and state the type of work for which such a machine can be used.
11. What information is necessary before a broach can be designed and constructed?
12. List five examples each of surface broaching and internal broaching.
13. How are helical grooves or splines cut by broaching?
14. What is meant by burnishing, and how is it done?
15. What determines the rake angle on a broaching tool? What should it be for mild steel?
16. Sketch a pull broach for cutting a keyway in a round hole.
17. How are broaching tools sharpened?

BIBLIOGRAPHY

BAUMBECK, W., "Development of Broach Rifling at Rock Island Arsenal," *Mechanical Engineering*, June 1943.

CADY, E. L., "Broaching of Machine Gun Barrels," *Metals & Alloys*, February 1943.

EINSTEIN, S., and ROMAINE, M., "Surface Broaching in High Production Industries," *Mechanical Engineering*, May 1937.

"Recent Developments in Broaching Automatic Parts," *Machinery*, February 1940.

"Rifle Parts Broached," *American Machinist*, November 1940.

GOTBERG, H. H., "Simplified Tools Designed for Broaching Spline Forms," *American Machinist*, October 25, 1945.

LIEBERT, HUGO W., "Broached Blades for Navy Turbines," *American Machinist*, August 5, 1943.

ROMAINE, M., "Broaching Cylinder Blocks and Heads in the Latest Engine Plant," *Machinery*, April 1938.

CHAPTER 20

TURRET AND AUTOMATIC LATHES

Turret and automatic lathes possess special designs and features which particularly adapt them to production work. The " skill of the worker " has been built into these machines, making it possible to reproduce identical parts with operators in charge who have little skill. In contrast to this, the engine lathe requires a highly skilled operator and more time to reproduce many parts which are dimensionally the same. The principal characteristic of this group of lathes is that the tools for consecutive operations can be continuously set up in readiness for use in the proper sequence. Although considerable skill is required to set and adjust the tools properly, once they are correct, little skill is required to operate them. Furthermore, many parts can be produced before adjustments are necessary. Eliminating the setup time between operations reduces the production time tremendously. The high development of turret and automatic lathes has made interchangeable manufacture what it is today.

Classification

The following classification of turret and automatic lathes, made according to single outstanding design characteristics, will serve as an outline of the discussion which is to follow.

1. Horizontal turret lathe
   (a) Chucking machine.
   (b) Bar machine.
   (c) Automatic machine.
2. Vertical turret lathe.
3. Vertical automatic multistation lathe.
4. Automatic screw machine.
5. Horizontal multispindle automatic
   (a) Bar machine.
   (b) Chucking machine.
6. Automatic lathes
   (a) Horizontal.
   (b) Vertical.

This classification may be further subdivided according to special design features, such as method of drive, method of feed, and type

Courtesy Warner & Swasey Manufacturing Company.

FIG. 1. No. 5 Universal Turret Lathe Equipped for Bar Work.

Courtesy Warner & Swasey Manufacturing Company.

FIG. 2. No. 5 Universal Turret Lathe Equipped with Chucking Tools.

of cross-tool slide. These details and other special features are explained in the discussion of the various types of machines.

Horizontal Turret Lathes

Turret lathes are listed according to the type of work they do. In appearance and general design the *bar* and *chucking* machines are much alike, as may be observed in Figures 1 and 2. Their principal difference is in the tools they use and in the manner the stock is held. Bar machines do not require the built-in rigidity that is required of chucking machines, as in most cases the bar tools can be made to

Courtesy Jones and Lamson Machine Company.

FIG. 3.   Saddle-Type Universal-Type Turret Lathe.

support the work. Chucking tools overhang and do not support the work, thus causing greater strain on both the work and tool support; hence, chucking machines must have the greatest possible rigidity.

Both the types of turret lathes previously mentioned may be made in either *ram* or *saddle* type. The ram type is so named because of the manner in which the turret is mounted. In these machines the turret is placed on a slide or ram which moves back and forth on a saddle clamped to the lathe bed. This arrangement permits quick and easy movement of the turret and is specially recommended for bar and light work. The saddle, although capable of adjustment, does not move in the operation of the turret. The saddle-type machine has the turret mounted directly on a saddle which moves back and forth with the turret in its operation. This design permits more rigid support of the tools, so necessary in heavy chucking work.

The stroke also is much longer, which is an advantage in long turning or boring cuts. Saddle-type turret lathes are illustrated in Figures 3 and 4.

Fig. 4. Gisholt 4L High-Production Saddle-Type Turret Lathe. Shown without Tooling.

Turret-Lathe Construction

Turret lathes have many features similar to those of modern engine lathes. The headstock in most cases is geared with provision for 6 to 12 spindle speeds. The various spindle speeds, as well as forward and reverse movement, are all controlled by levers extending from the head. The drive motor is usually located in the motor leg below the headstock and connected to the geared-head pulley by means of a belt. Early-design turret lathes were frequently belt-driven through step pulleys, but this arrangement is no longer used. Some few machines, designed for light work, are driven by a multiple-speed motor mounted directly on the spindle inside the headstock housing. High speeds, up to 3600 rpm, are possible on these machines.

The *cross-slide* unit, on which the tools are mounted for facing, forming, and cutting off, is somewhat different in construction from the tool-post and carriage arrangement used on lathes. It is made up of four principal parts: the cross slide, the square turret, the carriage, and the apron. These parts are readily discernible in the various turret-lathe illustrations. Some of the cross slides are supported entirely on the front and lower front ways, permitting more swing clearance for the work. This arrangement is frequently used on saddle-type machines which are to be used for large-diameter chucking jobs. The other arrangement for mounting has the cross slide riding on both upper bedways and further supported by a lower way. This is used on machines engaged in bar work and other applications where a large swing clearance is not necessary. An advantage of this type is the added tool post in the rear, frequently used in cutting-off operations.

On top of the cross slide is mounted a *square turret* capable of holding four tools in readiness for use. If several different tools are required, they are set up in sequence and can be quickly indexed and locked in correct working position. In order that cuts may be duplicated, the slide is provided with either a positive stop or a feed trip. Likewise the longitudinal position of the entire assembly may be accurately controlled by positive stops on the left side of the apron. Cuts may be taken with square-turret tools simultaneously with tools mounted on the hexagon turret.

The greatest difference between a turret lathe and an engine lathe is the use of a *hexagon turret* in place of the usual lathe tailstock. This turret, mounted either on the sliding ram or saddle, carries the tools for the various operations. The tools are mounted in proper sequence on the various faces of the turret so that, as it indexes around between operations, the proper tools are brought into position. For each set of tools there is provided a stop screw which controls the distance the tool will feed. When this distance is reached, an automatic trip lever stops further movement of the tool by disengaging the drive clutch.

Methods of Holding Stock

As the turret lathe is a production machine, special attention is given to methods of holding the work so that it can be done quickly and accurately. Since the operation is usually simple, extreme rigidity can be built into such equipment so that heavy cuts can be made. The usual devices for holding work are *collets, arbors, chucks*, and *special holding fixtures*.

Collets. Collets, commonly used for bar-stock material, are made with jaws of standard sizes to accommodate round, square, and hexagon stock. For large stock, collets of the parallel closing type are sometimes used, but in most cases collets of the spring type are recommended. These collets are solid at one end and split on the other end, which is tapered. The tapered end contacts with a similarly tapered hood or bushing, and, when the tapered end is forced into the hood, the jaws of the collet tighten around the stock. Spring collets are made in three designs: the *push-out type*, the *draw-in type*, and the *stationary type*. In each, however, the operation is similar to that just described. A cross section of a stationary type is shown in Figure 5. With this type there is no movement of the stock when

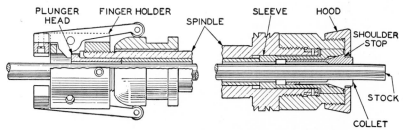

Courtesy Warner & Swasey Manufacturing Company.

FIG. 5. Stationary-Type Spring Collet.

it is tightened in the collet, since the latter is held in place against the hood. As the tapered surface of the plunger sleeve is pushed against a similar surface on the collet, the jaws are forced against the work. For work that must be accurately located endwise, this type of collet assembly is best. Push-out type collets are recommended for bar

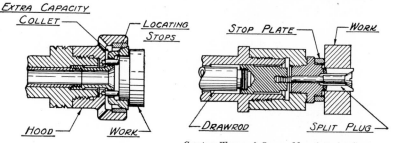

Courtesy Warner & Swasey Manufacturing Company.

FIG. 6. Draw-Back Extra-Capacity Collet.

FIG. 7. Expanding Plug-Type Arbor.

work, as the slight movement of the stock pushes it against the bar stop. This type is of the same construction as shown in Figure 5 except there is no hood to stop the forward movement of the bar stock. Draw-back-type collets are not widely used for bar stock, but are useful when the collets are of extra-capacity size and are used for holding short pieces, as shown in Figure 6. The slight back motion in closing forces the work against the locating stops.

Arbor. Expanding or threaded arbors are used to hold short pieces of stock that have a previously machined accurate hole in them. The action in holding the work is controlled by a mechanism very similar to that used with collets. An expanding plug-type arbor is shown in Figure 7. The work is placed on the arbor against the stop plate, and, as the draw rod is pulled, the tapered pin expands the partially split plug and grips the work. The threaded arbor operates in a similar fashion except that the work is screwed on the arbor by hand and is then forced back against a stop tube or flange.

FIG. 8. Two-Jaw Box Chuck.

Both collets and arbors may be power-operated by pneumatic, hydraulic, or electrical means located at the end of the spindle. Such an arrangement is frequently used on high-production work to provide quicker and easier operation.

Chucks. Chucks are used for holding large and irregularly shaped parts and, in general, are the same types as used in engine-lathe work. Illustrations of the *universal, independent,* and *combination* types are shown in Chapter 12. These chucks are either bolted or screwed to the spindle and have a very rigid mounting.

In addition to standard chucks, there are several other special types adapted for holding work of irregular shape. One of these, known as a *two-jaw box chuck*, is shown in Figure 8. Both jaws of this chuck move in and out together. Separate jaws which fit the work involved can be mounted on the master jaws, as shown in the illustration.

Courtesy Warner & Swasey Manufacturing Company.

FIG. 9. Revolving-Jaw Chuck:

Another special chuck, known as a *revolving-jaw chuck*, is illustrated in Figure 9. This is adapted for use in finishing small parts where several faces must be machined. After a cut is made, the work can be indexed to the next position, usually 90 degrees, and another cut can be taken. The revolving-jaw chuck is used principally for the machining of pipe fittings.

Power chucks operated by air, hydraulic means, or electricity to relieve the operator of the effort involved in tightening and loosening the work are available for production jobs. This is a distinct advantage if the work is large. An additional advantage of the power chuck is that it is quick acting.

It is difficult to mount all types of work on standard equipment; therefore, many special chuck jaws or holding fixtures must be devised. Standard face plates are frequently used for mounting such fixtures. The holding device is held to the face plate either by bolting or by means of the T slots on the face of the plate.

All these work-holding devices can be used equally well with both automatic lathes and turret lathes.

Differences between Turret and Engine Lathes

It has been stated previously in discussing turret-lathe construction that the turret lathe has several distinguishing design features not found in engine lathes. The principal difference, so far as appearance is concerned, is the turret for holding the tools. However, the main difference in these two machines is the fact that the turret lathe is adapted to quantity-production work, whereas the engine lathe is primarily used for miscellaneous jobbing, toolroom, or single-operation work. The essential features of a turret lathe which make it a quantity-production machine are these:

1. Tools may be permanently set up in the turret in the proper sequence of their use.
2. Each tool is provided with a stop or feed trip so that each cut of a tool is the same as its previous cut.
3. Combining cuts can be made — that is, tools on the cross slide can be used at the same time that tools on the turret are cutting.
4. Extreme rigidity in the holding of work and tools is built into the machine.

When a turret lathe is once set up for a certain job, many parts may be machined identically, without further adjustment of the tools. All types of work that can be done on an engine lathe can likewise be done on a turret lathe, and in many cases can be done quicker: bar and chuck work, thread cutting, taper turning, drilling, reaming, and many other similar operations. Although we now have many other production machines, the turret lathe has been largely responsible for the development of interchangeable manufacture as we know it today.

Tools and Tooling Principles

As has been stated, once a turret lathe is properly tooled, an experienced machinist is not required to operate it. However, skill is required in the proper selection and mounting of the tools. In small-lot production it is important that this work be done in as short a time as possible so as not to consume too much of the total production time. This time is made up of four factors: setup time, work-handling time, machine-handling time, and cutting time. Consideration must be given to all these factors in the original setup if the final production time is to be as short as possible.

UNIVERSAL BAR EQUIPMENT
PERMANENT SET-UP

COMBINATION STOCK STOP AND CENTER

SHORT FLANGED TOOL HOLDER

DIE HEAD

STARTING DRILL

DRILL CHUCK

COLLET AND BUSHING

SINGLE CUTTER TURNER

CENTER DRILLING TOOL

MULTIPLE CUTTER TURNER

LONG FLANGED TOOL HOLDER

SQUARE TURRET

COMBINATION END FACER AND TURNER

FLOATING TOOL HOLDER

CLUTCH TAP AND DIE HOLDER

ADJ. KNEE TOOL

Courtesy Warner & Swasey Manufacturing Company.

FIG. 10. Permanent Setup for Universal Bar Equipment.

Setup time can be reduced by having all necessary tools in condition and readily available. A thorough knowledge of the tools and the machine is also important. For short-run jobs a permanent setup of the usual tools on the turret is an excellent means of reducing time. In Figure 10 is shown a permanent setup for bar work with the tools mounted in the logical sequence for their use. The tools selected are standard tools and the ones most commonly used in this type of work. Permanently mounted,

SET ROLLS TO FOLLOW CUTTER

FEED

FEED

Courtesy The Warner & Swasey Company.

FIG. 11. "Combined Cuts" on Bar Work.

they may be quickly adjusted for various jobs. A similar setup can be prepared for chucking jobs.

The *work-handling time*, that time consumed in mounting or removing the work, is largely dependent on the type of work-holding devices used. For bar work this time is reduced to a minimum by having quick means for advancing the stock built into the machine.

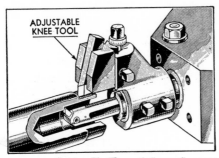

FIG. 12. "Multiple Cuts" from Hexagon Turret.

The time which it takes to bring the respective tools into cutting position is the *machine-handling time.* This can be reduced by having the tools in proper position and sequence for convenient use, and also by taking multiple or combined cuts whenever possible.

The actual *cutting time* for a given operation is largely controlled by the use of proper cutting tools, feeds, and speeds. However, additional time may often be saved by combining cuts as shown in Figure 11. *Combined cuts* refers to the simultaneous use of both slide and turret

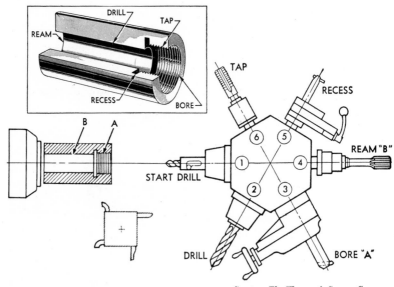

FIG. 13. Basic Hexagon-Turret Setup Illustrating the Correct Sequence of Operations to Handle Required Internal Cuts on Threaded Adapter Shown in Insert.

tools. In bar work combined cuts are especially desirable, as additional support is given to the work, thereby eliminating spring and chatter. In chucking work internal operations, such as drilling or boring, may frequently be combined with turning or facing cuts from the square turret. Time also may be saved by taking *multiple cuts* — that is having two or more tools mounted on one tool station. Figure 12 shows both boring and turning tools set up on one station of the turret.

To illustrate the method of tooling and sequence of operations for a given job, a basic hexagon-turret setup is shown in Figure 13 for making necessary internal cuts on a threaded adapter. Figure 14 shows the details of the internal cuts required to machine the adapter. With reference to the sequence shown in the figure, the various operations are as follows:

1. The bar stock is advanced against the combination stock stop and start drill and clamped in the collet. The start drill is then advanced in the combination tool, and the end of the work is centered.

2. The hole through the solid stock is drilled the required length.

3. The thread diameter is bored to correct size for the threads specified. A stub boring bar in a slide tool is used.

4. The drilled hole is reamed to size with the reamer supported in a floating holder.

5. A groove for thread clearance is recessed. For this operation a quick-acting slide tool is used with a recessing cutter mounted in a boring bar.

6. The thread is cut with a tap held in a clutch tap and die holder. For odd-size threads a single-point tool may be used. This operation is followed by a cutting-off operation not shown in the figure.

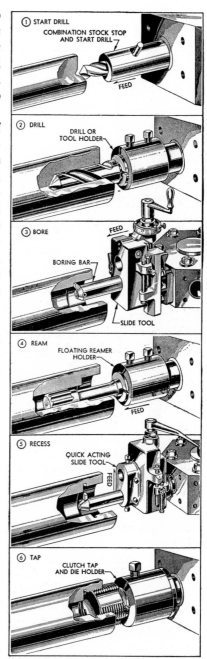

Courtesy The Warner & Swasey Company.

Fig. 14. Setup for Machining Internal Operations on Threaded Adapter.

Another example of tooling is illustrated in Figure 15. In this case the tool setup is for a shoulder stud shaft made of 2½-inch bar stock. The tooling shown is for a quantity of these parts and is

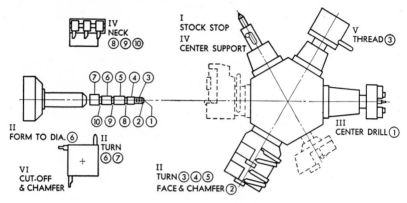

Fig. 15.   Tool Setup for Shoulder Stud Shaft.

slightly more complicated than a setup for producing only a few. With one exception all operations are external cuts. As shown in the figure, the tools used for the respective operations are as follows:

OPERATIONS AND TOOLS USED

Operation	Hexagon Turret	Square Turret
I	Feed stock to stop	
II	Turn (3), (4), and (5) diam. Face and chamfer (2)	Turn (6) and (7) to diameter
III	Center drill (1)	
IV	Support (1) with center	Neck cuts (8), (9), and (10) with back tools
V	Thread (3)	
VI		Cut-off and chamfer

Based on an output of 120 pieces the time per piece is estimated to be:

Setup time per piece	1 min.
Work-handling time per piece	½ min.
Machine-handling time per piece	½ min.
Cutting time per piece	5½ min.
Total production time	7½ min per piece

A few of the typical tools used in turret-lathe work have been illustrated in Figure 14. These tools are so designed that they may

be quickly mounted in the turret and adjusted for use. In addition
to the usual operations of drilling, boring, reaming, and internal
threading shown in the figure, various other threading, centering,
and turning tools are available. Internal threading is frequently
done with collapsible taps to facilitate quick removal of the tool.
For the same reason automatic die head cutters which open at the
end of the thread are used for external threads.

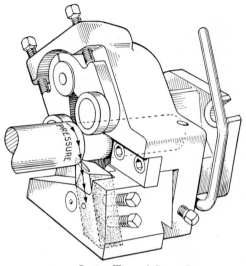

Courtesy Warner & Swasey Company.

Fig. 16.   Box Tool for Bar Stock.

For outside turning a *box tool,* shown in Figure 16, has been developed. As bar stock is supported only at the collet, additional
support must be provided in order for heavy cuts to be taken. This
is done by means of two rollers which contact the outside diameter
of the stock and take up the thrust of the cutting tool. Adjustment
of the rolls for varying diameter work is controlled by two setscrews
at the top of the holder. When the rolls are set slightly behind the
cutting tool, they tend to smooth out or burnish the surface. However, if the turned diameter must be concentric with the adjacent
surface, the rolls are set ahead and adjusted to its diameter. For
light cuts, a similar box tool which supports the work by means of
a V-back rest can be used. Cutters are held in position by two
setscrews, and the lever at the back of the assembly is for the purpose
of withdrawing the cutter from the work on the return stroke to
prevent marking.

Many small standard tools are also used in turret-lathe work. Drills, boring tools, reamers, and lathe-type turning tools can be rigidly supported in the various tool-holding devices. Cutting angles and speeds for these tools are the same as these described for lathe and other machine-tool operation.

Courtesy The Bullard Company.

FIG. 17. A 42-Inch Vertical Turret Lathe Machining Master Spindles. Table in Motion.

Vertical Turret Lathe

A vertical turret lathe is a machine tool resembling a vertical boring mill, but having the characteristic turret arrangement for holding the tools. It consists of a rotating chuck or table in horizontal position, with the turret mounted above on a cross slide. In addition, there is at least one side head provided with a square turret for holding tools. All tools mounted on the turret or side head have their respective stops set so that the length of cuts can be the same in successive machining cycles. It is, in effect the same as a turret lathe standing on the headstock end, and it has all the features necessary for the production of duplicate parts. This machine was developed to facilitate the mounting, holding, and machining of heavy parts. Only chucking work is done on this kind of machine.

In Figure 17 is shown a 42-inch master vertical turret lathe machining a large spindle. This machine is constructed with three heads — namely, the swiveling main turret head, the swiveling left-hand ram head, and the nonswiveling side head. Adjustable feed stops are provided for feed stop-off in all directions. A selector bar is built into the ram which automatically picks up different stops for each successive position of the turret. These prevent any possibility of overrunning the cut and after having been set for given dimensions are of further aid in the duplication of work sizes. The ram head at the left has vertical as well as horizontal feed screws and feed stops. This head may be swiveled to cut any angle up to 45 degrees and can operate simultaneously with the turret tools. The right-hand side head is also provided with feeding mechanism and stops. It provides for simultaneous machining adjacent to operations performed by the vertical turret, without interference. The table is supported rigidly by tapered roller bearings and driven by spiral-bevel gears. Two types of work tables are used, either the built-in three-jaw combination chuck or the four-jaw independent chuck. In addition, radial T slots are provided between the jaws to hold specially shaped parts. Vertical turret lathes have wide application in the quantity production of large and heavy parts.

Vertical Automatic Multistation Lathes

Machines of this type are designed for high production and are usually provided with either six or eight stations. In a six-station machine there would be five working spindles and a loading position.

All varieties of machining operations can be performed, including milling, drilling, facing, threading, tapping, reaming, and boring. The advantage of this type of machine is that all operations can be done simultaneously. In actual operation all tools are fed to the work, held in chucks or fixtures, at the respective loading stations.

Fig. 18.   Bullard 8-Inch Six-Spindle Type-D Mult-Au-Matic on Short-Run Work.

When all operations are complete, the tools or work move vertically out of the way, the work table indexes one station, and the operations are repeated. The time between indexing operations is controlled by the time of the longest single operation.

In Figure 18 is shown an 8-inch six-spindle Bullard Mult-Au-Matic, automatic production machine of this type. It is made in several sizes, accommodating work up to $17\frac{1}{2}$ inches in diameter, with either six or eight spindles. This machine is designed to accommodate all classes of machine work on castings, forgings, or cut-off bar stock. Several types of heads are provided for this machine. The plain vertical head of one-piece construction is combined with the saddle and permits only vertical movement with 8-inch strokes. The plain compound slide has a single slide mounted on the saddle, permitting a 4-inch horizontal movement. By using a standard universal head,

Courtesy The Bullard Company.

FIG. 19.   Close-up of Station 2 on Mult-Au-Matic Showing Cutting Lubricant in Action.

the tool may be fed vertically, horizontally, or in any angular direction. Standard double-purpose heads have two slides. The left slide operates only in a vertical direction, while simultaneously the right-hand slide moves left in a horizontal direction. With these motions available, cuts may be taken in any desired direction. Feed for the tools is controlled by cam action, and various spindle speeds may be obtained at each station. Parts to be machined, after being chucked, are indexed from one station to another, as previously described.

Figure 19 shows a close-up of station 2 of the previous figure with cutting lubricant in action. The completed piece at the unloading station can be seen at the right.

Eight-spindle machines can be equipped with either single or double indexing. With single indexing seven operations can be performed in the cycle; with double indexing only three operations are possible. In the second case two pieces are completed in each cycle.

An eight-spindle rotary "rigid-turning" machine, operating on a different cutting principle than the previous multiple-spindle machine,

Courtesy The Cleveland Hobbing Machine Company.

Fig. 20. Eight-Spindle Rotary Rigid-Turning Machine Setup for Shell Turning.

is shown set up for shell turning in Figure 20. In this machine, the work is rotated and fed past a stationary single-point tool, clearly shown in the insert next to the figure. This cutting tool is controlled by a profile cam which imparts the form to the work. Only turning operations are performed on this machine. Other production machines, similar in appearance to this one, are designed for spline cutting on axle shafts, gear hobbing, or mold turning.

Automatic Screw Machines

The automatic screw machine was invented by Christopher N. Spencer of the Billings & Spencer Company about 60 years ago. The principal feature of the invention was to provide a controlling movement for the turret so that tools could be fed into the work at

Courtesy Brown & Sharpe Manufacturing Company.

FIG. 21.   Automatic Screw Machine.

desired speeds, withdrawn, and indexed to the next position. This was all accomplished by means of a cylindrical or drum cam located beneath the turret. Another feature, also cam-controlled, was a mechanism for clamping the work in the collet, releasing it at the end of the cycle, and then feeding the bar stock up against the stop. These features are still used in about the same way as originally worked out.

An automatic screw machine is essentially a turret lathe designed to use only bar stock. It is so named, because the first machines of this type were used mainly for manufacturing bolts and screws. Since it can produce parts, one after the other, with little attention from the operator, it is naturally called automatic. Machines requiring an operator to load and unload the work during each cycle are properly termed semiautomatic machines. Most automatic screw machines are fully automatic and not only feed in an entire bar of stock but also are provided with a magazine so that several bars can be fed through the machine automatically.

Automatic screw machines may be classified according to the type of turret used or according to the number of spindles the machine has. Multispindle machines, however, are not usually spoken of as screw machines, but rather as multispindle automatics. The type of work which the two machines do is the same, although there is considerable difference in the design and production capacity.

In Figure 21 is illustrated an automatic screw machine designed for bar work of small diameter. This machine has a cross slide, capable of carrying tools both front and rear, and a turret mounted in a vertical position on a slide with longitudinal movement. The two disk cams controlling the cross slide are directly underneath and driven by the front drive shaft. Also mounted on this same drive shaft are three disk-shaped carriers, upon which are mounted dogs to engage various trip levers to control the operation of the machine. The one to the extreme right controls the indexing of the turret, the center one controls the collet and feeding of the stock, and the one to the left the rotation and speed of the spindle. The various tools used in the machine are mounted around the turret in a vertical plane in line with the spindle. All usual machining operations, such as turning, drilling, boring, and threading, can be done on these machines. The type of bar stock used, whether round, square, hexagon, or some special shape, is determined by the cross section preferred in the finished product. Collets for any commercial shape are available.

The machine shown in the figure is usually equipped with an automatic rod magazine to keep it supplied with material for a period of time. When a rod of material is completely used the machine stops and another rod is fed into the collet up to the stop. The machine then automatically resumes operation. In addition to attending several machines, the operator checks the work and tools and sees that the magazines are supplied with materials.

Multiple-Spindle Automatics

Multiple-spindle automatic machines are the fastest type of production machines for bar work. They are fully automatic in their operations and are made in a variety of models, with two, four, five, six, or eight spindles. In these machines all spindles operate simul-taneously, and one piece is completed each time the tools are with-drawn and the spindles indexed.

The general construction of a multiple-spindle automatic is shown in Figure 22. The spindles carrying the bar stock are all held and rotated in the spindle carrier. Opposite each spindle are mounted the necessary tools for the respective operations. Most of the tools are supported on the end tool slide, which is centrally located with reference to all spindles. This tool slide does not index or revolve with the spindle carrier, but slides forward and back on the stem shaft to carry the end working tools to and from contact with the revolving bars of stock. Both above and below the spindle carrier and end tool slide are two cross slides upon which side-cutting tools can be mounted. In six- and eight-spindle models there are two additional intermediate or side slides available. All slides are inde-pendently operated and are used in combination with end-slide tools for such operations as form turning, knurling, thread rolling, slotting, and cutting off.

Bars of stock are loaded into each spindle when it has been indexed to the first position. If automatic stock feeding is used, it is done in the lower spindle position at the rear of the machine. In operation, the spindle carrier is indexed by steps to bring the bar of stock in each of the work spindles successively in line with the various tools held on the tool slides. All tools in the successive positions are at work on different bars at the same time. The time to complete one part is equal to the time of the longest operation plus the time necessary for withdrawing the tools and indexing to next position. This time can frequently be reduced to a minimum by dividing the long cuts between two or more operations.

The drive for the multiple-spindle automatic is somewhat com-plicated, as all tool operations and machine movements are auto-matically controlled. The motor, mounted on the end of the machine opposite the spindles, operates the entire machine. The main drum shaft is located below the spindles and extends the full length of the machine. This shaft, with its several drum cams and gear connec-tions, controls all tool movements, indexing, stock feeding, and timing of operations.

In Figure 23 is shown the setup for making steel-tube end fronts on a four-spindle automatic bar machine. The illustration shows the operator's side of the machine. The machine automatically feeds the

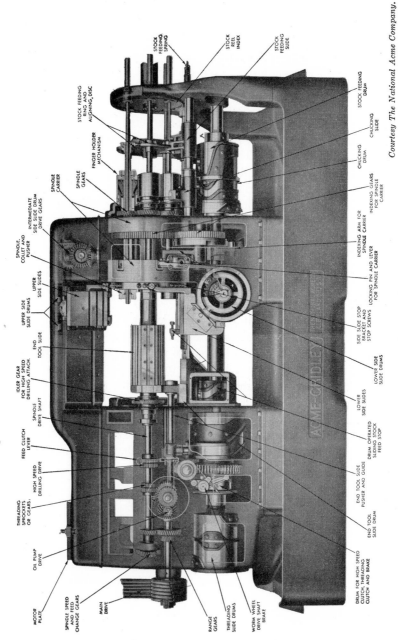

STOCK FEEDING SPRING

STOCK REEL INDEX

STOCK FEEDING SLIDE

STOCK FEEDING RING AND ALIGNING DISC

STOCK FEEDING DRUM

CHUCKING SLIDE

FINGER HOLDER MECHANISM

SPINDLE CARRIER

SPINDLE GEARS

CHUCKING DRUM

INTERMEDIATE SIDE SLIDE DRUM DRIVE GEARS

INDEXING GEARS FOR SPINDLE CARRIER

SPINDLE, COLLET AND PUSHER

INDEXING ARM FOR SPINDLE CARRIER

UPPER SIDE SLIDE DRUMS

UPPER SIDE SLIDES

LOCKING PIN AND LEVER FOR SPINDLE CARRIER

SIDE SLIDE STOP BRACKET AND STOP SCREWS

END TOOL SLIDE

IDLER GEAR FOR HIGH SPEED DRILLING ATTACH.

LOWER SIDE SLIDE

SPINDLE DRIVE SHAFT

LOWER SIDE SLIDES

FEED CLUTCH LEVER

HIGH SPEED DRILLING DRIVE

END TOOL SLIDE PUSHER AND GUIDE

DRUM OPERATED SLIDING STOCK FEED STOP

THREADING SPROCKETS OR GEARS

OIL PUMP DRIVE

END TOOL SLIDE DRUM

DRUM FOR HIGH SPEED CLUTCH THREADING CLUTCH AND BRAKE

MOTOR PLATE

SPINDLE SPEED AND FEED CHANGE GEARS

MAIN DRIVE

RANGE GEARS

THREADING SLIDE DRUMS

WORM WHEEL DRIVE SHAFT BRAKE

Courtesy The National Acme Company.

FIG. 22. Construction Features of a Six-Spindle Horizontal Automatic Bar Machine.

bar stock against the stop in the lower spindle position, and, after the collet has closed on the stock, the stop is automatically withdrawn. The rear piece is then rough formed while the hole is drilled and end-faced. At the second station the front part is formed and the hole bored and chamfered. At the third station, shown at upper rear position in the figure, the rear is finish-formed and the hole reamed. The piece is cut off at the fourth station by the upper front slide.

Courtesy The National Acme Company.

Fig. 23.   Four-Spindle Acme-Gridley Automatic Bar Machine Making Steel-Tube End Front.

As a practical example to show how a multiple-spindle automatic is set up, an operation sheet for the production of differential pinions is shown in Figure 24. Before a machine is set up for a new job, it is first necessary to make several preliminary calculations. These will include choice of tools, sequence of operations, spindle speeds and tool feeds for the stock and tools, selection of the proper gears and cams, and the estimation of the production. This preliminary calculation is usually a function of the production-planning department, and, when it is completed, a sheet, similar to the one in the figure, is turned over to the setup man for his guidance. In making

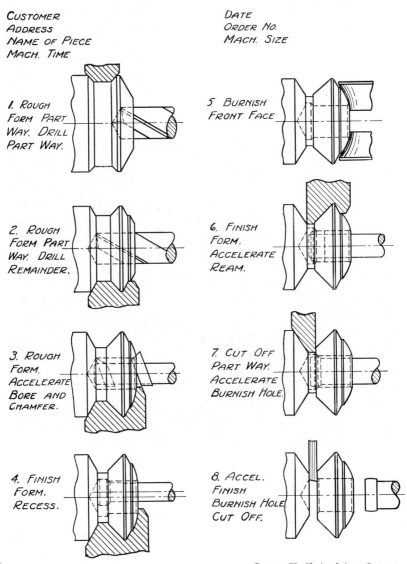

CUSTOMER
ADDRESS
NAME OF PIECE
MACH. TIME

DATE
ORDER NO.
MACH. SIZE

1. ROUGH FORM PART WAY. DRILL PART WAY.

5. BURNISH FRONT FACE

2. ROUGH FORM PART WAY. DRILL REMAINDER.

6. FINISH FORM. ACCELERATE REAM.

3. ROUGH FORM. ACCELERATE BORE AND CHAMFER.

7. CUT OFF PART WAY. ACCELERATE BURNISH HOLE.

4. FINISH FORM. RECESS.

8. ACCEL. FINISH BURNISH HOLE CUT OFF.

Courtesy The National Acme Company.

Fig. 24. Operation Sheet for Differential Pinion.

out such a sheet it should be kept in mind that the time to make one piece can never be less than the longest operation. However, if it is possible to split the longest operation between two or more stations, the time to produce one piece can be reduced. Reference to the figure will show that this procedure has been followed in the planning of this job. Space on the figure does not permit including all the information normally included on such a sheet. In addition to the data shown, there should be indicated the tools used for each operation, spindle speeds, tool feeds, change gear sizes, cam descriptions, and kind of material to be used.

Courtesy The National Acme Company.

Fig. 25. Eight-Spindle Acme-Gridley Automatic Chucking Machine.

Multiple-spindle automatics are not limited to bar stock, but may be provided with hydraulic or air-operated chucks for holding individual pieces. In some cases the chucks are loaded by the operator; in others magazine feeder are arranged to load the machine at one of the lower stations. Machines of this type are known as multiple-spindle automatic chucking machines and are similar to the bar machines except for the stock-holding equipment.

An eight-spindle automatic chucking machine is shown in Figure 25. The machine is making lifting plugs from steel forgings and is arranged for double indexing with duplicate tooling to give two pieces per cycle. Loading and unloading occur in the two center positions, where both the spindles are stopped. The first operation, taking place at the lower stations, consists of rough-facing shoulder, rough-facing end, and finish-turning thread diameter. The second operation, shown in the figure, is machining the threads. The final operation of finish-facing the end and shoulder is shown at the upper spindles.

A great many attachments to permit almost any type of machine operation are available for these machines. Both solid and self-opening dies and taps may be applied in position to suit the work. Taper turning, combined taper turning and taper boring, or recessing attachments are applied to the end tool slide. A spindle-stopping mechanism can be arranged for such operations as milling, slotting, and cross drilling. Many machines are provided with a small chip conveyor which picks up the chips beneath the tooling area and dumps them into a container at the end of the machine. To assist in production records, a chronolog can be used to count the production and record the idle time of the machine.

A great variety of parts can be produced in multiple-spindle automatics, the only limiting factor being the capacity of the machine. However, long-run jobs are necessary to offset the high initial investment, high maintenance, and expensive tooling costs. Both single-spindle automatics and hand-turret lathes have wide application and in short- and medium-run work prove to be economical in operation. Each machine is good in its field, but care must be taken in making the initial selection.

Automatic Lathes

Lathes which have their tools automatically fed to the work and withdrawn after the cycle is complete are known as *automatic* lathes. Most lathes of this type require that the operator place the part to be machined in the lathe and remove it after the work is complete, and so are perhaps incorrectly called automatic lathes. Lathes which are fully automatic are provided with a magazine feed so that a number of parts can be machined, one after the other, with little attention from the operator. Machines in this group differ principally in the manner of feeding the tools to the work. Most machines, especially those holding the work between centers, have front and rear tool slides. Others, adapted for chucking jobs, have an end tool slide located in the same position as the turret on the turret lathe. These machines may also have the two side-tool slides. Still another construction employs a flat table in front of the chucking spindle, upon which can be mounted tool slides at any angle or in any position. Each tool slide has individual feed and receives its power from individual drive shafts at the end of the machine. Several types of automatic lathes are described in the following paragraphs.

Courtesy Jones and Lamson Machine Company.

FIG. 26. A 12-Inch Fay Automatic Lathe.

A 12-inch Fay automatic lathe tooled for machining short shafts between centers is shown in Figure 26. This machine has both front and rear tool carriages mounted on heavy cylindrical bars. All turning tools are mounted on the front carriage and receive their longitudinal motion from the forward and reverse cams located beneath the headstock. These tools may be given a tilting motion, by means of the former slide in front of the machine, in order to turn tapers or relieve the tools at the end of the cut. In multiple tooling it is advisable to have some automatic means of relieving cutting

Courtesy Jones and Lamson Machine Company.

FIG. 27.   A 12-Inch Fay Automatic Lathe Tooling for Airplane-Engine Pistons.

Courtesy The Lodge and Shipley Machine Tool Company.

FIG. 28.   No. 3A Duomatic Lathe Setup for Rough-Turning 155-Mm Shell.

pressure of the tools at the points where the cuts are matched. This is obtained by moving the former slide down the angle of the guide. This action relieves the tool pressure at the desired rate to insure definite matching of all cuts without undercutting or marking on the return stroke. All facing, forming, and chamfering tools are carried on the rear carriage. These tools move into the work in a transverse plane controlled by a rear former slide which is moved by cylindrical cams beneath the machine.

This same machine is shown in Figure 27, tooled for airplane-engine pistons. The ring grooves are formed and faced by the back arm while the front carriage turns the outside diameter and chamfers the ring grooves. The faces of the ring grooves are held absolutely parallel to each other, their width held to 0.0002 inch. The closed end is faced with a tool held on the rear arm which is automatically relieved at the end of the stroke to eliminate marking the work on the return stroke.

Figure 28 shows a no. 3A Duomatic Lathe set up for rough-turning 155-mm shell forgings. This machine also has two complete carriages, one in front and one in the rear. Both carriages have cross tool slides which can be swiveled to any desired angle for power feed. Each carriage has an independent lead screw to provide power feed for its respective carriages during turning or boring operations. Both carriages can be used simultaneously in turning operations, as is the case in the machining operation shown in the figure. Either one or both may be used in facing or grooving operations by using the tool slides mounted on the carriages. The independent control boxes, which regulate tool feeds and stop and return tools to the commencement of cycle, are located at front and rear of the headstock end of the lathe. When once set up, the lathe is entirely automatic from the time of loading to the end of the machining cycle.

The setup for the finish turning of a 155-mm shell is shown in Figure 29. The straight diameter is turned by the tools mounted in the rear carriage slide, while the tapered portion is turned by the tool in the front slide. The method controlling the movement of the tool slides is clearly illustrated in the figure.

An automatic lathe, known as a platen-type Simplimatic, is shown in Figure 30. This machine is provided with front, rear, and center tool slides mounted on the platen or the table end of the machine. The machine, as illustrated, is tooled for machining flywheels. In operation all tools feed simultaneously, performing all turning, facing,

Courtesy The Lodge and Shipley Machine Tool Company.

FIG: 29;   Duomatic Lathe Setup for Finish-Turning 155-Mm Shells.

Courtesy Gishott Machine Company.

FIG. 30.   Platen-Type Simplimatic Lathe Setup for Flywheel Work.

taper boring, and chamfering operations at one time. Power for feeding the tools comes to the tool slides from one end through universal couplings and drive shafts. The entire operation of the

Courtesy Gisholt Machine Company.

FIG. 31. Machining Heavy Steel Pipe Flanges on a Gisholt Simplimatic Lathe.

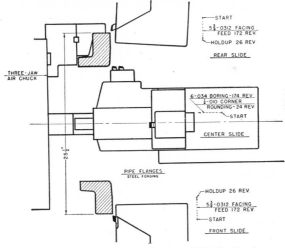

Courtesy Gisholt Machine Company.

FIG. 32. Method of Setting Up Tools for Machining Steel Pipe Flanges.

machine is automatic, and the operator has only to load and unload the machine. One operator may attend two or three machines.

Figures 31 and 32 show the method of tool setup for machining heavy steel flanges. The line diagram in Figure 32 shows the motion and direction of each of the three tool slides. The front slide holding one tool faces the flange. The rear slide mounted at the proper angle faces concave from the mid-point of the face to the inside. Rough and finish boring, as well as the rounding of the corner, is done by the center slide. The tool holder is provided with cam control which guides the boring tools on the proper taper as they are fed into the work by the slide. The actual details of this arrangement are shown in the close-up photograph in Figure 31. The construction of the slides and the manner in which the tools are fed are clearly illustrated. Platen-type machines have considerable flexibility in the mounting of the various tool slides.

Courtesy Gisholt Machine Company.

FIG. 33. Machining Flywheels on Simplimatic Vertical-Head Automatic Lathe.

For the machining of automotive flywheels, a radial- or vertical-head Simplimatic lathe is shown in Figure 33. This machine is a development from the platen-type Simplimatic and is especially designed for large work requiring end facing cuts. The tools are placed on the face of the radial head around the work. They are brought up close to the work and are applied with a minimum of overhang. All tools cut simultaneously and at high speed. Cam segments, on a single master drum, actuate and feed the tool slides at their proper rate.

A fully automatic lathe developed for supporting work at both ends is shown in Figure 34. This lathe is designed to take care of those jobs which cannot be fed through a hollow spindle or, on account of length, can be satisfactorily turned only on centers. The lathe has both a headstock and a tailstock, with centers and all the usual features of an engine lathe. In addition, it has a magazine for holding the work blanks and a fully automatic work-handling and control mechanism. This mechanism places the work on centers, adjusts and clamps the tailstock spindle, grips the work by means of the rotating chuck, starts the feed of the tool carriage, releases the work at the end of the cut, and finally returns the carriage to

Pratt and Whitney Company.

FIG. 34. Fully Automatic Lathe.

the starting position. These operations are repeated automatically so long as there are pieces in the magazine. The machine shown will handle work $\frac{1}{2}$ to $1\frac{1}{8}$ inches in diameter and from $3\frac{1}{2}$ to 18 inches in length. For ordinary turning work, one or more tools are mounted in the front slide. A tool block for grooving, necking, or chamfering may be mounted in the rear. Taper turning is accomplished by using a tool block with a former. The number of machines an operator can run depends upon his ability to keep the magazines loaded and the tools in cutting condition. Automatic lathes of this type have wide application for the quantity production of small machine shafts, reamer blanks, valve stems, spindles, and similar parts.

A single-spindle vertical machine known as a Rigidturner is shown in Figure 35. This machine, using a different method of turning

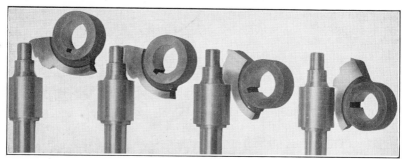

Courtesy The Cleveland Hobbing Machine Company.

FIG. 35. (Top). Single-spindle Rigidturner Machine Adapted to Turning Multiple Diameters on a Wide Variety of Work.

(Bottom). Sequence of Operations on Mold Turning Illustrating Type of Cutter Used.

from the usual single-point tool method, is specially adapted to turning multiple diameters on a wide variety of production work. A special form cutter is supported on a short horizontal shaft, and the work is held between centers in a vertical position. The type of tool used in mold cutting and the method employed in its operation are illustrated in the insert above the figure of the machine. The work advances past the cutter, which is rotated in a variable timed relation, thus producing the various diameters. The entire operation is automatic except for loading and unloading the work from the machine.

Each of the machines described has its field of usefulness, although often the selection of the proper machine is difficult to make. In general, the more automatic a machine is the greater will be its initial cost. However, in deciding upon a machine, this cost must be weighed against unit production cost, cost of maintenance and tooling, accuracy, adaptability of the machine to the product, and numerous other factors. Rate of production and the quantity to be produced are also influencing factors, as they have considerable bearing on the production cost. Turret and automatic machines are essentially production machines and require long runs for most efficient operation.

REVIEW QUESTIONS

1. List the various types of turret and automatic lathes.
2. How does a turret lathe differ from an engine lathe?
3. Distinguish between the ram and saddle types of turret lathes.
4. What are the principal parts of a turret-lathe cross-slide assembly?
5. List the various devices for holding stock on a turret lathe.
6. What type of spring collet is recommended for bar work?
7. How does a revolving jaw chuck operate?
8. What controls the actual cutting time of a given turret-lathe operation?
9. Explain the term " combined cuts " as applied to turret-lathe operation.
10. How is bar stock supported so that it will not deflect when being cut?
11. Show by sketch the tool setup for making 1/2 by $2\frac{1}{2}$ hexagon-head machine bolts.
12. Prepare an operation sheet for the bolt described in Problem 11, indicating the tools used for each operation.
13. What is the purpose of the following: (a) Box tool, (b) automatic die head, (c) square turret, (d) spring collet, (e) bar stop?
14. What type of work is done on a vertical turret lathe?
15. Explain the operation of a vertical automatic multistation lathe.
16. What is the cutting principle used on a multispindle rotary Rigidturning machine?
17. How does an automatic screw machine differ from a turret lathe?

18. What is an automatic lathe?

19. How is the tool feed accomplished on a platen-type Simplimatic lathe?

20. Explain the operation of mold cutting on a vertical-turning machine known as a Rigidturner.

BIBLIOGRAPHY

How to Machine Parts on a Turret Lathe, Warner & Swasey Company, 1944.

LONGSTREET and BAILEY, *Turret Lathe Operators Manual,* Operators' Service Bureau, Warner & Swasey Company, 1940.

CHAPTER 21

ABRASIVES, GRINDING WHEELS, AND
GRINDING MACHINES

Grinding

To grind means "to abrade, to wear away by friction or to sharpen." As applied to machine-shop practice, it refers to the removal of metal by means of a rotating abrasive wheel. The action of a grinding wheel is very similar to that of a milling cutter. It is made up of many small abrasive grains bonded together, each one acting as a small cutting tool. Definite elongated metal chips may be seen clearly by examining the material removed under a microscope.

The grinding process is one of extreme importance in production work. It possesses certain advantages that are not found in other cutting processes:

1. It is the only method of cutting such materials as hardened steel. Parts requiring hard surfaces are first machined to shape while the metal is in an annealed state, only a small amount of excess material being necessary for the grinding operation. The amount of this allowance depends on the size, shape, and tendency of the part to warp during the heat-treating operation. The sharpening of hand cutting tools is an important use of this process.

2. It produces finishes that are extremely smooth and hence very desirable at contact and bearing surfaces. This is due to the many small cutting edges on the wheel. As the wheel has considerable width there are no marks as a result of feeding it across the work.

3. This process can finish work to very accurate dimensions in a short time. Since only a small amount of material is removed, the grinding machines require a close regulation of the wheel, and it is possible to hold work to a fraction of a thousandth of an inch with considerable ease.

4. Very little pressure is required in this process, thus permitting its use on very light work that would otherwise tend to spring away from the tool. This characteristic permits the use of magnetic chucks for holding the work in many applications of grinding.

Abrasives

An *abrasive* is a hard material which can be used to cut or wear away other materials. Theoretically, any material can act as an

577

abrasive to other materials that are softer. However, certain few materials are known which have suitable characteristics for grinding work, and it is these materials that we have in mind when we speak of abrasives. A brief classification of the common abrasive materials used for grinding wheels is given here:

ABRASIVES FOR GRINDING WHEELS

1. Natural
   (a) Sandstone or solid quartz.
   (b) Emery, 50–60% crystalline Al_2O_3 plus iron oxide.
   (c) Corundum, 75–90% crystalline Al_2O_3 plus iron oxide.
   (d) Diamonds.
2. Manufactured
   (a) Silicon carbide, SiC.
   (b) Aluminum oxide, Al_2O_3.

For many years it was necessary to rely on natural abrasives in the manufacture of grinding wheels. *Sandstone* wheels are still used to some extent for hand-operated grindstones. While they are cut from high-grade quartz or sandstone, they have the disadvantage that they frequently do not wear evenly in use because of the variations in the natural bond. Most wheels of this type are made in Ohio, where suitable deposits of sandstone are found.

Corundum and *emery* have long been used for grinding purposes. Both are made up of crystalline aluminum oxide in combination with iron oxide and other impurities. In the United States corundum is found in Tennessee, Georgia, and South Carolina. Emery first came from Greece and Asia Minor, but is now mined in New York and Massachusetts. As is the case with sandstone, these minerals also lack a uniform bond and consequently are not suitable for high-speed grinding work. Before the discovery of artificial abrasives, these abrasives were crushed and bonded with various materials in the manufacture of grinding wheels. The best results were obtained by using the vitrified process. Although they were a great improvement over natural stones, these wheels still lacked uniform structure because of the impurities associated with the emery and corundum.

Diamond wheels, made with a resinoid bond, are especially useful in sharpening cemented-carbide tools. In spite of high initial cost, they have proved economical because of their rapid cutting ability, slow wear, and free cutting action. Very little heat is generated with their use, which is an added advantage in tool grinding.

Manufactured or electric-furnace abrasives were not known until the latter part of the 19th century. *Silicon carbide* was first discovered by E. G. Acheson of Monongahela City, Pa., in 1891, while he was attempting to manufacture precious gems in an electric furnace. The hardness of this material, according to the Mohs's scale,* is slightly over 9.5, which approaches the hardness of a diamond. Realizing the possibilities of this hard crystalline material as an abrasive, the Carborundum Company developed the process for its manufacture on a commercial scale. The raw materials now used are silica sand, petroleum coke, sawdust, and salt. The furnace employed

"Lecture Course on Coated Abrasives" Behr-Manning.

Fig. 1. Silicon Carbide Furnace Charged with Coke, Sand, and Sawdust under Operating Conditions. Note inflammable gases escaping at the sides.

(see Figure 1) is quite long and of the resistance type. The raw materials are piled around the carbon electrode and walled up on each side with loose brick. The purpose of the sawdust is to give porosity to the mixture and to permit the escape of the carbon monoxide gas. The furnace is heated to around 4200 F and held there for a considerable period of time. The product consists of a mass of crystals surrounded by partially unconverted raw material. After cooling, the material is broken up, graded, and then crushed

* Mohs's scale of hardness: (1) Talc, (2) gypsum, (3) calc spar, (4) fluor spar, (5) apatite, (6) feldspar, (7) quartz, (8) topaz, (9) sapphire, (10) diamond.

to the desired grain size. Silicon carbide crystals are very sharp and extremely hard, but their use as an abrasive is limited because of brittleness.

The development of *aluminum oxide* occurred a few years after the discovery of silicon carbide through experiments made by C. B. Jacobs of the Ampere Electro-Chemical Company of Ampere, N. J. The raw material for this process is the claylike mineral *bauxite* (mined in Arkansas), which is the principal source of the metal aluminum. Bauxite consists principally of aluminum oxide in combination with water and various impurities. In brief, the process consists in first driving off the excess moisture by heating the ore. Small amounts of coke and iron filings are added to the ore to act as reducing and purifying agents, and it is then put into the electric furnance of the arc type. The furnace consists of an unlined conical shell which is placed on a carbon base. Two carbon electrodes hang down inside the shell, and the bauxite is charged from above, filling the space around the electrodes. The current arcs from one electrode to the mass and then into the other electrode producing intense heat which melts the mass and eliminates the impurities. The finished product is a large pig, weighing several tons, which is broken up and graded. The purest material is at the center of the mass. Aluminum oxide is slightly softer than silicon carbide, but it is much tougher. Most manufactured wheels are made of aluminum oxide for this reason.

Manufacture of Grinding Wheels

The process of making a grinding wheel is the same for both the aluminum oxide and silicon carbide materials. In brief, the procedure is:

1. The material is first reduced to small sizes by being run through roll and jaw crushers. Between crushing operations the fines are removed by passing the material over screens.

2. All material is passed through magnetic separators to remove iron compounds.

3. A washing process removes all dust and foreign material.

4. The grains are graded by being passed over vibrating standard screens. (A standard no. 30-mesh screen has 30 meshes per inch or 900 openings per square inch. No. 30-size material is that which passes through a no. 30 screen and is retained on the next finer size, which in this case is a no. 36.)

5. Grains are mixed with bonding material, molded or cut to proper

shape, and heated.   The heating or burning procedure varies considerably, according to the type of bond used.

6. The wheels are finally bushed, trued, tested, and given a final inspection.

Bonding Processes

1. Vitrified process.   The abrasive grains are mixed with claylike ingredients which are changed to glass up being burned at a high temperature.   In the puddling process sufficient water is added to form a thick smooth mixture.   It is then poured into a steel mold and allowed to dry for several days in a room with controlled temperature.   The dry-press process requires the addition of little water. In this case the wheels are shaped in metal molds under a hydraulic press.   Wheels made this way are dense and are accurately shaped. The time for burning varies with the wheel size, being anywhere from 2 to 14 days.   The process is similar to burning tile or pottery.

The advantages of vitrified wheels are that they are porous, strong, and unaffected by water, acids, oils, and climatic or temperature conditions.   About 75% of all wheels are made by this process.   The recommended speed for these wheels is 5500 feet per minute with a maximum speed of 6500 feet per minute.

2. Silicate process.   In this process silicate of soda is mixed with the abrasive grains and the mixture is tamped in metal molds.   After drying several hours, the wheels are baked at 500 F 1 to 3 days.

Silicate wheels are milder-acting than those made by other procesess and wear away more rapidly.   This type of wheel is suitable for grinding edge tools where the heat must be kept to a minimum.   This process is also to be recommended for very large wheels, as they have little tendency to crack or warp in the baking process.   The hardness of the wheel is controlled by the amount of silicate of soda used and the amount of tamping given the material in the mold.

3. Shellac process.   The abrasive grains are first coated with shellac by being mixed in a steam-heated mixer.   The material is then placed in heated steel molds and rolled or pressed.   Finally, the wheels are baked a few hours at a temperature around 300 F.

This type of bond is adapted to thin wheels, as it is very strong and has some elasticity.   Shellac-bonded wheels are also used for grinding camshafts and other parts, where a high polish is desired.   Other applications are sharpening large saws, cutting-off operations, and finishing large rolls.

4. Rubber process. Pure rubber with sulfur as a vulcanizing agent is mixed with the abrasive by running the material between heated mixing rolls. After it is finally rolled to desired thickness, the wheels are cut out with proper-shaped dies and then vulcanized under pressure. Very thin wheels can be made by this process because of the elasticity of the material. Wheels having this bond are used for high-speed grinding (9000–16,000 feet per minute), since they afford rapid removal of the stock. They are used a great deal as snagging wheels in a foundry and also for cutting-off wheels.

5. Bakelite or resinoid process. The abrasive grains in this process are mixed with a synthetic resin powder and a liquid solvent. This plastic mixture is then molded to proper shape and baked in an electric oven at 312 F 1/2 to 3 days. This bond is very hard and strong, and wheels made by this process can be operated at speeds around 9500 to 16,000 feet per minute. These wheels are used for general-purpose grinding and are widely used in foundries and billet shops for snagging purposes because of their ability to remove metal rapidly.

Grinding-Wheel Selection

The proper selection of a grinding wheel for a definite purpose is important. There is a great variation in the wheels from which one may choose, and the selection is somewhat difficult because of the many factors involved. The factors to be considered in ordering a wheel are:

1. Size and shape of wheel. The principal grinding wheel shapes have been standardized by the United States Department of Commerce and the Grinding Wheel Manufacturers Association. *Standard shapes* which are available are shown in Figure 2, each having its own type number. These types may be obtained from any wheel manufacturer. In addition, all principal dimensions must be given. Grinding wheels of the straight wheel type have also been standardized according to *wheel face* as shown in Figure 3. These wheels are used for grinding special contours, sharpening saws, and other special applications, and are designated by a letter.

2. Kind of abrasive. A decision as to whether to use silicon carbide or aluminum oxide is largely dependent on the physical properties of the material to be ground. Silicon carbide wheels are recommended for materials of low tensile strength, such as cast iron, brass, stone, rubber, leather, and cemented carbides. The aluminum oxide wheels are best used on materials of high tensile strength, as hardened steel, high-speed steel, alloy steel, and malleable iron.

3. Grain size of abrasive particles. In general, coarse wheels are used for fast removal of materials. Fine-grained wheels are used where finish is an important consideration. Coarse wheels may be used for soft materials, but generally a fine grain should be used for hard and brittle materials.

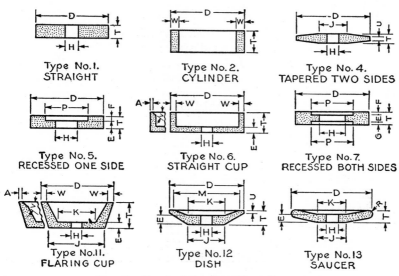

Fig. 2. Standard Grinding-Wheel Shapes.

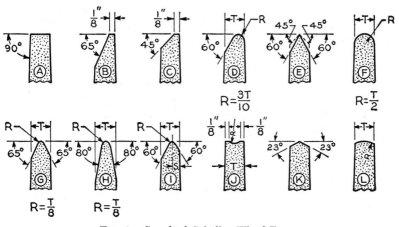

Fig. 3. Standard Grinding-Wheel Faces.

Grain size is specified according to standard screen sizes. The Norton Company classes abrasives from no. 6 to no. 24 coarse, no. 30 to no. 60 medium, no. 70 to no. 120 fine, and no. 150 to no. 240 very fine. Flour sizes run as high as no. 600 mesh.

4. Grade or strength of bond. The grade depends on the kind and hardness of the bonding material used. If the bond is very strong and capable of holding the abrasive grains against the force tending to pry them loose, it is said to be hard. If only a small force is needed to release the grains, the wheel is said to be soft. Most companies indicate the grade of the wheel by a letter. Although company standards differ, in general the grade letters increase in hardness from D to Z. Hard wheels are recommended for soft materials, and soft wheels for hard materials.

5. Structure or grain spacing. The structure refers to the number of cutting edges per unit area of wheel face as well as to the number and size of void spaces between grains. The structure to use depends principally on the physical properties of the material to be ground and the type of finish desired. Soft, ductile materials require a wide spacing. A fine finish requires a wheel with a close spacing of the abrasive particles.

6. Kind of bond material. The vitrified bond is most commonly used, but, where thin wheels are required or high operating speed or high finish is necessary, the selection of other types is advantageous.

7. Function of grinding wheel.* The use or purpose for which a grinding wheel is to be employed is a definite factor in wheel selection. Following are listed the basic functions of grinding wheels:

(*a*) Generation of size or grinding to close tolerance.

(*b*) Generation of surface finishes or effects which may or may not involve close tolerances.

(*c*) Removal of a large amount of stock, as in snagging.

(*d*) Cutting-off operations.

(*e*) Production of sharp edges or points as in knife grinding.

(*f*) Reduction of material to particle form.

TABLE 11

RECOMMENDED GRINDING-WHEEL SPEEDS

Type of Grinding	Wheel Speed, Surface Ft per min
Internal	2000–6000
Hemming cylinders	2100–5000[1]
Machine knives	3500–4500
Surface	4000–5000
Cutlery — large wheels offhand	4000–5000
Wet tool	5000–6000
Cylindrical	5500–6500
Snagging — vitrified bond	5000–6000
Snagging — resinoid and rubber bond	7000–9500
Cutoff — rubber, resinoid, and shellac bond	9000–16,000[1]

[1] Recommended only where bearings, protection devices, and machine rigidity are adequate (Abrasive Company).

* "Grinding Facts," The Carborundum Company, 1944.

8. Other factors that must be given some consideration are the wheel speed, speed of work, materials to be ground, and general condition of the machine. Table 11 lists recommended grinding-wheel speeds.

STANDARD MARKING SYSTEM CHART

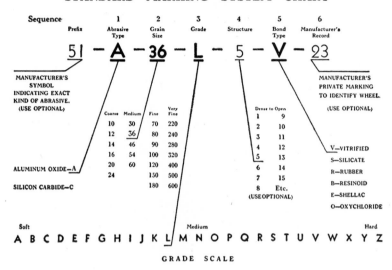

American Standard B 17—1943

A standard system* of marking grinding wheels, recently adopted by the American Standards Association, is shown in the accompanying chart. Although the standard greatly facilitates ordering from the standpoint of uniform marking of all wheels, there is no assurance that competitors' wheels marked alike will cut the same. Provision for each manufacturer to incorporate into the system such symbols as further describe the wheel and qualify the standard markings are stated in the first and last symbols of the identification marking.

Grinding Machines

Grinding machines are designed principally to finish parts having cylindrical, flat, or internal surfaces. The type of surface machined largely determines the type of grinding machine used; thus, a machine grinding cylindrical surfaces is called a cylindrical grinder. Machines designed for some special function such as tool grinding or cutting off, are designated according to the type of operation they perform.

* American Standard B5.17 — 1943.

A classification of grinding machines according to type of surface generated or work done is as follows:

CLASSIFICATION OF GRINDING MACHINES

1. Cylindrical grinder
   (a) Work between centers.
   (b) Centerless.
   (c) Tool post.
   (d) Crankshaft and other special applications.
2. Internal grinder
   (a) Work rotated in chuck.
   (b) Work rotated and held by rolls.
   (c) Work stationary.
3. Surface grinder
   (a) Planer type (reciprocating table)
       (1) Horizontal spindle.
       (2) Vertical spindle.
   (b) Rotating table
       (1) Horizontal spindle.
       (2) Vertical spindle.
4. Tool grinder
   (a) Universal.
   (b) Special
       (1) Drill.
       (2) Tool bit.
       (3) Cutter.
       (4) Pedestal, etc.
5. Special grinding machines
   (a) Swinging frame — snagging.
   (b) Cutting off — sawing.
   (c) Portable — offhand grinding.
   (d) Honing and lapping — accurate finishing.
   (e) Superfinishing.
   (f) Flexible shaft — general purpose.
6. Surface finishing
   (a) Disk.
   (b) Flexible band.
   (c) Two-wheel polishing or buffing machine.

Cylindrical Grinders

As the name implies, this type of machine is used primarily for grinding cylindrical surfaces, although tapered and simple formed surfaces may also be ground on most cylindrical grinders. This type of machine may be further classified according to the method of supporting the work. Schematic diagrams illustrating the essential difference in supporting the work between centers and the centerless grinder are shown in Figure 4. In the centerless type, the work is supported by the arrangement of the work rest, a regulating wheel, and the grinding wheel itself. Both types use plain grinding wheels with the grinding face as the outside diameter.

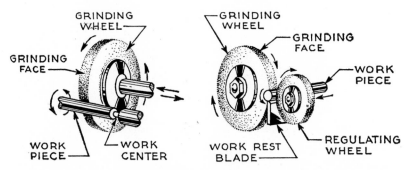

FIG. 4. Illustrating the Methods of Supporting Work in the Center and Centerless Type of Cylindrical Grinding.

Two illustrations of center-type grinding machines are shown in Figures 5 and 6: one adapted for the grinding of large rolls, and the other for smaller cylindrical parts. In the design of these machines there are incorporated three movements that are necessary in the operation of cylindrical grinding:

1. Rapid rotation of the grinding wheel at the proper grinding speed, usually from 5500 to 6500 surface feet per minute.

2. Slow rotation of the work against the grinding wheel at a speed to give best performance (this varies from 60 to 100 surface feet per minute in the grinding of steel cylinders).

3. Horizontal traverse of the work back and forth along the grinding wheel so as to grind the entire surface of a long piece.

In some machines the work remains stationary except for its rotation, and the wheel is slowly fed back and forth across the work. The narrower the face of the wheel, the slower must be the traverse and the faster should be the work revolution. For most cases the work should be traversed nearly the entire width of the wheel during each revolution of the work. In finishing, the traverse may be reduced to one-half the width of the wheel.

The depth of the cut is controlled by feeding the wheel into the work. Roughing cuts around 0.002 inch may be made, but for

Courtesy Cincinnati Milling and Grinding Machines, Inc.

FIG. 5.   A 16-Inch Roll Grinding Machine.

Courtesy Landis Tool Company.

FIG. 6.   A 14x48-Inch Cylindrical Grinder.

finishing the feed should be reduced to about 0.00025 inch. In selecting the amount of infeed, consideration must be given to the size and rigidity of the work, the finish desired, and whether or not a coolant is used.

Where the face of the wheel is wider than the part to be ground, it is not necessary to traverse the work. This is known as " plunge-cut " grinding and is common practice in the grinding of crankshafts (see Figure 7). Cuts up to 9 inches wide may be made in this

Courtesy Landis Tool Company.

FIG. 7. Crankshaft Grinder.

manner if the work is properly supported. Grinders for crankshaft work are usually especially built for that purpose, owing to the special features necessary for supporting and driving the crankshaft. Another special machine of this general type is the camshaft grinder. In order for the cams to be ground to proper shape, the movement of the work to and away from the wheel is controlled by master cams at the end of the shaft.

The *tool-post grinder* is used for miscellaneous and small grinding work on a lathe. It is held on the tool post and fed across the work, the regular longitudinal or compound rest feed being used. A common application of this grinder is the truing up of lathe centers.

Centerless grinders are designed so that they support and feed the work by using two wheels and a work rest, as illustrated diagrammatically in Figure 8. The large wheel is the grinding wheel and the smaller one the pressure or regulating wheel. The regulating wheel is a rubber bonded abrasive wheel having the proper frictional characteristics to rotate the work at its own rotational speed. The speed of this wheel may be controlled and varies from 50 to 200 surface feet per minute. Both wheels are rotated in the same direction. The slide assists in supporting the work while it is being ground and is extended on both sides to direct the work travel to and from the wheels.

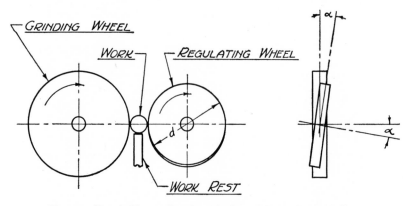

Fig. 8.   Sketch Illustrating the Principle of Centerless Grinding.

The axial movement of the work past the grinding wheel is obtained by tilting the wheel at a slight angle from horizontal. An angular adjustment of zero to 8 or 10 degrees is provided in the machine for this purpose. The actual feed can be calculated by this formula:

$$F = \pi \, d \, N \sin \alpha$$

where

$F =$ feed in inches per minute
$N =$ revolutions per minute
$d =$ diameter of regulating wheel in inches
$\alpha =$ angle of inclination of wheel

The formula assumes no slippage, and in actual practice the error is slight. This type of grinding may be applied to any cylindrical parts of one diameter, as shown in Figure 9. In production work on such parts as piston pins, a magazine feed is arranged, and the parts may go through several machines before completion, each grinder removing from 0.0005- to 0.002-inch stock.

Where parts are not uniformly of the same diameter, or where they require form grinding as a ball bearing (see Figure 10), the *infeed* type of centerless grinding must be used. The method of operation corresponds to the plunge-cut form of grinding, and the length of

Courtesy Cincinnati Milling and Grinding Machines, Inc.

FIG. 9. Tube Grinding in Centerless Grinder.

the section to be ground is limited to the width of the grinding wheel. The part is placed on the work rest and is moved against the grinding wheel with the regulating wheel. Upon completion, the

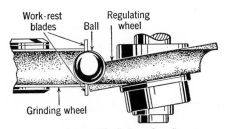

Courtesy The Carborundum Company.

FIG. 10. Centerless Grinding of Ball Bearings.

gap between the wheels is increased either manually or automatically and the work is ejected from between the wheels.

A third type of centerless grinding called *end feed* has been devised for use only on short taper work. Both wheels are dressed to the correct taper, and the work is automatically fed in from one side to a fixed stop.

The advantages of centerless grinding are:

1. Less skill is required in the operation of the machine.

2. No chucking or mounting of the work on mandrels or other holding devices is required.

3. The work is rigidly supported, and there is no tendency for chatter or deflection of the work.

4. The process is rapid and especially adapted for production work. Idle machine time is negligible.

5. The size of the work is easily controlled.

6. As a true floating condition exists during the grinding process, less grinding stock is required.

Some disadvantages are:

1. Work with flats and keyways cannot be ground.

2. In hollow work there is no assurance that the outside diameter will be concentric with the inside diameter.

3. Work having several diameters is not easily handled in this type of machine.

Internal Grinders

The work done on an internal grinder is diagrammatically shown in Figure 11. Tapered holes, or those having more than one diameter, may also be accurately finished in this manner. Although especially

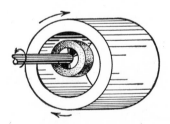

FIG. 11.   Sizing to close Tolerance by Internal Grinding.

adapted for heat-treated parts, internal grinding is frequently used on production parts that have not been heat-treated to save on reamer cost and maintenance.

According to general construction there are several types of internal grinders.

1. The wheel is rotated in a fixed position while the work is slowly rotated and traversed back and forth. The usual setup for this type of work is shown in Figure 12. The cam is fastened to the slowly rotating chuck while the grinder wheel is rotated at high speed against one side of the hole.

2. The wheel is rotated and at the same time reciprocated back and forth through the length of the hole. The work is rotated slowly but otherwise has no movement.

Courtesy Landis Tool Company.

FIG. 12. Internal Grinder.

3. The work remains stationary, and the rotating wheel spindle is given an eccentric motion, according to the diameter of hole to be ground. This type of grinder is frequently called the planetary type, and it is used for work that is difficult to rotate. In actual construction the wheel spindle is adjusted eccentrically in a larger one that rotates about a fixed axis. The wheel spindle is driven at high speed and at the same time rotates about the axis of the large spindle.

4. In another type of grinder which embodies the principle of centerless grinding, the work is rotated on the outside diameter by driven rolls, thus making it possible to grind the base absolutely concentric with the outside diameter. This arrangement lends itself to production work, as loading is simplified and magazine feed may be used if desired.

There is another variation in internal-grinding-machine design, involving the method of supporting the grinding-wheel head assembly. In most machines the wheel assembly is supported on a cross slide similar to the carriage on a lathe. Another type has the wheel assembly on a swinging arm supported from above, as shown in Figure 13. This machine is designed for a double purpose, having one spindle for internal grinding and one for face grinding. Thus two surfaces may be accurately finished at one chucking. This machine is used for grinding such parts as pistons, ring gears, drawing dies, bearing races, and brake drums.

Courtesy Bryant Chucking Grinder Company.

FIG. 13.   Hydraulic Hole and Face Grinder.

Since internal-grinding wheels are small in diameter, the spindle speed is much higher than for cylindrical grinding in order to attain surface speeds up to 6000 feet per minute. Most toolroom grinding is done dry, but common practice on production work is to grind steel wet and to grind bronze, brass, and cast iron dry. The amount of metal to be allowed for internal grinding depends on the size of the hole to be ground; in most cases this allowance is around 0.010 inch.

Surface Grinding

The grinding of flat or plane surfaces is known as *surface grinding*. Two general types of machines have been developed for this purpose: those of the planer type with a reciprocating table and those having a rotating work table. Each type of machine has the possible variation of having the grinding-wheel spindle in either a horizontal or vertical position. The four possibilities of construction are diagrammatically illustrated in Figure 14.

A planer-type surface grinder with a horizontal spindle is shown in Figure 15. Straight or recessed wheels (types 1, 5, and 7) grinding on the outside face or circumference are used on a machine of this type.

This grinder has hydraulic control of the table movement with possible speeds up to 100 feet per minute. Likewise, a hydraulic

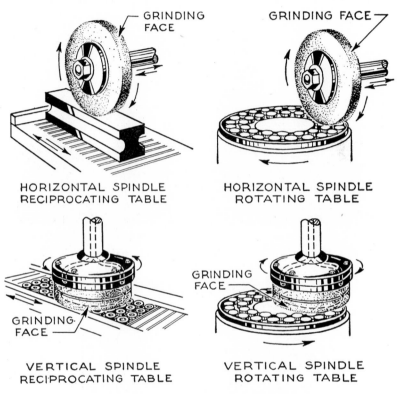

HORIZONTAL SPINDLE
RECIPROCATING TABLE

HORIZONTAL SPINDLE
ROTATING TABLE

VERTICAL SPINDLE
RECIPROCATING TABLE

VERTICAL SPINDLE
ROTATING TABLE

FIG. 14.   Types of Surface-Grinding Machines.

cross-feed is used which may be varied up to one-half the wheel face or more if desired. This type of grinder is widely used for reconditioning dies, as the large-diameter wheels make possible this operation without the guide pins being removed. Other applications include the grinding of grooves, ways on machine tools, and other long surfaces.

Another type of construction for planer-table grinders is the vertical spindle design, the grinding being done by a large-diameter ring-shaped wheel.

Several types of these wheels are shown in Figures 16 and 17. The wheel is a hollow cylinder which cuts on its end, and may be made up as a *plain cylindrical* wheel, a *sectored* wheel, or a *segment* wheel. All three of these wheels are illustrated in the first figure.

Courtesy Thompson Grinder Company.

FIG. 15.   Surface Grinder.   Planer Type with Horizontal Spindle.

Courtesy The Blanchard Machine Company.

FIG. 16.   A Sectored Wheel, a Plain Cylindrical Wheel, and a Segment Wheel, Used in Vertical-Spindle Surface Grinding.

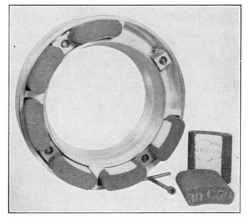

Courtesy The Blanchard Machine Company.

FIG. 17. Method of Mounting Wheel Segment in Chuck.

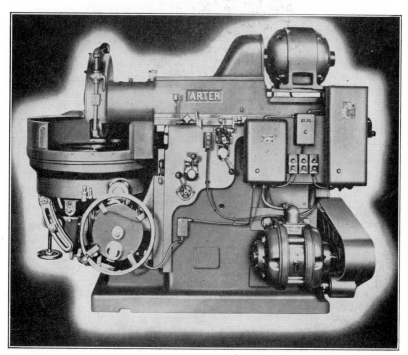

Courtesy Arter Grinding Machine Company.

FIG. 18. Rotary Surface Grinder.

The cylinder and sectored wheels are set with sulfur into a cast-iron ring which fastens to the face plate of the spindle. The segment wheel, shown also in Figure 17, uses a chuck secured to the face plate in which the segment blocks are clamped. Because of the large area of work in contact with these wheels, they are especially adapted to grinding large surfaces. It is also possible to place many small parts on a magnetic chuck and grind them with equal effectiveness.

Courtesy The Blanchard Machine Company.

FIG. 19.   Surface Grinder with Vertical Spindle and Adjustable Worktable.

Two other surface grinders with rotating tables, but with horizontal and vertical wheel spindles, are shown in Figures 18 and 19, respectively. In the Arter machine, shown in the first figure, the work is held on a magnetic chuck and slowly rotated under the grinding wheel. The wheel and spindle assembly is given a reciprocating movement during the grinding operation. This type of surface grinder is adapted to circular work, such as milling cutters, piston rings, saws, and valves. By tilting the work-table bevels, concave surfaces or short tapers may be ground. The Blanchard grinder, shown in the other figure, uses a cylindrical-type wheel in connection with the vertical spindle. The work to be ground is placed on a rotary magnetic chuck which rests on a table body that may be moved out from under the wheel. After the table is loaded, body and chuck are moved along the base to bring the center of the chuck

Courtesy Bergram Mechanical Engineering Company.

FIG. 20. Production Surface Grinder with Rotating Table.

just under the near edge of the wheel. In that position the work is rotated continuously in one direction, and the wheel head fed gradually downwards until the desired amount of metal has been removed. By changing the position of the chuck, pieces with central projections may be ground as well as small pieces laid radially. Examples of work ground on this machine are thrust washers, small gears, small blanking dies, ball-bearing housings, pump housings, and plier forgings.

The Bergram grinder, shown in Figure 20, is a production surface grinder similar to the last one described, but with a larger work table. It is designed to fill the need of a grinder for comparatively small parts which must be ground to close tolerance. The parts are held in a series of small fixtures which pass slowly under the two grinding wheels, the work being completed by one pass under the wheels. Diamonds for dressing the wheels to maintain uniform finished heights of work are mounted on the work table.

Magnetic Chucks

Work can be held successfuly on surface grinders and other machine tools by means of magnetic chucks. This method of holding has the advantage of being both simple and rapid. Parts to be held are placed on the chuck, and the chuck is energized by the turning of a switch.

The two types of magnetic chucks used are those which are magnetized by means of a *direct current* and the *permanent-magnet*

Courtesy O. S. Walker Company, Inc.

Fig. 21.   Universal Rectangular Magnetic Chuck.

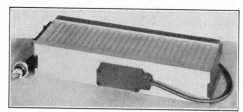

Courtesy O. S. Walker Company, Inc.

Fig. 22.   Standard-Type Rectangular Magnetic Chuck.

type. Figures 21 and 22 illustrate two d-c rectangular magnetic chucks as used on reciprocating table grinders or for light milling work. One of these chucks is known as a universal type and may

be swiveled in either direction. In addition, it is provided with a
T slot and two clamps to assist in holding work under unusual
conditions. The pulling power of these chucks is about 125 to 130
pounds per square inch. D-c chucks are also made in rotary styles,
as shown in Figure 23, for use on rotating table grinders and lathes.
Getting the current to the rotating chuck presents a problem, but
this is overcome by the use of collector rings and a brush unit
mounted at the back end of chuck or spindle. The pulling power

Courtesy O. S. Walker Company, Inc.

FIG. 23. Concentric-Gap and Radial-Pole Rotary Chucks.

varies according to the type of winding used and may be as high as
165 pounds per square inch. The equipment for furnishing the direct
current consists of a motor–generator set and demagnetizing switch.

All parts held on a magnetic chuck should be demagnetized after
the work is finished. Several types of demagnetizers are available,
operating on either alternating or direct current, which successfully
remove the residual magnetism from knives, bearing races, blades,
and many other parts.

Permanent-magnet chucks do not require any electric equipment,
and work can be held on these chucks as long as desired without
damage to work or chuck. A rectangular model of this type of chuck
is shown in Figure 24 set up for a surface grinding job. The operation
of this chuck is by means of the lever shown on the right-hand end.
The one internal movable member is made of alternating magnet and
conductor bars with nonmagnetic separators between them. Figure

25 shows what takes place when the operating lever is shifted. In the " off " position the conductor bars and separator are shifted in

Courtesy Brown and Sharpe Manufacturing Co.

FIG. 24. Rectangular Permanent-Magnetic Chuck in Use on Surface Grinder.

such a way that the magnetic flux passes through the top plate and is short-circuited from the work. When the handle is turned to the

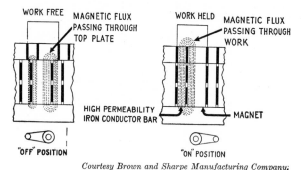

Courtesy Brown and Sharpe Manufacturing Company.

FIG. 25. Diagrammatic Sketch Showing How Work Is Held on Permanent-Magnet Chuck.

" on " position the conductor bars and nonmagnetic separators line up so that magnetic flux, in following the line of least resistance, goes through the work in completing the circuit. The holding power, obtained by the magnetic flux passing through, is sufficient to withstand the action of grinding wheels and other light machining operations. Both this type of chuck and the d-c chucks may be used for either wet or dry operations.

The permanent-magnet arrangement, as used in the magnetic chuck, has several other useful tool applications. V blocks and parallels are used in this fashion, both of which offer convenient means for holding special shapes which are not adapted to the regular-type chuck. Another application is the base for dial-test indicators, which permits holding the dial in any desired position by contact with the machine frame.

Courtesy Oliver Machinery Company.

FIG. 26.   Grinding Stand.

Tool Grinders

In grinding tools by hand (known as offhand grinding) a bench- or pedestal-type grinder is used similar to the one shown in Figure 26. The tool is held by hand and moved across the face of the wheel continually to avoid excessive grinding in one spot. This type of

Courtesy Cincinnati Milling and Grinding Machines, Inc.

FIG. 27.   No. 2 Cincinnati Cutter Grinder.

Courtesy Landis Tool Company.

FIG. 28.   Sharpening the Teeth on a Large Face Mill.

grinding is used to a large extent on single-point tools and is dependent upon the skill of the operator for good results.

In large production plants much of this type of grinding is done on special single-purpose grinders. Special drill or tool-bit grinders are justified by the large amount of grinding work necessary to keep production tools in proper cutting condition. In addition, tools can be ground uniformly and with accurate cutting angles.

For the sharpening of miscellaneous cutters a universal-type grinder as shown in Figure 27 is used. It is equipped with a universal head, vise, headstock and tailstock, and numerous other attachments for holding tools and cutters. Although essentially designed for cutter sharpening, it can also be used for cylindrical, taper, internal, and surface grinding.

A typical setup for grinding the inserts on a large face mill is shown in Figure 28. In the operation shown, the respective teeth are held in place by a flexible finger against the back of the insert while the faces of the teeth are ground by the wheel. In sharpening form cutters, a dish-shaped wheel is used, and the grinding is done only on the face of the tooth.

Honing

Honing is a special grinding process in which very little material is removed. It is used primarily to remove the grinding marks left by previous grinding or machining operations. Most of such work is done by hand with honing sticks or stones to improve the edges of cutting tools. India oil stones and fine-grained manufactured abrasives are used for this work.

For the honing of cylinders of internal-combustion engines, many special machines have been developed. These machines employ a holding device for the honing stones which has either an adjusting feature or some expanding mechanism to maintain uniform pressure against the sides of the cylinders. Four or six honing stones are uniformly spaced around the circumference of the holder. As the hone rotates at a peripheral speed of about 250 feet per minute, it is slowly reciprocated. Kerosene is the usual coolant employed for cast iron or steel. Very accurate dimensions can be maintained by honing, as the amount of metal to be removed seldom exceeds 0.002 inch.

Lapping

The purpose of lapping is to produce a very smooth surface, to correct minor surface imperfections, or to provide a very close fit between two contact surfaces. Although it is a material-removing operation, it is not an economical one. The amount of material removed is usually less than 0.001 inch. It is accomplished by using a lap charged with abrasive particles or by mixing the abrasive particles with oil and applying the mixture between the lap and the part being worked on. Originally lapping was done with a soft metal lap into which was rolled some hard abrasive, but more effective lapping is obtained by embodying the abrasive particles in a suitable carrying vehicle and placing the mixture between the metal lap and work. The metal lap should be either porous or soft to support and hold the abrasive. This causes the greatest wear to occur on the hard surface being worked on. Open-grained cast iron is frequently used in this work, although some applications require softer metals, such as copper and lead. The vehicle to hold the abrasive may be oil, grease, or kerosene.

Hardened gears are frequently run in or lapped to improve surface conditions and eliminate excessive wear. Valves and valve seats are lapped to insure tight contact when closed. This process also is used for finishing gages, dies, and carbide cutting tools. Crankshafts, bearing races, and surface plates are other applications which employ this process for proper surface condition.

Superfinishing

The commercial preparation of suitable smooth surfaces is of great importance in present-day high-speed machinery. All machining operations as well as the usual grinding processes leave a surface coated with fragmented, noncrystalline, or smear metal which is easily removed by sliding contact even though lubricated. The result is excessive wear, increased clearances, nosier operation, and lubrication difficulties. A recent process, known as *superfinishing,* has been developed by the Chrysler Corporation. It removes this undesirable surface metal, leaving a base of solid crystalline metal. This process, which is primarily a finishing process and not a dimensional one, can be superimposed on any commercial finishing operation such as turning, grinding, lapping, or honing. A bonded abrasive stone operating at a low abrasive speed and pressure is given a combination of multiple and random motions. A lubricant of proper viscosity is used to carry away the minute particles abraded from the surface by the short abrasive stone stroke. These short strokes, varying from 1/16 to 1/4 inch in length, have 300 to 3000 reversals per minute. Having both a primary and a secondary reciprocating

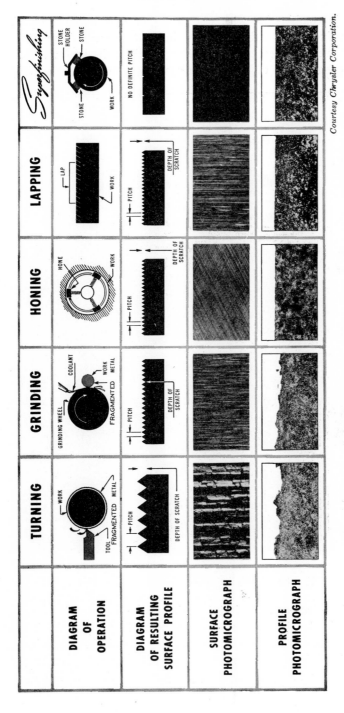

Courtesy Chrysler Corporation.

Fig. 29.   Diagram Illustrating the Surface Finishes Obtained by Various Manufacturing Processes.

motion in combination with work rotation gives a random path to the abrasive stone. The action is similar to a scrubbing movement and removes all excess and defective metal on the surface. Microscopic examination of the surface reveals clean metal and a solid

Courtesy Chyrsler Corporation.

FIG. 30.   General-Purpose Superfinishing Machine.

crystalline structure. It is interesting to note that this is accomplished with low abrasive stone pressure. The pressure for internal surfaces ranges from 3 to 10 pounds per square inch. Experimental work with lubricants of various viscosities demonstrated that the lubricant would support the stone pressure after the surface had been worn sufficiently smooth. Any further stone action had little or no effect, owing to the oil film between the stone and the metal surface. With proper equipment it takes only 5 to 30 seconds to finish a single bearing area. Figure 29 is an interesting diagram illustrating the types of surfaces obtained by superfinishing, as well as several other machining operations.

A general-purpose superfinishing machine is shown in Figure 30. This machine is designed for cylindrical surfaces, but may be

FIG. 31. Crankshaft Superfinisher.

equipped with an attachment which permits the finishing of flat surfaces. The machine is flexible and may be quickly adapted to many different jobs, either short-run or in production quantities. A special superfinisher which has been designed for finishing the main bearings of a crankshaft is shown in Figure 31. The time for completing the operation varies from 15 to 25 seconds, depending on the initial condition of the surface. Numerous other variations of the superfinishing process have been worked out for the many other bearing surfaces found in high-speed machinery.

Coated Abrasives

When abrasive particles are glued to paper or other flexible backings, as illustrated in Figure 32, they are known as coated abrasives. Any of the abrasives used in wheel manufacture may be applied in

Courtesy The Carborundum Company.

FIG. 32.   Various Types of Metal-Working Coated Abrasives.

this way. The most common type is the ordinary "sandpaper," which name is frequently applied to this entire group. The abrasive in this case is a flint quartz, which is mined in large lumps and then crushed to size and graded. Another important natural abrasive is the red mineral garnet. Of the several kinds of garnet known, the one called almandite is the best for abrasive coatings. It is much harder and sharper than flint and, when broken down, breaks into crystals with many cutting edges. The best garnet comes from the Adirondack Mountains in New York State. Other natural abrasives used are emery and corundum. The two manufactured abrasives which have wide application in this work are silicon carbide and aluminum oxide. Figure 33 is a photomicrograph of three natural abrasive grains compared with common sea sand, showing outstanding differences in shape and structure.

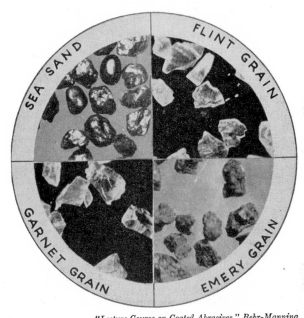

"Lecture Course on Coated Abrasives," Behr-Manning.

FIG. 33. Photomicrographs of Three Natural Abrasive Grains.

The three types of backing used are paper, cloth, and a combination of the two. Paper-coated abrasives are cheapest but lack flexibility. For applications requiring both strength and flexibility, a cloth backing is used. Combination backings are used where a backing stronger than paper but not with the extreme flexibility of cloth is needed.

An important phase in the manufacture of coated abrasives is the application of the abrasive particles on the backing material. The abrasive grains must be securely held in a manner to give the best cutting action for the type of work to be done. For severe service, closed coating is recommended, the grains completely covering the surface of the backing. Where increased flexibility is desired, with no tendency for the particles to become loaded or clogged, an open coating is used. Particles are separated at predetermined distances, leaving the base surface of the backing exposed. A recent method of coating is the electrocoating process, as illustrated in Figure 34. This method is based on the scientific fact that particles oppositely charged attract one another. Referring to the figure, we see the abrasive particles being carried between two electrodes, across which a high electrostatic field is built up. As the particles enter the field, they first stand on end, aligning themselves in the direction of the flow of the electric force, and then they are attracted to the glue-

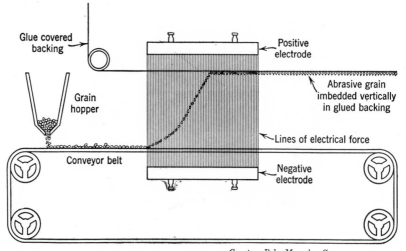

Courtesy Behr-Manning Company.

FIG. 34.   Electrocoating Method of Applying Abrasive Grain to Glue-Covered Backing.

covered backing which is traveling in the same direction as the other belt. The abrasive grains are imbedded on end and are equally spaced on the backing. By this process the spacing of the abrasive particles is controlled.

The principal application of coated abrasives is the finishing of surfaces. All forms of woodwork, such as mill work, patterns, floors, and furniture, are finished in this manner. Leather goods, felt hats, metallurgical specimens, and many metal parts also rely on coated abrasives in manufacturing operations. The selection of the proper coated abrasive for a given job depends on the finish desired, amount of stock to be removed, speed and pressure used by the sanding equipment, and the kind of surface to be sanded. These factors should be considered for each application, as they influence the selection of the grit number, type of abrasive, the abrasive spacing, and the backing material. Unfortunately, all coated abrasives are not graded in the same manner. Flint paper and emery cloth each have their own system while garnet and manufactured abrasives are graded by another system. Table 12 will be of assistance in the selection of the proper grades to use.

TABLE 12

GRADING CHART FOR COATED ABRASIVES

Garnet Silicon Carbide Aluminum Oxide	Flint Paper	Emery Cloth	Description
500			
400 or 10/0			
360			
320 or 9/0			Very fine
280 or 8/0			
240 or 7/0	5/0		
220 or 6/0	4/0		
	3/0		
180 or 5/0		3/0	
150 or 4/0		2/0	Fine
	2/0		
120 or 3/0			
	0	0	
100 or 2/0			
	$\frac{1}{2}$	$\frac{1}{2}$	
80 or 0		1	
	1	$1\frac{1}{2}$	Medium
60 or $\frac{1}{2}$			
		2	
50 or 1	$1\frac{1}{2}$		
40 or $1\frac{1}{2}$	2	$2\frac{1}{2}$	
	$2\frac{1}{2}$		Coarse
36 or 2			
30 or $2\frac{1}{2}$	3	3	
24 or 3	$3\frac{1}{2}$		
20 or $3\frac{1}{2}$			
16 or 4			Very coarse
12 or $4\frac{1}{2}$			

Table prepared by Behr-Manning Company.

REVIEW QUESTIONS

1. What advantage does the grinding process have over other cutting processes?

2. List the abrasives used for grinding wheels.

3. Why are natural abrasives inferior to manufactured abrasives?

4. How is silicon carbide made?

5. State briefly how vitrified wheels are made.

6. What bonding materials are used in wheel manufacture?

7. In purchasing a wheel for a specific purpose, what factors must be considered?

8. What are the basic functions of a grinding wheel?

9. A vitrified wheel 12 inches in diameter should run at what rpm for cylindrical grinding?

10. Classify grinding machines according to the type of work they will do.

11. What are two methods of grinding cylindrical work pieces?

12. What holds the work in centerless grinding?

13. Illustrate with line sketches four types or designs of surface grinders.

14. What are the three fundamental machine movements necessary in all cylindrical grinding?

15. What is " plunge-cut " grinding, and for what type of work is it used?

16. What elements in wheel construction are designated in the standard wheel-marking symbol?

17. What are the advantages and limitations of centerless grinding?

18. Name three types of centerless grinding.

19. What variations in construction are used in internal-grinding machines?

20. Describe the type of wheel used on vertical spindle surface grinders.

21. What type of grinding wheel is best adapted for grinding hardened steel?

22. What procedure should be followed for sharpening a form cutter such as a gear cutter?

23. Describe the process of honing an engine cylinder.

24. What is superfinishing and how is it done?

25. What abrasives are used in the manufacture of coated abrasives?

BIBLIOGRAPHY

BURGHARDT, H. D., *Machine Tool Operation*, Part II, McGraw-Hill Book Company, 1922.

BOSTON, O. W., *Metal Processing*, Chapter XV, John Wiley & Sons, 1941.

COLVIN, F. H., and STANLEY, F. A., *Grinding Practice*, 1st edition, McGraw-Hill Book Company, 1937.

Principles of Tool Room Grinding, Carborundum Company, 1944.

HEYWOOD, J., *Grinding Wheels and Their Uses*, Penton Publishing Company, 1938.

KLINE, J. E., "Desired Characteristics of Surface Finishes," *Mechanical Engineering*, Vol. 57, no. 12, 1935.

ROSE, KENNETH, " Ultra-Fine Surfaces on Metals," *Metals & Alloys*, July 1945.

SHOEMAKER, S. S., " Selecting the Correct Speeds and Feeds for Cylindrical Grinding," *Machinery*, February 1945.

SWIGERT, A. M., *The Story of Superfinish*, Lynn Publishing Company, 1940.

TROWBRIDGE, T., "Use of Coated Abrasives in Woodworking Industries," *Mechancal Engineering*, June 1945.

VICTORY, FREDERIC C., "Small Hole Grinding with Diamond Charged Mandrels," *American Machinist*, February 15, 1945.

TEXTBOOKS ON MACHINE-SHOP PRACTICE

BARRITT, J. W., *Machine Shop Operations*, American Technical Society, 1937.

BURGHARDT, H. D., *Machine Tool Operation*, Parts I and II, 1st edition, McGraw-Hill Book Company, 1919 and 1922.

BOSTON, O. W., *Engineering Shop Practice*, II, John Wiley & Sons, 1935.

BOSTON, O. W., *Metal Processing*, John Wiley & Sons, 1941.

CLAPP and CLARK, *Engineering Materials and Processes*, 1st edition, International Textbook Company, 1938.

COLE, C. B., *Tool Making*, American Technical Society, 1939.

COLVIN, F. H., and STANLEY, F. A., *Running a Machine Shop*, McGraw-Hill Book Company, 1941.

COLVIN, F. H., and STANLEY, F. A., *Drilling and Surfacing Practice*, 1st edition, McGraw-Hill Book Company, 1936.

COLVIN, F. H., and STANLEY, F. A., *American Machinist Handbook*, 8th edition, McGraw-Hill Book Company, 1945.

FREY, C. J., and KOGUT, S. S., *Metal Forming by Flexible Tools*, Pitman Publishing Corporation, 1943.

Henry Ford Trade School Shop Theory, Henry Ford Trade School, Dearborn, Michigan, 1934.

HESSE, H. C., *Engineering Tools and Processes*, D. Van Nostrand Company, 1941.

HEYWOOD, J., *Grinding Wheels and Their Uses*, Penton Publishing Company, 1938.

JONES, E. J. H., *Production Engineering*, Chemical Publishing Company, 1941.

JONES, F. D., *Machine Shop Training Course*, Vols. 1 and 2, Industrial Press, 1940.

JONES, M. M., and AXELROD, A., *Introductory Shopwork*, McGraw-Hill Book Company, 1943.

Kent's Mechanical Engineering Handbook, Vol. III, John Wiley & Sons, 1938.

Manual on Cutting Metals, ASME Committee on Cutting Metals, American Society of Mechanical Engineers, 1939.

Precision Measurement in the Metalworking Industry, educational department, International Business Machines Corporation, 1941.

ROE, J. W., and LYTLE, C. W., *Factory Equipment*, 2d edition, International Textbook Company, 1937.

SMITH, R. H., *Advanced Machine Work*, 7th edition, Industrial Education Book Company, 1940.

TURNER, F. W., and PERRIGO, O. E., *Machine Shop Work*, American Technical Society, 1941.

TURNER, W. P., and OWEN, H. F., *Machine Tool Work*, McGraw-Hill Book Company, 1945.

YOUNG, J. F., *Materials and Processes*, John Wiley & Sons, 1944.

INDEX